Springer Optimization and Its Applications

VOLUME 124

Aims and Scope
Optimization has been expanding in all directions at an astonishing rate during the last few decades. New algorithmic and theoretical techniques have been developed, the diffusion into other disciplines has proceeded at a rapid pace, and our knowledge of all aspects of the field has grown even more profound. At the same time, one of the most striking trends in optimization is the constantly increasing emphasis on the interdisciplinary nature of the field. Optimization has been a basic tool in all areas of applied mathematics, engineering, medicine, economics and other sciences.

The series *Springer Optimization and Its Applications* publishes undergraduate and graduate textbooks, monographs and state-of-the-art expository works that focus on algorithms for solving optimization problems and also study applications involving such problems. Some of the topics covered include nonlinear optimization (convex and nonconvex), network flow problems, stochastic optimization, optimal control, discrete optimization, multi-objective programming, description of software packages, approximation techniques and heuristic approaches.

More information about this series at http://www.springer.com/series/7393

Janusz Brzdęk • Krzysztof Ciepliński
Themistocles M. Rassias
Editors

Developments in Functional Equations and Related Topics

Springer

Editors
Janusz Brzdęk
Department of Mathematics
Pedagogical University of Cracow
Kraków, Poland

Krzysztof Ciepliński
Faculty of Applied Mathematics
AGH University of Science and Technology
Kraków, Poland

Themistocles M. Rassias
Department of Mathematics
National Technical University of Athens
Athens, Greece

ISSN 1931-6828 ISSN 1931-6836 (electronic)
Springer Optimization and Its Applications
ISBN 978-3-319-61731-2 ISBN 978-3-319-61732-9 (eBook)
DOI 10.1007/978-3-319-61732-9

Library of Congress Control Number: 2017948656

Mathematics Subject Classification (2010): 39-XX, 43-XX, 41-XX, 46-XX, 51-XX, 47-XX, 60-XX, 65-XX, 52-XX

Printed on acid-free paper

This Springer imprint is published by Springer Nature
The registered company is Springer International Publishing AG
The registered company address is: Gewerbestrasse 11, 6330 Cham, Switzerland

Dedicated to the memory of
Stanisław Marcin Ulam (1909–1984)
who more than 75 years ago
posed a problem concerning
approximate homomorphisms of groups
which stimulated
a long-lasting interest
in stability of functional equations

Preface

The present book has been composed on the occasion of the 16th International Conference on Functional Equations and Inequalities (ICFEI) that took place in the Mathematical Research and Conference Center in Będlewo, Poland, on May 17–23, 2015. We dedicate it to the memory of Stanisław Marcin Ulam (April 13, 1909–May 13, 1984), who 77 years ago posed a problem concerning approximate homomorphisms of groups that stimulated long-lasting research into stability of functional equations and inequalities (FEI). Several papers featured in this volume have been devoted fully or partly to Ulam's stability problem.

The book consists of articles written by eminent scientists from the international mathematical community, who present important research works in the field of FEI as well as related subjects. These works provide an insight into the progress achieved on the study of various problems of nonlinear flavor and present up-to-date developments of selected topics of FEI as well as of related fields of mathematics. Both old and new results are presented in expository and research papers written by 17 authors from 8 countries who have been intensively involved in those areas of investigations. Special emphasis has been placed on a variety of topics applying methods and techniques involving or originating from FEI.

Several of these results have been influenced and inspired by the work of S.M. Ulam, the well-known mathematician and physicist. Emphasis is placed on those questions, concerning approximate homomorphisms, that he posed in 1940.

We aim for this publication to serve as a kind of guidebook for mathematicians and other researchers, whose works are somewhat connected or related to the fields of FEI and in particular to Ulam's type stability.

Subjects which have been treated in this book include (in order of appearance in the volume):

- Some quasi-means and the behavior of their difference
- The isometric approximation problem in bounded sets and some applications of the results related to it to the extension problems for bilipschitz and quasisymmetric maps
- A mathematization method of social choice

- One-parameter subgroups (iteration groups) of the group of all invertible power series in one indeterminate x over $\mathbb{C}$ and a description of their construction
- The Fischer-Muszély equation, its pexiderization, and Hyers-Ulam stability, as well as two inequalities related to it
- The "alienation phenomenon" for functional equations (and inequalities)
- Haar meager sets and Haar null sets and some analogies between them
- Different types of stability of a system of two equations related to one-dimensional dynamical systems
- The role of functional equations in the asymptotic analysis needed to elicit the characterization of various laws in probability theory
- The translation equation and its stability
- Stochastic convex ordering and some applications of the results related to it to the Hermite-Hadamard type inequalities
- Two constructions of the field of reals closely related to functional equations and their stability
- The generalized Dhombres functional equation and a classification of its possible solutions as well as a description of the structure of periodic points contained in the range of the solutions
- Functional equations as well as their stability and superstability on hypergroups
- The nonstandard analysis approach to some systems of functional equations and their stability in the compact-open topology

It is a pleasure to express our deepest thanks to all of the mathematicians who, through their works, participated in this volume. We would also wish to acknowledge the support of the reviewers and the superb assistance that the staff of Springer has provided for this publication.

Kraków, Poland — Janusz Brzdęk
Kraków, Poland — Krzysztof Ciepliński
Athens, Greece — Themistocles M. Rassias
March 2017

Contents

List of Contributors

Shoshana Abramovich Department of Mathematics, University of Haifa, Haifa, Israel

Pekka Alestalo Department of Mathematics and Systems Analysis, Aalto University, Aalto, Finland

Anna Bahyrycz Faculty of Applied Mathematics, AGH University of Science and Technology, Mickiewicza, Kraków, Poland

Harald Fripertinger Institut für Mathematik und wissenschaftliches Rechnen, NAWI-Graz, Karl-Franzens-Universität Graz, Graz, Austria

Roman Ger Institute of Mathematics, University of Silesia, Bankowa, Katowice, Poland

Eliza Jabłońska Department of Discrete Mathematics, Rzeszów University of Technology, Rzeszów, Poland

Zenon Moszner Institute of Mathematics, Pedagogical University, Podchorążych, Kraków, Poland

Adam J. Ostaszewski Department of Mathematics, London School of Economics, London, UK

Barbara Przebieracz Institute of Mathematics, University of Silesia, Bankowa, Katowice, Poland

Teresa Rajba Department of Mathematics, University of Bielsko-Biala, Willowa, Bielsko-Biala, Poland

Ludwig Reich Institut für Mathematik und wissenschaftliches Rechnen, NAWI-Graz, Karl-Franzens-Universität Graz, Graz, Austria

Maciej Sablik Institute of Mathematics, University of Silesia, Bankowa, Katowice, Poland

Jens Schwaiger Institute for Mathematics and Scientific Computing, University of Graz, Graz, Austria

Jaroslav Smítal Mathematical Institute, Silesian University, Opava, Czech Republic

Marta Štefánková Mathematical Institute, Silesian University, Opava, Czech Republic

László Székelyhidi Department of Mathematics, Institute of Mathematics, University of Debrecen, Debrecen, Hungary

Pavol Zlatoš Faculty of Mathematics, Physics and Informatics, Comenius University, Mlynská dolina, Bratislava, Slovakia

Chapter 1
The Behavior of the Difference Between Two Means

Shoshana Abramovich

Abstract The inequalities derived in this article are related to the two quasi-arithmetic means $W_p(x,\lambda)$ and $M_q(x,\lambda)$. Here we extend some results about the difference $W_p(x,\lambda) - M_q(x,\lambda)$ for several sets of the values p and q.

Keywords Quasi-arithmetic means • Power means • Convex functions • Jensen's inequality • Subquadratic functions

Mathematics Subject Classification (2010) Primary 26D15, 47A63, 47A64; Secondary 26A51

1.1 Introduction

In this paper we extend the results proved in [2]. We discuss the behavior of the quasi-means

$$W_f(x,\lambda) = f^{-1}\left(f\left(\sum_{r=1}^{n}\lambda_r x_r\right) + \sum_{r=1}^{n}\lambda_r f\left(\left|x_r - \sum_{i=1}^{n}\lambda_i x_i\right|\right)\right), \tag{1.1}$$

$$\sum_{r=1}^{n}\lambda_r = 1, \qquad \lambda_r \ge 0, \qquad x_r \ge 0, \qquad r = 1,\dots,n,$$

in relation to the quasi-arithmetic means

$$M_g(x,\lambda) = g^{-1}\left(\sum_{r=1}^{n}\lambda_r g(x_r)\right) \tag{1.2}$$

S. Abramovich (✉)
Department of Mathematics, University of Haifa, Haifa 31905, Israel
e-mail: abramos@math.haifa.ac.il

J. Brzdęk et al. (eds.), *Developments in Functional Equations and Related Topics*, Springer Optimization and Its Applications 124,
DOI 10.1007/978-3-319-61732-9_1

where $x = (x_1, \ldots, x_n)$, $\boldsymbol{\lambda} = (\lambda_1, \ldots, \lambda_n)$ and the function f in (1.1) and g in (1.2) are increasing convex functions on $x \geq 0$, satisfying $f(0) = g(0) = 0$.

In particular we deal with $f(x) = x^p$, $p \geq 1$, $x \geq 0$ and $g(x) = x^q$, $q \geq 1$, $x \geq 0$. In these cases (1.1) and (1.2) are

$$W_p(x, \lambda) = \left(\left(\sum_{r=1}^{n} \lambda_r x_r \right)^p + \sum_{r=1}^{n} \lambda_r \left| x_r - \sum_{i=1}^{n} \lambda_i x_i \right|^p \right)^{\frac{1}{p}} \tag{1.3}$$

and

$$M_q(x, \lambda) = \left(\sum_{r=1}^{n} \lambda_r x_r^q \right)^{\frac{1}{q}}. \tag{1.4}$$

The identity

$$W_2(x, \lambda) - M_2(x, \lambda) = 0 \tag{1.5}$$

leads to the question about what can be said about the difference

$$W_f(x, \lambda) - M_g(x, \lambda)$$

and in particular about the difference

$$W_p(x, \lambda) - M_q(x, \lambda), \quad p \neq q.$$

In [7] and in [4] it was proved that

$$W_p(x, \lambda) - M_q(x, \lambda) \leq 0, \quad p = q \geq 2,\, x \geq 0, \tag{1.6}$$

holds, and

$$W_p(x, \lambda) - M_q(x, \lambda) \geq 0, \quad 1 \leq p = q \leq 2,\, x \geq 0. \tag{1.7}$$

In [2] it was proved that $W_p(x, \lambda)$ is decreasing in p when $0 \leq x_i \leq 2\overline{x}$, $i = 1, \ldots, n$, where

$$\overline{x} = \sum_{i=1}^{n} \lambda_i x_i.$$

This, together with (1.5)–(1.7) leads to Theorem 3 in [2] that part of which we quote here:

Theorem 1.1 *Let*

$$x_i \geq 0, \quad \lambda_i \geq 0, \quad x_i \leq 2 \sum_{j=1}^{n} \lambda_j x_j, \quad i = 1, \ldots, n, \quad \sum_{i=1}^{n} \lambda_i = 1.$$

If

$$1 \le s \le t \le 2,$$

then

$$\begin{aligned} M_1(x,\boldsymbol{\lambda}) &\le M_s(x,\boldsymbol{\lambda}) \le M_t(x,\boldsymbol{\lambda}) \le M_2(x,\boldsymbol{\lambda}) = W_2(x,\boldsymbol{\lambda}) \\ &\le W_t(x,\boldsymbol{\lambda}) \le W_s(x,\boldsymbol{\lambda}) \le 2^{1/s} M_1(x,\boldsymbol{\lambda}). \end{aligned}$$

If

$$s \ge t \ge 2,$$

then

$$\begin{aligned} M_s(x,\boldsymbol{\lambda}) &\ge M_t(x,\boldsymbol{\lambda}) \ge M_2(x,\boldsymbol{\lambda}) = W_2(x,\boldsymbol{\lambda}) \\ &\ge W_t(x,\boldsymbol{\lambda}) \ge W_s(x,\boldsymbol{\lambda}) \ge M_1(x,\boldsymbol{\lambda}), \end{aligned}$$

where $W_p(x,\boldsymbol{\lambda})$ *and* $M_p(x,\boldsymbol{\lambda})$ *are as in (1.3) and (1.4).*
Moreover, the difference

$$M_s(x,\boldsymbol{\lambda}) - W_s(x,\boldsymbol{\lambda})$$

increases when $s \ge 1$, *is negative when* $1 \le s < 2$, *positive when* $s > 2$, *and is equal to zero when* $s = 2$.

In Section 1.2 we prove inequalities related to the difference

$$W_p(x,\lambda) - M_q(x,\lambda)$$

for

$$\lambda = \left(\frac{1}{2}, \frac{1}{2}\right), \qquad p \ge 2, \qquad \frac{1}{p} + \frac{1}{q} = 1$$

and show an example for

$$\lambda = (\lambda_1, \lambda_2), \qquad 0 \le \lambda_1 \le \frac{1}{2} \le \lambda_2 \le 1.$$

In Section 1.3 we discuss the general quasi-means (1.1) and (1.2).

1.2 The Behavior of $M_p(x_1, x_2) - W_q(x_1, x_2)$

In this section we investigate the difference

$$\Delta(p,q) = M_p(x_1,x_2) - W_q(x_1,x_2) \tag{1.8}$$
$$= \left(\frac{x_1^p + x_2^p}{2}\right)^{\frac{1}{p}} - \left(\left(\frac{x_1+x_2}{2}\right)^q + \left(\frac{|x_2-x_1|}{2}\right)^q\right)^{\frac{1}{q}}, \quad 0 \le x_1, x_2,$$

for different values of $p, q \ge 1$.

We know that $M_p(x_1,x_2)$ is increasing with p, $p \ge 0$ and that $W_q(x_1,x_2)$ decreases with q, $q \ge 0$. Hence the difference $\Delta(p,q)$ as defined in (1.8) increases when p and q increase. In particular

$$\Delta(p,p) \ge \Delta(2,2) = 0 \ge \Delta(q,q), \quad 1 \le q \le 2 \le p, \tag{1.9}$$

and

$$\Delta(p_2,q_2) \ge \Delta(p,p) \ge \Delta(2,p) \ge \Delta(2,2) = 0 \tag{1.10}$$
$$\ge \Delta(q,2) \ge \Delta(q,q) \ge \Delta(p_1,q_1),$$
$$1 \le p_1,\ q_1 \le q \le 2 \le p \le p_2, q_2.$$

Inequalities (1.9) and (1.10) lead to the following questions:

Question 1 For what p and q

$$\Delta(q,p) = \left(\frac{x_1^q + x_2^q}{2}\right)^{\frac{1}{q}} - \left[\left(\frac{x_1+x_2}{2}\right)^p + \left(\frac{x_2-x_1}{2}\right)^p\right]^{\frac{1}{p}} \ge 0, \tag{1.11}$$
$$x_2 \ge x_1 \ge 0, \quad 1 \le q \le 2 \le p,$$

holds?

Remark 1.1 It is obvious that (1.11) holds for $q = 2$ and $p \ge 2$ and for $q = 1$ and $p \ge 2$ this is not the case. This means that $\Delta(q,p)$ does not hold for every $1 \le q \le 2 \le p$.

Question 2 For what p and q

$$\Delta(p,q) = \left(\frac{x_1^p + x_2^p}{2}\right)^{\frac{1}{p}} - \left[\left(\frac{x_1+x_2}{2}\right)^q + \left(\frac{x_2-x_1}{2}\right)^q\right]^{\frac{1}{q}} \le 0, \tag{1.12}$$
$$x_2 \ge x_1 \ge 0, \quad 1 \le q \le 2 \le p,$$

holds.

Remark 1.2 It is obvious that (1.12) is valid for $1 \leq q < 2 = p$ and does not hold for $q = 2 < p$.

First we will see that these two questions are related:

Theorem 1.2 *Let p and q be given and* $p \geq 2 \geq q \geq 1,\ x_2 \geq x_1 \geq 0$. *Then:*

Case a. *When* $1/p + 1/q = 1$, *inequality (1.11) holds iff inequality (1.12) holds.*
Case b. *When* $1/p + 1/q > 1$, *if inequality (1.11) holds, then inequality (1.12) holds.*
Case c. *When* $1/p + 1/q < 1$, *if inequality (1.12) holds, then inequality (1.11) holds.*

Proof For $p \geq 2 \geq q \geq 1,\ x_2 \geq x_1 \geq 0$, substituting

$$y_1 = \frac{x_2 - x_1}{2}, \qquad y_2 = \frac{x_2 + x_1}{2}$$

in $\Delta(q,p)$ we get

$$-2^{-\frac{1}{q}}\,\Delta(p,q) = \left(\frac{y_1^q + y_2^q}{2}\right)^{\frac{1}{q}} - 2^{1-\left(\frac{1}{p}+\frac{1}{q}\right)}\left[\left(\frac{y_1 + y_2}{2}\right)^p + \left(\frac{y_2 - y_1}{2}\right)^p\right]^{\frac{1}{p}}. \tag{1.13}$$

Then from (1.13) we deduce the following.

Case a: $1/p + 1/q = 1$. From

$$-2^{-\frac{1}{q}}\,\Delta(p,q) = \Delta(q,p),$$

we get that the proof of Case a is complete.

Case b: $1/p + 1/q > 1$. Note that if $\Delta(q,p) \geq 0$, then (1.11) holds, because

$$2^{1-\left(\frac{1}{p}+\frac{1}{q}\right)} < 1.$$

Hence we get from (1.13) that also

$$-2^{-\frac{1}{q}}\,\Delta(q,p) \geq 0$$

and therefore (1.12) holds. The proof of Case b is complete.

Case c: follows similarly. □

Conclusion Once we have p_0 and q_0, $1 < q_0 \leq 2 \leq p_0$ that satisfy $\Delta(q_0, p_0) \geq 0$, then because of the monotonicity with respect to q of the power means it is clear that if $1 < q_0 \leq q_1$ Inequality (1.11) holds too for q_1 and p_0. It is also obvious that there is an interval $[1, q_2]$, $q_2 < q_0$ such that Inequality (1.11) is not satisfied because for $q = 1$

$$\frac{x_1 + x_2}{2} < \left[\left(\frac{x_1 + x_2}{2}\right)^p + \left(\frac{x_2 - x_1}{2}\right)^p\right]^{\frac{1}{p}},$$
$$x_2 > x_1 > 0, \quad 1 = q, \quad 2 \le p.$$

Similarly, let p_0 and q_0 satisfy (1.12). Then if $p_1 < p_0$ (1.12) is also satisfied for p_1 and q_0. On the other hand, there is $p_2 > p_0$ for which Inequality (1.12) for p_2 and q_0 does not hold because

$$\left[\left(\frac{x_1 + x_2}{2}\right)^q + \left(\frac{x_2 - x_1}{2}\right)^q\right]^{\frac{1}{q}}, \qquad x_2 \ge x_1 \ge 0$$

is decreasing in q and

$$\left[\left(\frac{x_1 + x_2}{2}\right)^q + \left(\frac{x_2 - x_1}{2}\right)^q\right]^{\frac{1}{q}} > \left[\left(\frac{x_1 + x_2}{2}\right)^2 + \left(\frac{x_2 - x_1}{2}\right)^2\right]^{\frac{1}{2}}$$
$$= \left(\frac{x_1^2 + x_2^2}{2}\right)^{\frac{1}{2}}.$$

Therefore for p_2 big enough and $q < 2$ but close enough to 2 the Inequality (1.12) for these q and p_2 is not satisfied.

In the following theorem we demonstrate sufficient answers to Questions 1 and 2 under the conditions of Case a of Theorem 1.2.

Theorem 1.3 *Let $p \ge 2$. Then for $x_2 \ge x_1 \ge 0$ and an integer p*

$$\left(\frac{x_1^p + x_2^p}{2}\right)^{\frac{1}{p}} \le \left[\left(\frac{x_1 + x_2}{2}\right)^{\frac{p}{p-1}} + \left(\frac{x_2 - x_1}{2}\right)^{\frac{p}{p-1}}\right]^{\frac{p-1}{p}} \tag{1.14}$$

and

$$\left(\frac{x_1^{\frac{p}{p-1}} + x_2^{\frac{p}{p-1}}}{2}\right)^{\frac{p-1}{p}} \ge \left[\left(\frac{x_1 + x_2}{2}\right)^p + \left(\frac{x_2 - x_1}{2}\right)^p\right]^{\frac{1}{p}} \tag{1.15}$$

hold.

Proof As

$$1 < \frac{p}{p-1} \le 2,$$

and

$$f(x) = x^{p/(p-1)}, \qquad x \geq 0,$$

is convex, we get from (1.7) that

$$\left(\frac{x_1^{\frac{p}{p-1}} + x_2^{\frac{p}{p-1}}}{2}\right)^{\frac{p-1}{p}} \leq \left[\left(\frac{x_1+x_2}{2}\right)^{\frac{p}{p-1}} + \left(\frac{x_2-x_1}{2}\right)^{\frac{p}{p-1}}\right]^{\frac{p-1}{p}}$$

$$x_2 \geq x_1 \geq 0.$$

As

$$\frac{p}{p-1} \leq p$$

we get the inequality satisfied by power means

$$\left(\frac{x_1^{\frac{p}{p-1}} + x_2^{\frac{p}{p-1}}}{2}\right)^{\frac{p-1}{p}} \leq \left(\frac{x_1^p + x_2^p}{2}\right)^{\frac{1}{p}}.$$

We will show that when $p \geq 2$ is an integer and $x_2 \geq x_1 \geq 0$

$$\left(\frac{x_1^p + x_2^p}{2}\right)^{\frac{1}{p}} \leq \left[\left(\frac{x_1+x_2}{2}\right)^{\frac{p}{p-1}} + \left(\frac{x_2-x_1}{2}\right)^{\frac{p}{p-1}}\right]^{\frac{p-1}{p}}$$

holds. Moreover as $W_q(x_1, x_2)$ decreases for

$$q_1 \leq q = \frac{p}{p-1}$$

and as $M_p(x_1, x_2)$ increases in p, therefore

$$\Delta(p_1, q_1) \leq 0$$

for $p_1 \leq p$ and

$$q_1 \leq \frac{p}{p-1}.$$

By the change of variables

$$y_1 = \frac{x_2 - x_1}{2}, \qquad y_2 = \frac{x_2 + x_1}{2}$$

instead of (1.14) we have to prove

$$(y_2 + y_1)^p + (y_2 - y_1)^p \le 2\left(y_2^{\frac{p}{p-1}} + y_1^{\frac{p}{p-1}}\right)^{p-1}. \tag{1.16}$$

Equality holds in (1.16) for $y_1 = 0$ and for $y_1 = y_2$.

By the change of variables

$$z_1 = y_1^{\frac{1}{p-1}}, \qquad z_2 = y_2^{\frac{1}{p-1}},$$

we get from (1.16) that we need to prove the inequality

$$\left(z_1^{p-1} + z_2^{p-1}\right)^p + \left(z_2^{p-1} - z_1^{p-1}\right)^p \le 2\left(z_1^p + z_2^p\right)^{p-1} \tag{1.17}$$

for $z_2 \ge z_1 \ge 0,\ p \ge 2$. Equality holds in (1.17) when $z_1 = 0$ and for $z_1 = z_2$.

With no loss of generality, when $z_1 > 0$, we may assume that $z_1 = 1$ and we get from (1.17) that we want to prove the inequality

$$F(z) = 2\left(z^p + 1\right)^{p-1} - \left[\left(z^{p-1} + 1\right)^p + \left(z^{p-1} - 1\right)^p\right] \ge 0, \qquad z \ge 1. \tag{1.18}$$

Note that

$$F(1) = 0.$$

For an integer $p,\ p \ge 2$ we will show that Inequality (1.18) holds.

We will use Newton Binomial Expansion of

$$(1 + z^p)^{p-1}$$

and for

$$(1 + z^{p-1})^p + (z^{p-1} - 1)^p.$$

It is easy to verify that in the expansion of (1.18) we get

$$F(z) = 2\sum_{k=0}^{p-1}\binom{p-1}{k}(z^p)^k - \left[\sum_{j=0}^{p}\binom{p}{j}(z^{p-1})^j + \sum_{j=0}^{p}\binom{p}{j}(z^{p-1})^j(-1)^{(p-j)}\right] \tag{1.19}$$

and that there are $3(m-1)$ powers with non-zero coefficients of z^i, $i = 0, \ldots, p(p-1)$ when $p = 2m$, when m is an integer. This is because in

$$2\sum_{k=0}^{p-1}\binom{p-1}{k}(z^p)^k$$

there are $p = 2m$ non-zero terms and in

$$\sum_{j=0}^{p}\binom{p}{j}(z^{p-1})^j + \sum_{j=0}^{p}\binom{p}{j}(z^{p-1})^j(-1)^{(p-j)}$$

there are $m+1$ non-zero terms, and also the coefficients of z^0 and of $z^{p(p-1)}$ in

$$2\sum_{k=0}^{p-1}\binom{p-1}{k}(z^p)^k$$

and in

$$\sum_{j=0}^{p}\binom{p}{j}(z^{p-1})^j + \sum_{j=0}^{p}\binom{p}{j}(z^{p-1})^j(-1)^{(p-j)}$$

of $F(z)$ in (1.19) cancel each other.

The same number of powers with non-zero coefficients is when $p = 2m+1$.

We prove our result for odd integers. We get the same result when p is an even integer.

For $p = 2m+1$, $m \geq 1$, when m is an integer, we will see that the sum of the following kth three consecutive term in the expansion (1.19) satisfies

$$W_k(z) = \binom{2m}{2k+1} z^{(2m+1)(2k+1)} - \binom{2m+1}{2k+1} z^{2m(2k+1)} + \binom{2m}{2k} z^{(2m+1)2k} \geq 0$$

for $k = 0, \ldots, m-1$.

The vector

$$\left(\binom{2m}{2k+1}, \binom{2m+1}{2k+1}, \binom{2m}{2k}\right)$$

can be rewritten as

$$\frac{(2m)!}{(2k+1)!\,(2m-2k)!}\,(2m-2k, 2m+1, 2k+1)\,.$$

Therefore

$$W_k(z) = \frac{(2m)!}{(2k+1)!\,(2m-2k)!} \times \left[(2m-2k)\,z^{(2m+1)(2k+1)} - (2m+1)\,z^{2m(2k+1)} + (2k+1)\,z^{2k(2m+1)}\right]$$

or

$$W_k(z) = \frac{(2m)!}{(2k+1)!\,(2m-2k)!}\,z^{(2m+1)2k} \times \left[(2m-2k)\,z^{2m+1} - (2m+1)\,z^{2m-2k} + (2k+1)\right],$$

$$W_k(1) = W_k(0) = 0.$$

Define

$$R_k(z) = (2m-2k)\,z^{2m+1} - (2m+1)\,z^{2m-2k} + (2k+1), \quad k = 0, \ldots, m-1$$

The derivative of $R(z)$ is

$$\begin{aligned} R_k'(z) &= (2m-2k)\,z^{2m}(2m+1) - (2m+1)\,z^{2m-2k-1} \cdot (2m-2k) \\ &= (2m+1)\left[z^{2k+1} - 1\right] \end{aligned}$$

and we get that

$$R_k'(1) = 0$$

and

$$R_k'(z) \geq 0$$

because $z \geq 1,\ k \geq 0,\ m > k$.

Therefore as $R_k'(1) = 0$ and $R_k'(z) > 0,\ R_k(z) > 0,\ k = 0, \ldots, m-1$ hence

$$W_k(z) \geq 0, \qquad k = 0, \ldots, m-1.$$

Therefore for $p = 2m + 1$

$$\sum_{i=0}^{m-1} W_i(z) = F(z) \tag{1.20}$$

$$= 2\sum_{k=0}^{p-1}\binom{p-1}{k}(z^p)^k - \left[\sum_{j=0}^{p}\binom{p}{j}(z^{p-1})^j + \sum_{j=0}^{p}\binom{p}{j}(z^{p-1})^j(-1)^{(p-1)(p-j)}\right] \geq 0,$$

and by using (1.20) and (1.19) we get that Inequality (1.18) holds when $x_2 \geq x_1 \geq 0$ and $p = 2m + 1$, $p \geq 2$. Hence (1.14) holds. Therefore Inequality (1.12) holds for integers $p \geq 2$ and for $q = p/(p-1)$ and (1.15) follows from Theorem 1.2, Case a as

$$\frac{1}{p} + \frac{1}{q} = 1.$$

This completes the proof of the theorem. □

We give now an example of

$$\left(\alpha x_1^q + \beta x_2^q\right)^{\frac{1}{q}} \geq \left((\alpha x_1 + \beta x_2)^p + (\alpha\beta^p + \alpha^p\beta)(x_2 - x_1)^p\right)^{\frac{1}{p}}$$

when $0 \leq \alpha \leq \frac{1}{2} \leq \beta \leq 1$, $\alpha + \beta = 1$, $q = \frac{3}{2}$, $p = 3$, which probably is valid also for $p = n$, $n \geq 2$ is an integer and $1/p + 1/q = 1$.

Example 1.1 In this example we show that when

$$\alpha + \beta = 1, \qquad 0 \leq \alpha \leq \frac{1}{2} \leq \beta, \qquad 0 \leq x_1 \leq x_2,$$

the inequality

$$\left(\alpha x_1^{\frac{3}{2}} + \beta x_2^{\frac{3}{2}}\right)^{\frac{2}{3}} \geq \left((\alpha x_1 + \beta x_2)^3 + \alpha\beta\left(\alpha^2 + \beta^2\right)(x_2 - x_1)^3\right)^{\frac{1}{3}}$$

holds.

Instead, with no loss of generality, we show that

$$\left(\alpha + \beta y^3\right)^{\frac{2}{3}} \geq \left(\left(\alpha + \beta y^2\right)^3 + \alpha\beta\left(\alpha^2 + \beta^2\right)\left(y^2 - 1\right)^3\right)^{\frac{1}{3}}.$$

It is clear that for $y = 1$ we get an equality.

The last inequaltity is equivalent to

$$F = \left(\alpha + \beta y^3\right)^2 - \left(\left(\alpha + \beta y^2\right)^3 + \alpha\beta\left(\alpha^2 + \beta^2\right)\left(y^2 - 1\right)^3\right) \geq 0,$$

from which we get that

$$\begin{aligned} F &= \alpha\beta\left[\alpha(\beta - \alpha) y^6 - 3y^4\alpha(\beta - \alpha) + 2y^3\right. \\ &\quad \left. -3(\alpha + 1 - 2\alpha\beta) y^2 + (\alpha + 1 - 2\alpha\beta)\right] \\ &= \alpha\beta (y - 1)^2 \left[\alpha(\beta - \alpha) y^4 + 2\alpha(\beta - \alpha) y^3\right. \\ &\quad \left. +2(1 - \alpha(\beta - \alpha)) y + (1 - \alpha(\beta - \alpha))\right]. \end{aligned}$$

Therefore under our conditions as $y \geq 1, \alpha + \beta = 1, 0 \leq \alpha \leq \frac{1}{2} \leq \beta \leq 1$ we get that $F \geq 0$ which means that

$$\left(\alpha x_1^{\frac{3}{2}} + \beta x_2^{\frac{3}{2}}\right)^{\frac{2}{3}} \geq \left((\alpha x_1 + \beta x_2)^3 + \alpha\beta\left(\alpha^2 + \beta^2\right)(x_2 - x_1)^3\right)^{\frac{1}{3}}$$

when $0 \leq x_1 \leq x_2$ and $\alpha + \beta = 1, 0 \leq \alpha \leq \frac{1}{2} \leq \beta \leq 1$.

In the special case we get that when $\alpha = \beta = 1/2$

$$F = \frac{1}{4}(y-1)^2(2y+1) \geq 0.$$

1.3 Inequalities for Quasi-Arithmetic Means and Subquadracity

In this section we state a result related to the more general quasi-means M_f and W_g defined in (1.1) and in (1.2). Indeed, Theorem 1.1 is a special case of Theorem 1.4 below. In a future paper we intend to generalize Theorems 1.2 and 1.3 to M_f and W_g so that Theorems 1.2 and 1.3 become a special case of a theorem that deals with the more general quasi-means M_f and W_g.

First we quote a definition and results that appear, for instance, in [1, 2, 4, 5] and [6] and their references.

Definition A A function f is defined on an interval $I = [0, b)$ or $[0, \infty)$ is subquadratic if for each x in I, there exists a real number $C(x)$ such that

$$f(y) - f(x) \leq f(|y - x|) + C(x)(y - x) \tag{1.21}$$

for all $y \in I$. The function f is superquadratic if $-f$ is subquadratic.

From (1.21) it is easy to verify that:

Lemma A *Let f be subquadratic on $[0, b)$, $0 < b \leq \infty$. Let $\lambda_r \geq 0$, $x_r \in [0, b)$, $r = 1, \ldots, n$ and*

$$\sum_{r=1}^{n} \lambda_r = 1.$$

Then

$$\sum_{r=1}^{n} \lambda_r f(x_r) \leq f\left(\sum_{i=1}^{n} \lambda_i x_i\right) + \sum_{r=1}^{n} \lambda_r f\left(\left|x_r - \sum_{i=1}^{n} \lambda_i x_i\right|\right). \tag{1.22}$$

If f is superquadratic, the reverse of Inequality (1.22) holds.

When f is superquadratic and nonnegative, f is also convex increasing and

$$f(0) = f'(0) = 0.$$

The functions

$$f(x) = x^p, \qquad 1 \le p \le 2,\ x \ge 0,$$

and

$$f(x) = 3x^2 - 2x^2 \log(x), \qquad 0 \le x \le 1,$$

are examples of subquadratic increasing functions which are also convex (see [1]). The functions

$$f(x) = x^p, \qquad p \ge 2,\ x \ge 0,$$

and $f(x) = x^2 \log(x)$ are examples of superquadratic functions.

The following theorem deals with the quasi-means M_f and W_g and was proved in [2], by using the properties of subquadratic functions and by using the results of [3].

Theorem 1.4 *Let* $x_i,\ \lambda_i \ge 0,$

$$x_i \le 2\sum_{j=1}^{n} \lambda_j x_j, \qquad i = 1, \ldots, n,$$

$$\sum_{i=1}^{n} \lambda_i = 1.$$

Let F and G be nonnegative strictly increasing functions on $[0, \infty]$ *satisfying*

$$F(0) = G(0) = 0.$$

Let $\varphi = G \circ F^{-1}$ be convex function.

Case I: If F and G are subquadratic functions, then

$$\begin{aligned} M_f &= F^{-1}\left(\sum_{j=1}^{n} \lambda_j F(x_j)\right) \le G^{-1}\left(\sum_{j=1}^{n} \lambda_j G(x_j)\right) \\ &\le G^{-1}\left(G\left(\sum_{j=1}^{n} \lambda_j x_j\right) + \sum_{i=1}^{n} \lambda_i G\left(\left|x_i - \sum_{j=1}^{n} \lambda_j x_j\right|\right)\right) \\ &\le F^{-1}\left(F\left(\sum_{j=1}^{n} \lambda_j x_j\right) + \sum_{i=1}^{n} \lambda_i F\left(\left|x_i - \sum_{j=1}^{n} \lambda_j x_j\right|\right)\right) = W_f. \end{aligned}$$

Case II: If G and F are superquadratic functions, then

$$\begin{aligned} G^{-1}\left(\sum_{j=1}^{n}\lambda_j G\left(x_j\right)\right) &\geq F^{-1}\left(\sum_{j=1}^{n}\lambda_j F\left(x_j\right)\right) \\ &\geq F^{-1}\left(F\left(\sum_{j=1}^{n}\lambda_j x_j\right)+\sum_{j=1}^{n}\lambda_i F\left(\left|x_i-\sum_{j=1}^{n}\lambda_j x_j\right|\right)\right) \\ &\geq G^{-1}\left(G\left(\sum_{j=1}^{n}\lambda_j x_j\right)+\sum_{j=1}^{n}\lambda_i G\left(\left|x_i-\sum_{j=1}^{n}\lambda_j x_j\right|\right)\right). \end{aligned}$$

Example 1.2 Let $\lambda_i \geq 0$ for $i = 1, \ldots, n$,

$$\sum_{i=1}^{n}\lambda_i = 1,$$

and

$$F(x) = x^{(a+b)/a}, \qquad G(x) = x^{a+b}, \qquad x \geq 0.$$

(a) If $1 \leq a \leq b$, we get that $\varphi(x) = G\left(F^{-1}(x)\right) = x^a$ is a convex function and as well as F and G.

As F and G are superquadratic too, we get according to Theorem 1.4

$$\begin{aligned} \left(\sum_{i=1}^{n}\lambda_i x_i^{a+b}\right)^{1/(a+b)} &\geq \left(\sum_{i=1}^{n}\lambda_i x_i^{(a+b)/a}\right)^{a/(a+b)} \\ &\geq \left(\left(\sum_{j=1}^{n}\lambda_j x_j\right)^{(a+b)/a}+\sum_{i=1}^{n}\lambda_i\left(\left|x_i-\sum_{j=1}^{n}\lambda_j x_j\right|\right)^{(a+b)/a}\right)^{a/(a+b)} \\ &\geq \left(\left(\sum_{j=1}^{n}\lambda_j x_j\right)^{a+b}+\sum_{i=1}^{n}\lambda_i\left(\left|x_i-\sum_{j=1}^{n}\lambda_j x_j\right|\right)^{a+b}\right)^{1/(a+b)}. \end{aligned}$$

(b) If $a \geq 1,\ b \geq 0,\ a+b \leq 2,\ F$ and G are subquadratic and we get that

$$\left(\sum_{j=1}^{n}\lambda_j x_j\right)^{a+b} \le \left(\sum_{i=1}^{n}\lambda_i x_i^{(a+b)/a}\right)^{a} \le \sum_{j=1}^{n}\lambda_j x_j^{a+b}$$
$$\le \left(\sum_{j=1}^{n}\lambda_j x_j\right)^{a+b} + \sum_{i=1}^{n}\lambda_i \left(\left|x_i - \sum_{j=1}^{n}\lambda_j x_j\right|\right)^{a+b}$$
$$\le \left(\sum_{j=1}^{n}\lambda_j x_j\right)^{(a+b)/a} + \sum_{i=1}^{n}\lambda_i \left(\left|x_i - \sum_{j=1}^{n}\lambda_j x_j\right|\right)^{(a+b)/a}$$

is satisfied.

References

1. Abramovich, S.: On superquadracity. J. Math. Inequal. **3**, 329–239 (2009)
2. Abramovich, S.: Quasiarithmetic means and subquadracity. J. Math. Inequal. **9**, 1157–1168 (2015)
3. Abramovich, S., Pečarić, J., Varošaneć, S.: Comparison theorems between several quasi-arithmetic means. Math. Inequal. Appl. **7**, 1–6 (2004)
4. Abramovich, S., Jameson, G., Sinnamon, G.: Refining Jensen's Inequality. Bull. Math. Soc. Math. Roum. (Novel Series) **47**(95), 3–14 (2004)
5. Bibi, R., Bohner, M., Pečarić, J.: Cauchy-type means and exponential and logarithmic convexity for superquadratic functions on time scales. Ann. Funct. Anal. **6**, 59–83 (2015)
6. Oguntuase, J.A., Persson, L.E.: Time scale Hardy type inequalities via superquadracity. Ann. Funct. Anal. **5**, 61–73 (2014)
7. Sinnamon, G.: Refining the Hölder and Minkowski inequality. J. Inequal. Appl. **6**, 633–640 (2001)

Chapter 2
Isometric Approximation in Bounded Sets and Its Applications

Pekka Alestalo

Abstract We give a review of results related to the isometric approximation problem in bounded sets, and their application in the extension problems for bilipschitz and quasisymmetric maps. We also list several recent articles dealing with the approximation problem for mappings defined in the whole space.

Keywords Nearisometry • Quasisymmetric • Bilipschitz • Extension

Mathematics Subject Classification (2010) Primary 30C65; Secondary 46B20

2.1 Introduction

Definition 2.1 Let X and Y be metric spaces with distance written (in the Polish notation) as $|x - y|$, and let $\varepsilon \geq 0$. A mapping $f: X \to Y$ is an ε-nearisometry if

$$\big||f(x) - f(y)| - |x - y|\big| \leq \varepsilon$$

for all $x, y \in X$.

We remark that these mappings are often called ε-isometries, ε-quasi-isometries, etc., and that the condition is equivalent to

$$|x - y| - \varepsilon \leq |f(x) - f(y)| \leq |x - y| + \varepsilon.$$

The nearisometry condition does not imply continuity (unless $\varepsilon = 0$), but these maps are closely related to $(1 + \varepsilon)$-bilipschitz maps that satisfy

$$|x - y|/(1 + \varepsilon) \leq |f(x) - f(y)| \leq (1 + \varepsilon)|x - y|$$

P. Alestalo (✉)
Department of Mathematics and Systems Analysis, Aalto University, FI-00076 Aalto, Finland
e-mail: pekka.alestalo@aalto.fi

J. Brzdęk et al. (eds.), *Developments in Functional Equations and Related Topics*, Springer Optimization and Its Applications 124,
DOI 10.1007/978-3-319-61732-9_2

for all $x, y \in X$. In particular, if the diameter $d(X)$ is finite, then every $(1+\varepsilon)$-bilipschitz map $f: X \to Y$ is a $d(X)\varepsilon$-nearisometry.

Our starting point is the following theorem from [14] and [19].

Theorem 2.1 *Let E and F be real normed spaces and let $f: E \to F$ be a surjective ε-nearisometry with $f(0) = 0$. Then there is a surjective linear isometry $T: E \to F$ satisfying*

$$\|T - f\|_E \equiv \sup\{|Tx - f(x)| \mid x \in E\} \leq 2\varepsilon.$$

The original proof in [14] was for Hilbert spaces only, and with a constant 10ε. The bound 2ε, obtained in [19], is the best universal one, but it can be improved to $J(E)\varepsilon$ for Hilbert spaces, cf. [13]. Here $J(E)$ is the Jung's constant of the space E.

A comprehensive history of these developments, some counterexamples demonstrating the sharpness of the constants, and a survey of further progress up to c. 2002 can be found in the article [25], which I recommend to the interested reader. See also [22] for some updates. Additional surveys of these problems in [11] and [21] are also useful. Furthermore, many of the original proofs are reproduced in Chapter 13 of the monograph [16], which contains also other closely related material.

However, some important counterexamples related to the approximation problem in bounded subsets were discovered only after Väisälä's survey article appeared. In the following sections I will describe these developments and present applications of the results to extension problems for mappings that are, in a certain sense, close to either an isometry or a similarity.

To close this introduction, I remark that also the case of mappings defined in the whole space, but without the surjectivity assumption, has attracted a lot of interest and new results in the last couple of years. Since this is not my area of speciality and I want to concentrate in the approximation problem for bounded sets, I will only list here some of these references: [7–10, 12, 20, 27, 28].

2.2 Isometric Approximation in Bounded Sets

We start with the approximation of nearisometries in the closed unit ball $\mathbf{B}^n \subset \mathbb{R}^n$.

Theorem 2.2 *There is a universal constant $C > 0$ such that every ε-nearisometry $f: \mathbf{B}^n \to \mathbb{R}^n$ has an isometric approximation $T: \mathbb{R}^n \to \mathbb{R}^n$ satisfying*

$$\|T - f\|_{\mathbf{B}^n} \equiv \sup\{|Tx - f(x)| \mid x \in \mathbf{B}^n\} \leq C \log(n+1) \cdot \varepsilon.$$

History A similar result was proved by John [15] already in 1961, but with an error term $10n^{3/2}\,\varepsilon$. A more general formulation of this can also be found in the book [6, Theorem 14.11.]. The logarithmic upper bound was found in 2003 by Kalton in [17], whereas Matouvsková [18] constructed already in 2002 examples, where the error grows logarithmically. It follows that the logarithmic dependence on the dimension n is optimal.

We consider next the case, where the subset $A \subset \mathbf{B}^n$ is otherwise arbitrary, but contains the points $\overline{0}, \mathbf{e}_1, \ldots, \mathbf{e}_n \in A$.

Theorem 2.3 *There is a universal constant $C > 0$ such that every ε-nearisometry $f: A \to \mathbb{R}^n$ has an isometric approximation $T: \mathbb{R}^n \to \mathbb{R}^n$ satisfying*

$$\|T - f\|_A \equiv \sup\{|Tx - f(x)| \mid x \in A\} \le C n \cdot \varepsilon.$$

History An upper bound $C n^{3/2} \varepsilon$ was obtained in [4, 3.12] using John's idea (see [6, Chapter 14]). In 2005, Vestfid [26] found the linear bound $C n \varepsilon$ and showed that the linear growth is optimal in n.

Before more general versions, we need a definition.

Definition 2.2 The **thickness** $\theta(A)$ of a set $A \subset \mathbb{R}^n$ is the infimum of numbers $t > 0$ such that A lies between two parallel hyperplanes with mutual distance t.

The inequality

$$0 \le \theta(A) \le d(A)$$

is always true, but the thickness $\theta(A)$ can be very small even if the diameter $d(A)$ is large. In particular, $\theta(A) = 0$ if and only if A is contained in some hyperplane.

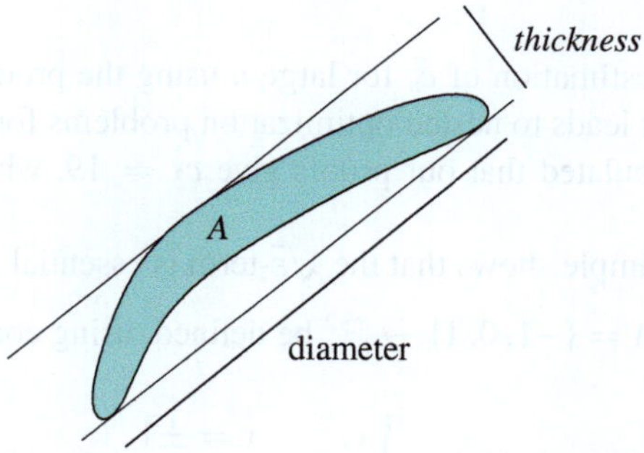

The following theorem is from [4, 3.3].

Theorem 2.4 *Let $A \subset \mathbb{R}^n$ be a compact set such that*

$$\theta(A) \ge \frac{d(A)}{t}$$

for some $t \ge 1$, and let $f: A \to \mathbb{R}^n$ be an ε-nearisometry. Then there is an isometry $T: \mathbb{R}^n \to \mathbb{R}^n$ such that

$$\|T - f\|_A \equiv \sup\{|Tx - f(x)| \mid x \in A\} \le C_n t \, \varepsilon.$$

Remark The upper bound is sharp with respect to the parameter t, but the asymptotic behaviour of C_n is unknown. Vestfrid's examples show that the growth of C_n is at least linear in n. On the other hand, an upper bound for C_n can be derived from the proof in [4]. The proof proceeds by induction on n, and the growth of C_n can be analysed from a system of recursion formulas. Numerical experiments for $n \leq 50$ by the author (unpublished) indicate that

$$\lim_{n\to\infty} \frac{\log C_n}{3^n} \approx 0.5756.$$

(Curiously, this number seems to be equal to $(1/4)\log 10$ up to at least 10 decimal places, which I discovered by accident.) It follows from this that

$$C_n \lesssim 1.778^{3^n},$$

so there seems to be a huge gap between upper and lower estimates.

Without any restrictions on the geometry of the set A we obtained the following result in [4, 2.2].

Theorem 2.5 *Let $A \subset \mathbb{R}^n$ be a compact set and let $f: A \to \mathbb{R}^n$ be an $\varepsilon d(A)$-nearisometry. Then there is an isometry $T: \mathbb{R}^n \to \mathbb{R}^n$ such that*

$$\|T - f\|_A \equiv \sup\{|Tx - f(x)| \mid x \in A\} \leq c_n d(A)\sqrt{\varepsilon}.$$

Remark Numerical estimation of c_n for large n using the proof seems difficult but should be possible: it leads to nested optimization problems for recursion formulas. However, I have calculated that our proofs give $c_3 = 19$, whereas $C_3 = 10^7$ for thick sets.

The following example shows that the $\sqrt{\varepsilon}$-term is essential in general.

Example 2.1 Let $f: A = \{-1, 0, 1\} \to \mathbb{R}^2$ be defined, using complex notation, by

$$f(x) = \begin{cases} x, & x = \pm 1 \\ i\sqrt{\varepsilon}, & x = 0. \end{cases}$$

Then f is $(1 + \varepsilon)$-bilipschitz and hence a 2ε-nearisometry, but for all isometric approximations $T: \mathbb{R}^2 \to \mathbb{R}^2$ the error is at least $\sqrt{\varepsilon}/2$. This follows easily by minimizing the distance from the set fA to the line $T\mathbb{R}$.

2.3 From Approximation to Bilipschitz Extension

In this and the following section we give some examples of extension results that can be proven by using the approximation results for bounded sets. The main problem is to extend a mapping $f: A \to \mathbb{R}^n$ to a mapping $F: \mathbb{R}^n \to \mathbb{R}^n$ having similar properties as the original f.

The easiest case for bilipschitz extension occurs if the set A has thickness in all scales.

Definition 2.3 Let $c \geq 1$. A set $A \subset \mathbb{R}^n$ is c-uniformly thick if

$$\theta(A \cap B(a,r)) \geq 2r/c$$

for all $a \in A$ and $r > 0$.

Uniform thickness does not allow isolated points, but, on the other hand, extending a map from an isolated point (at least to its neighbourhood) is very easy. In order to obtain the most general setting for extension, we need a more general definition that does not rule out isolated points if there is enough thickness around them, in a larger scale related to the distance from an isolated point of A to the rest of A.

Definition 2.4 Let $A \subset \mathbb{R}^n$. For $a \in A$ we set $s(a) = d(a, A \setminus \{a\})$.

Then $s(a) > 0$ if and only if a is isolated in A.

Definition 2.5 Let $c \geq 1$. We say that the set $A \subset \mathbb{R}^n$ is c-sturdy if

(1) $\theta(A \cap B(a,r)) \geq 2r/c$ whenever $a \in A,\ r \geq cs(a),\ A \not\subset B(a,r)$,
(2) $\theta(A) \geq d(A)/c$.

If A is unbounded, we omit (2), and the condition $A \not\subset B(a,r)$ of (1) is unnecessary.

Examples of sturdy sets are $\mathbb{Z}^n \subset \mathbb{R}^n$, the Koch snowflake curve in the plane, bounded Lipschitz domains, and all uniformly thick sets.

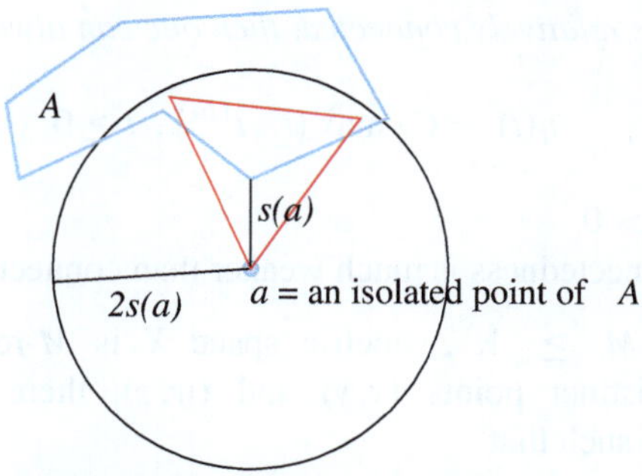

intersection of A and $B(a,r)$ contains vertices of a thick triangle

The following extension theorem is from [5].

Theorem 2.6 *Let $A \subset \mathbb{R}^n$ be c-sturdy. Then there are $\delta = \delta(c,n)$ and $C = C(c,n)$ such that every $(1+\varepsilon)$-bilipschitz map $f: A \to \mathbb{R}^n$, with $\varepsilon \leq \delta$, extends to a $(1+C\varepsilon)$-bilipschitz map $F: \mathbb{R}^n \to \mathbb{R}^n$.*

The proof is based on approximation results for nearisometries and will be sketched in a more general setting in the next section.

Remark The converse result in $\mathbb{R}^2$ was proven in [1]: If a set $A \subset \mathbb{R}^2$ has the above extension property for $(1+\varepsilon)$-bilipschitz maps with a small ε, then it is sturdy, and there are quantitative relations between all constants involved.

2.4 From Approximation to Quasisymmetric Extension

In this section we consider a more general class of mappings, the quasisymmetric ones, and present the main extension result from [3].

Let $\eta\colon [0,\infty) \to [0,\infty)$ be a homeomorphism, called a growth function.

Definition 2.6 An injective map $f\colon A \to \mathbb{R}^n$ is η-quasisymmetric if the ratios of distances are changed in a controlled way:

$$\frac{|f(x)-f(y)|}{|f(x)-f(z)|} \leq \eta\left(\frac{|x-y|}{|x-z|}\right)$$

for all distinct $x, y, z \in A$. If $\eta(t) = t$, then f is a similarity.

An L-bilipschitz map is quasisymmetric with $\eta(t) = L^2 t$, $t \geq 0$. Conversely, a quasisymmetric map with a linear growth $\eta(t) = Ct$ is always bilipschitz. However, since there is no general bound for the Lipschitz-constant of a similarity, we cannot say anything about the constant in this converse part.

It was proven in [24, 3.12] and [23, 6.5] that one can often replace the growth function η with a power form.

Theorem 2.7 *If A is relatively connected, then one can always choose*

$$\eta(t) = C \cdot \max\left(t^{\alpha}, t^{1/\alpha}\right),\ t \geq 0,$$

where $C \geq 1$ and $\alpha > 0$.

Here relative connectedness is much weaker than connectedness.

Definition 2.7 Let $M \geq 1$. A metric space X is M-relatively connected if, for all pairs of distinct points (x, y) and (w, z), there is a finite sequence $(x_0, x_1, \ldots, x_{k-1}, x_k)$ such that

$$x_0 = x,\ x_1 = y,\ x_{k-1} = w,\ x_k = z$$

and

$$\frac{1}{M} \leq \frac{|x_{j+1} - x_j|}{|x_j - x_{j-1}|} \leq M$$

for all $1 \leq j \leq k-1$.

Examples of relatively connected spaces include all connected ones, the Cantor middle-third set, etc.

In this light, the following condition seems natural and turns out to be the best way to measure how close a quasisymmetric mapping is from a similarity. Some problems with other possible approaches are considered in [2].

Definition 2.8 A mapping $f: A \to \mathbb{R}^n$ is ε-power-quasisymmetric if it is η-quasisymmetric with

$$\eta(t) = (1+\varepsilon) \cdot \max(t^{1+\epsilon}, t^{1/(1+\epsilon)}).$$

We remark that suitable radial stretching maps in $\mathbb{R}^n$ will satisfy this condition, but they are not bilipschitz.

Theorem 2.8 *Let $A \subset \mathbb{R}^n$ be c-sturdy. Then there are $\delta = \delta(c,n)$ and $C = C(c,n)$ such that every ε-power-quasisymmetric map $f: A \to \mathbb{R}^n$, with $\varepsilon \leq \delta$, extends to a $C\varepsilon$-power-quasisymmetric map $F: \mathbb{R}^n \to \mathbb{R}^n$.*

Main Steps of the Proof

- Show that ε-power-quasisymmetric maps can be well approximated by similarities in balls $A \cap B(a,r)$ if r is suitably chosen. This follows from sturdiness and the isometric approximation results of Alestalo et al. [4] by scaling.
- Show that one may assume A to be unbounded, so that sturdiness is easier to handle. This a rather easy, but a very technical part.
- Decompose $\mathbb{R}^n \setminus A$ into Whitney cubes and define the extension in the vertices v by using suitable approximating similarities of f in sets of the type $A \cap B(v,r)$. Here the radius r must be carefully chosen in order to guarantee the thickness of this intersection.
- Triangulate the Whitney cubes and extend affinely to each simplex.
- The result will be a continuous map $F: \mathbb{R}^n \to \mathbb{R}^n$.
- The final step is based on showing that the assumptions for the following theorem from [3, 3.7] are satisfied if ε is small enough. Indeed, if a mapping has been extended in a suitable way using approximating similarities, it should not be surprising that it can be well approximated by similarities. However, the details of the proof are again quite technical.

Theorem 2.9 *Let $F: \mathbb{R}^n \to \mathbb{R}^n$ satisfy the following condition for some $\varepsilon \leq 1/100$: For every ball $B = B(x,r)$ there is a similarity $S = S_{x,r}$ such that*

$$\|S \circ F - \mathrm{id}\|_B \leq \varepsilon r.$$

Then F is 50ε-power-quasisymmetric.

References

1. Alestalo, P., Trotsenko, D.A.: Plane sets allowing bilipschitz extension. Math. Scand. **105**, 134–146 (2009)
2. Alestalo, P., Trotsenko, D.A.: On mappings that are close to a similarity. Math. Rep. (Bucur.) **15**(4), 313–318 (2013)
3. Alestalo, P., Trotsenko, D.A.: On the extension of quasisymmetric maps. Ann. Acad. Sci. Fenn. Math. **41**, 881–896 (2016). doi:10.5186/aasfm.2016.4154
4. Alestalo, P., Trotsenko, D.A., Väisälä, J.: Isometric approximation. Isr. J. Math. **125**, 61–82 (2001)
5. Alestalo, P., Trotsenko, D.A., Väisälä, J.: Linear bilipschitz extension property. Sibirsk. Mat. Zh. **44**, 1226–1238 (2003). Translation in Sib. Math. J. **44**, 959–968 (2003)
6. Benyamini, Y., Lindenstrauss, J.: Geometric Nonlinear Functional Analysis I. AMS Colloquium Publications, vol. 48. American Mathematical Society, Providence (2000)
7. Cheng, L., Zhou, Y.: On perturbed metric-preserved mappings and their stability characterizations. J. Funct. Anal. **266**, 4995–5015 (2014)
8. Cheng, L., Dong, Yu., Zhang, W.: On stability of nonlinear non-surjective ε-isometries on Banach spaces. J. Funct. Anal. **264**, 713–734 (2013)
9. Cheng, L., Dai, D., Dong, Y, Zhou, Y.: Universal stability of Banach spaces for ε-isometries. Stud. Math. **221**, 141–149 (2014)
10. Cheng, L., Cheng, Q., Tu, K., Zhang, J.: A universal theorem for stability of ε-isometries of Banach spaces. J. Funct. Anal. **269**, 199–214 (2015)
11. Ding, G.: A survey on the problems of isometries. Southeast Asian Bull. Math. **29**, 485–492 (2005)
12. Dong, Y.: A note on the Hyers-Ulam problem. Colloq. Math. **138**, 233–239 (2015)
13. Huuskonen, T., Väisälä, J.: Hyers-Ulam constants of Hilbert spaces. Stud. Math. **153**, 31–40 (2002)
14. Hyers, D.H., Ulam, S.M.: On approximate isometries. Bull. Am. Math. Soc. **51**, 288–292 (1945)
15. John, F.: Rotation and strain. Commun. Pure Appl. Math. **14**, 391–413 (1961)
16. Jung, S.-M.: Hyers-Ulam-Rassias Stability of Functional Equations in Nonlinear Analysis. Springer Optimization and Its Applications, vol. 48. Springer, New York (2011)
17. Kalton, N.J.: A remark on quasi-isometries. Proc. Am. Math. Soc. **131**, 1225–1231 (2003)
18. Matoušková, E.: Almost isometries of balls. J. Funct. Anal. **190**, 507–525 (2002)
19. Omladič, M., Šemrl, P.: On non linear perturbations of isometries. Math. Ann. **303**, 617–628 (1995)
20. Protasov, V.Y.: On stability of isometries in Banach spaces. In: Rassias, T.M., Brzdęk, J. (eds.) Functional Equations in Mathematical Analysis, pp. 273–285. Springer, New York (2012)
21. Rassias, T.M.: Isometries and approximate isometries. Int. J. Math. Math. Sci. **25**, 73–91 (2001)
22. Šemrl, P., Väisälä, J.: Nonsurjective nearisometries of Banach spaces. J. Funct. Anal. **198**, 269–278 (2003)
23. Trotsenko, D.A., Väisälä, J.: Upper sets and quasisymmetric maps. Ann. Acad. Sci. Fenn. Math. **24**, 465–488 (1999)
24. Tukia, P., Väisälä, J.: Quasisymmetric embeddings of metric spaces. Ann. Acad. Sci. Fenn. Math. **5**, 97–114 (1980)
25. Väisälä, J.: A survey of nearisometries. Report. Univ. Jyväskylä **83**, 305–315 (2001). Electronic version with an addendum (2002). http://arxiv.org/abs/math/0201098.Cited20Sep2016
26. Vestfrid, I.A.: ϵ-isometries in Euclidean spaces. Nonlinear Anal. **63**, 1191–1198 (2005)
27. Vestfrid, I.A.: Stability of almost surjective ε-isometries of Banach spaces. J. Funct. Anal. **269**, 2165–2170 (2015)
28. Zhou, Y., Zhang, Z., Liu, Ch.: On linear isometries and ε-isometries between Banach spaces. J. Math. Anal. Appl. **435**, 754–764 (2016)

Chapter 3
On the Indicator Plurality Function

Anna Bahyrycz

Abstract This survey paper is dedicated to certain mathematization method of social choice, given by Roberts, and its generalizations.

Keywords Social choice • Plurality function • Indicator plurality function • Consistency condition • Cone • Additive function

Mathematics Subject Classification (2010) Primary 39B22; Secondary 39B52, 39B72, 39B82

3.1 Introduction

There is a long history in the theory of social choice of finding axioms that characterize a particular group consensus function and of finding all group consensus functions that satisfy certain axioms. Much of the early history of this theory has been concerned with impossible theorems, which show that under certain reasonable axioms there is no social choice function that merges individual judgements into a consensus judgement (see [8]). From 1974 there have been a variety of positive results. The first concerned an axiomatization of Borda's rule [21], next outcomes, among others, of social choice scoring functions [22], of the plurality rule [15], and of the plurality function [16].

Much of the literature of social choice functions falls into the following setting. Let A be a set of alternatives, for instance, alternative strategies, alternative new technologies, alternative diagnoses, or alternative candidates and let B be a set of individuals (voters or experts), who are expressing opinions about the alternatives in the set A. A social choice function is a function which, based on the opinions

---

A. Bahyrycz (✉)
Faculty of Applied Mathematics, AGH University of Science and Technology,
Mickiewicza 30, 30-059 Kraków, Poland
e-mail: bahyrycz@agh.edu.pl

J. Brzdęk et al. (eds.), *Developments in Functional Equations and Related Topics*, Springer Optimization and Its Applications 124,
DOI 10.1007/978-3-319-61732-9_3

of the individuals in B, picks a "consensus." In many contexts, the opinions of the individuals in B are given as rankings or linear orders of the elements of A. In other contexts, the opinions are simply a choice of the best or most preferred alternative in A. The "consensus" can be either a single element of A, a subset of alternatives in A, a ranking of elements of A, or a set of such rankings. In the situation where the opinions are rankings and the consensus is a subset of A, a well-known social choice function is the plurality rule, which chooses for the consensus all those alternatives which receive the greatest number of first place rankings. In [15], Richelson was able to characterize the plurality rule by giving five simple axioms, which are based on some earlier axioms of Young (see [22]).

In the situation where each of individual gives only the first choice from A (the opinions are elements of the set A) and the consensus is a subset of A a social choice function is a consensus function. The plurality function is that consensus function which chooses as consensus all alternatives which receive the largest number of first choices. The axioms that characterize the plurality function were introduced by Roberts [16]. Mathematical theory of this approach was developed by Bahyrycz [2–7], Forti and Paganoni [9, 10], Moszner [7, 11–14], Roberts [17, 18], and Rosenbaum [20]. In this paper we present some of these results.

In the second section we present the definitions and characterizations of the plurality function and the indicator plurality function and we give their election interpretation. In the Section 3.3 we determine all functions which are consistent. In the next section we describe a way of construction of all m-elements consistent system related to the indicator plurality function. In the last section we consider the systems of equations with unknown multifunctions related to the indicator plurality function.

3.2 The Plurality Function and the Indicator Plurality Function

In this section we start by recalling the results of Roberts [16, 17], which are based on some earlier axioms of Richelson [15].

Suppose A is a set of alternatives and each voter provides us with a first choice from A. *The plurality function* is the function F from $\cup_{n=1}^{\infty} A^n$ into 2^A, where $F(x_1, \ldots, x_n)$ is the set of all those y in A so that no z in A appears more often in $(x_1, \ldots, x_n)$ than y.

To state the characterization of the plurality function, we introduce the following definitions.

Anonymity For all permutations π of $\{1, \ldots, n\}$

$$F(x_{\pi(1)}, \ldots, x_{\pi(n)}) = F(x_1, \ldots, x_n)$$

for all sequences of alternatives $(x_1, \ldots, x_n)$.

The election interpretation of this property is the following. The election result does not depend on the order of inserting the votes into the ballot box.

Neutrality For all permutations σ of A

$$F(\sigma(x_1), \ldots, \sigma(x_n)) = \sigma[F(x_1, \ldots, x_n)]$$

where $\sigma(X) = \{\sigma(x) : x \in X\}$.

The election result does not depend on the order of placing the candidates on the list.

Consistency If $F(x_1, \ldots, x_n) \cap F(y_1, \ldots, y_m) \neq \emptyset$, then

$$F(x_1, \ldots, x_n, y_1, \ldots, y_m) = F(x_1, \ldots, x_n) \cap F(y_1, \ldots, y_m). \tag{3.1}$$

It expresses the following. If $(x_1, \ldots, x_n)$ and $(y_1, \ldots, y_m)$ are two vectors representing the votes of two different groups of voters among the same set of candidates and some candidate is chosen by both groups, then a candidate x is chosen by the combined group if and only if this candidate x is chosen by both groups, separately. The combined group is represented by the vector $(x_1, \ldots, x_n, y_1, \ldots, y_m)$.

The assumption that at least one candidate won the election in both groups is important, otherwise the equality (3.1) could not take place, the set on the right side would be empty and the set on the left side could never be empty.

Faithfulness $F(x) = \{x\}$ for all $x \in A$.

If we have one voter and this voter gives his or her first choice on a candidate x, then this candidate is chosen.

In [16] was given a following characterization of the plurality function.

Theorem 3.1 *Suppose* $F : \cup_{n=1}^{\infty} A^n \to 2^A$ *and* $F(x) \neq \emptyset$ *for any* $x \in A$. *Then the following are equivalent:*

(1) F is the plurality function.
(2) F is anonymous, neutral, consistent, and faithful.

For more characterization of the plurality function see [16].

Now, suppose that A is a finite set $\{v_1, v_2, \ldots, v_m\}$ and F is a plurality function. We may rewrite any vector $(x_1, \ldots, x_n)$ from A^n, after possibly permuting the subscripts, in the form

$$\underbrace{v_1, \ldots, v_1}_{c_1}, \underbrace{v_2, \ldots, v_2}_{c_2}, \ldots, \underbrace{v_m, \ldots, v_m}_{c_m},$$

where v_i occurs c_i times. If v_j doesn't occur in the vector $(x_1, \ldots, x_n)$, then $c_j = 0$ and all of the c_i are non-negative integers and at least one of them is positive.

The vector $(x_1, \ldots, x_n)$ we can write in the following way: $(c_1 v_1, \ldots, c_m v_m)$. Since the function F is anonymous we have

$$F(x_1, \ldots, x_n) = F(c_1 v_1, \ldots, c_m v_m).$$

We can define a new function $f = (f_1, \ldots, f_m) : \mathbb{Z}(m) \to 0(m)$, where $\mathbb{Z}(m)$ is the set of all m-vectors of non-negative integer numbers, except the vector $\underline{0} := (0, \ldots, 0)$ and $0(m)$ is a subset of $\mathbb{Z}(m)$ in which each component is 0 or 1 and

$$f_k(c_1, \ldots, c_m) = 1 \Leftrightarrow v_k \in F(c_1 v_1, \ldots, c_m v_m) \quad \text{for } k \in \{1, \ldots, m\}.$$

We will think of f as the indicator function corresponding to the plurality function F.

From the above considerations follows that we may define the function f independently of the plurality function in the following way:

$$f_k(c_1, \ldots, c_m) = 1 \Leftrightarrow c_k \geq c_j \quad \text{for } j \in \{1, \ldots, m\}. \tag{3.2}$$

It expresses the following. If we vote for m candidates and $c_1, \ldots, c_m$ is a description of this vote (c_i is the number of votes which received the ith candidate on the list), then 1 in the kth position in $f(c_1, \ldots, c_m)$ means that the kth candidate on the list received at least as many votes as the other and he won the election maybe simultaneously with other candidates.

More generally, if we allow fractional votes or vote splitting, the domain of f would consist of the set of all m-vectors of non-negative rational numbers, except $\underline{0}$ $(\mathbb{Q}(m))$ or even the set of all m-vectors of non-negative real numbers, except $\underline{0}$ $(\mathbb{R}(m))$.

A function $f = (f_1, \ldots, f_m) : U \to 0(m)$, where $U \subset \mathbb{R}(m)$ is called *the indicator plurality function on U* if f satisfies (3.2) for all $(c_1, \ldots, c_m) \in U$.

From now on, we assume that $U \in \{\mathbb{Z}(m), \mathbb{Q}(m), \mathbb{R}(m)\}$.

The indicator plurality function on U has analogous properties to those defined above for the plurality function F. The anonymity was used to define the indicator plurality function and other properties have the following form.

Neutrality For all $(c_1, \ldots, c_m) \in U$ and all permutations π of $\{1, \ldots, m\}$

$$f_k(c_{\pi(1)}, \ldots, c_{\pi(m)}) = f_{\pi(k)}(c_1, \ldots, c_m) \qquad \text{for } k \in \{1, \ldots, m\}.$$

Consistency For all $c, d \in U$

$$f(c) \cdot f(d) \neq \underline{0} \Rightarrow f(c + d) = f(c) \cdot f(d), \tag{3.3}$$

where $x + y := (x_1 + y_1, \ldots, x_m + y_m)$ and $x \cdot y := (x_1 \cdot y_1, \ldots, x_m \cdot y_m)$ for $x = (x_1, \ldots, x_m), y = (y_1, \ldots, y_m) \in U$.

Faithfulness For all $i \in \{1, \ldots, m\}$

$$f(e_i) = e_i$$

where e_i denotes the vector $(0, \ldots, 0, 1, 0, \ldots, 0)$ with a 1 in the ith position.

We have the following characterization of the indicator plurality function on $\mathbb{Z}(m)$ and $\mathbb{Q}(m)$ (see [17]).

Theorem 3.2 *Suppose that* $V \in \{\mathbb{Z}(m), \mathbb{Q}(m)\}$ *and* $f : V \to 0(m)$*. Then the following are equivalent:*

(i) f *is the indicator plurality function on* V*.*
(ii) f *is neutral, consistent, and faithful.*

The original indicator plurality function is homogeneous, because the election result does not depend on the number of votes, but on the proportion of the votes for individual candidates. One can consider the following properties:

Weak Homogeneity For all $c \in U$

$$f(2c) = f(c).$$

Homogeneity For all positive real number r, all $c \in \mathbb{R}(m)$ and $U = \mathbb{R}(m)$

$$f(rc) = f(c),$$

where $rc := (rc_1, \ldots, rc_m)$ for $c = (c_1, \ldots, c_m) \in \mathbb{R}(m)$.

Homogeneity Faithful For all positive real number r, all $j \in \{1, \ldots, m\}$ and $U = \mathbb{R}(m)$

$$f(re_j) = f(e_j).$$

In the case, when we weaken the assumption in Theorem 3.2 that the range of f is contained in $O(m)$ we have the following (see [17]):

Theorem 3.3 *Suppose that* $V = \mathbb{Z}(m)$ *or* $V = \mathbb{Q}(m)$ *and* $f : V \to \mathbb{R}(m)$*. Then the following are equivalent:*

(i) f *is the indicator plurality function on* V*.*
(ii) f *is neutral, consistent, faithful, and weakly homogeneous.*

If we consider the function $f : \mathbb{R}(m) \to 0(m)$, then there exist the functions which are neutral, consistent, and faithful but different from the indicator plurality function which shows the following example ([20], see also [11]).

Let b_0 be a positive irrational, b_1 be a non-zero rational, and H be a Hamel base of the space $\mathbb{R}$ over a field $\mathbb{Q}$ such that $b_0, b_1 \in H$. Every $x \in \mathbb{R}$ has a representation, unique up to terms with coefficients zero

$$x = \sum_{l=0}^{n} q_l b_l,$$

where $q_l \in \mathbb{Q}$, $b_l \in H$ for $l \in \{0, \ldots, n\}$. We put $\gamma := q_0$ and $\delta := \sum_{l=1}^{n} q_l b_l$, then $x = \gamma b_0 + \delta$. We define a function $g = (g_1, \ldots, g_m) : \mathbb{R}^m \to 0(m)$ by

$$g_j(x_1, \ldots, x_m) = \begin{cases} 1 & \text{if } x_j \geq x_k \text{ for } k = 1, \ldots, m, \\ 0 & \text{otherwise} . \end{cases}$$

Now, we define a function $f : \mathbb{R}(m) \to 0(m)$ in the following way:

$$f(x_1, \ldots, x_m) = g(\delta_1 - \gamma_1, \ldots, \delta_m - \gamma_m),$$

where $x_j := \gamma_j b_0 + \delta_j$ for $j \in \{1, \ldots, m\}$.

The function f is neutral, consistent, and faithful but different from the indicator plurality function because

$$f(b_0, 0, \ldots, 0) = g(-1, 0, \ldots, 0) = (0, 1, \ldots, 1).$$

On the other hand, we have the following theorems (see [17]):

Theorem 3.4 *Let $f : \mathbb{R}(m) \to 0(m)$ be an arbitrary function. Then the following are equivalent:*

(i) f is the indicator plurality function on $\mathbb{R}(m)$.
(ii) f is neutral, consistent, and homogeneous faithful.

Theorem 3.5 *Let $f : \mathbb{R}(m) \to \mathbb{R}(m)$ be an arbitrary function. Then the following are equivalent:*

(i) f is the indicator plurality function on $\mathbb{R}(m)$.
(ii) f is neutral, consistent, faithful, and homogeneous.

If we consider the function $f : \mathbb{R}(m) \to \mathbb{R}(m)$, then there exist the functions which are neutral, consistent, and homogeneous faithful but different from the indicator plurality function on $\mathbb{R}(m)$, which shows the following example (see [2, 12]).

We define a function $f = (f_1, \ldots, f_m) : \mathbb{R}(m) \to \mathbb{R}(m)$ as follows:

$$f_i(x) = \begin{cases} \exp(x_1 + \cdots + x_m - x_i) & \text{for } x \in Z_i, \\ 0 & \text{for } x \in \mathbb{R}(m) \setminus Z_i, \end{cases}$$

where $Z_i := \{x \in \mathbb{R}(m) : x_i \geq x_j \text{ for } j \in \{1, \ldots, m\}\}$ and $i \in \{1, \ldots, m\}$.

The function f is neutral, consistent, and homogeneity faithful but f is not the indicator plurality function on $\mathbb{R}(m)$ because the range of f is not contained in $0(m)$, for example,

$$f(1, 1, 0 \ldots, 0) = (e, e, 0 \ldots, 0).$$

In the original problem of the social choice the function $f(x_1, \ldots, x_m)$ is defined on the set $\mathbb{Z}(m)$ and $x_1 + \cdots + x_m$ is the sum of the votes cast. In practice this sum is limited, for example, by the number $c > 0$. This begs the idea of replacing the property consistency by the condition

$$x_1+\cdots+x_m+y_1+\cdots+y_m \leq c \ \wedge \ f(x)f(y) \neq \underline{0} \Rightarrow f(x+y)=f(x)f(y) \qquad (3.4)$$

and a problem arises if each function f satisfying Equation (3.4) can be uniquely extended to the solution of Equation (3.3). We have the following (see [12, 13]):

Theorem 3.6 *Every function* $f : E := \{(x_1, \ldots, x_m) \in \mathbb{R}(m) : x_1 + \cdots + x_m \leq c\} \to \mathbb{R}(m)$ *which is the solution of Equation* (3.4) *can be uniquely extended to the solution of Equation* (3.3).

Every solution of Equation (3.4) *can be obtained by restricting the solution of Equation* (3.3) *to the set E.*

This result shows that the generalization of the considerations about the election of the case of natural numbers to the case of real numbers is not good for the description of the election, because the outcome of the election on a small population determines the result for the whole population. Note that this anomaly does not take place if the real numbers replace integers, because in this case an analogue of Theorem 3.6 is not true. Indeed, we consider the function f defined as follows (see [13]):

$$f(1,0) = f(2,0) = f(1,1) = (1,0) \quad \text{and} \quad f(0,1) = f(0,2) = (0,1).$$

This function f satisfies the condition (3.4) with $c = 2$ and $m = 2$ and can be extended onto $\mathbb{Z}(2)$ at least two different ways

$$f_1(x_1,x_2) = \begin{cases} (1,0) & \text{for } x_1 \geq x_2, \\ (0,1) & \text{for } x_1 < x_2 \end{cases} \qquad (3.5)$$

and

$$f_2(x_1,x_2) = \begin{cases} (1,0) & \text{for } x_1 \neq 0, \\ (0,1) & \text{for } x_1 = 0. \end{cases}$$

These and other characterizations of the indicator plurality function one may be found in [2, 11–13, 17, 18].

3.3 On the Functions Which Are Consistent

In [19] Roberts stated that it is of interest in the theory of social choice to determine all functions $f : \mathbb{R}(m) \to \mathbb{R}(m)$ which are consistent, i.e., satisfy the conditional functional equation

$$f(x) \cdot f(y) \neq \underline{0} \Rightarrow f(x+y) = f(x) \cdot f(y).$$

As a generalization, one may consider functions $f : \mathbb{R}(n) \to \mathbb{R}(m)$ (where n, m are arbitrary natural numbers, independent of each other) satisfying the condition

$$\forall_{x,y\in\mathbb{R}(n)}:\ f(x)\cdot f(y)\neq \underline{0} \Rightarrow f(x+y)=f(x)\cdot f(y). \tag{3.6}$$

It may be shown that in such a case the description of all the solutions $f = (f_1, \ldots, f_m)$ of Equation (3.6) takes the following form (see [4] and [11] for $n = m$):

$$f_\nu(x) = \begin{cases} \exp a_\nu(x) & \text{for } x \in Z_\nu, \\ 0 & \text{for } x \in \mathbb{R}(n) \setminus Z_\nu, \end{cases} \tag{3.7}$$

where $a_\nu : \mathbb{R}^n \to \mathbb{R}$ are additive functions for $\nu = 1, \ldots, m$, whereas the sets Z_ν satisfy the conditions

$$Z_1 \cup \cdots \cup Z_m = \mathbb{R}(n), \tag{3.8}$$

$$ij \neq 0_m \Rightarrow Z_1^{i_1} \cap \cdots \cap Z_m^{i_m} + Z_1^{j_1} \cap \cdots \cap Z_m^{j_m} \subset Z_1^{i_1 j_1} \cap \cdots \cap Z_m^{i_m j_m}, \tag{3.9}$$

for every $i = (i_1, \ldots, i_m),\ j = (j_1, \ldots, j_m) \in 0(m),\ E_1 + E_2 := \{x + y :\ x \in E_1,\ y \in E_2\}$ for $E_1, E_2 \subset \mathbb{R}^n,\ E^1 := E,\ E^0 := \mathbb{R}(n) \setminus E$ for $E \subset \mathbb{R}(n)$.

Let us observe that if sets $Z_1, \ldots, Z_m$ satisfy condition (3.8), then for every $a \in \mathbb{R}(n)$ and every $k \in \{1, \ldots, m\}$ there exists a unique $i_k \in \{0, 1\}$ such that $a \in Z_k^{i_k}$.

The parameters determining the solutions of Equation (3.6) are systems of sets $Z_1, \ldots, Z_m$ satisfying conditions (3.8) and (3.9), as well as additive functions $a_\nu : \mathbb{R}^n \to \mathbb{R}$. Additionally condition (3.9) has a complicated form. For this reason, it is interesting to find conditions equivalent to condition (3.9) under the assumption of condition (3.8) which are of simpler form than the ones obtained from (3.9). We have the following theorem (see [4]).

Theorem 3.7 *Assume that sets $Z_1, \ldots, Z_m$ satisfy condition* (3.8). *The following conditions are equivalent:*

(i) condition (3.9)*;*

(ii) the sets $Z_1, \ldots, Z_m$ are cones over $\mathbb{Q}$ for which

$$Z_l^1 + Z_l^1 \cap Z_k^0 \subset Z_l^1 \cap Z_k^0 \tag{3.10}$$

for all $k, l \in \{1, \ldots, m\}$ such that $k \neq l$;

(iii) the sets $Z_1, \ldots, Z_m$ satisfy the conditions

$$Z_k^1 + Z_k^1 \subset Z_k^1 \tag{3.11}$$

for every $k \in \{1, \ldots, m\}$ and condition (3.10) *for all $k, l \in \{1, \ldots, m\}$ such that $k \neq l$;*

(iv) *for all* $x, y \in \mathbb{R}(n)$ *if there exists* $\nu \in \{1, \ldots, m\}$ *such that* $x \in Z_\nu$ *and* $y \in Z_\nu$, *then*

$$\forall k \in \{1, \ldots, m\} : \quad x + y \in Z_k \Leftrightarrow x \in Z_k \text{ and } y \in Z_k;$$

(v) *for all* $k, l \in \{1, \ldots, m\}$ *the following implication holds:*

$$ij \neq \underline{0} \Rightarrow Z_k^{i_k} \cap Z_l^{i_l} + Z_k^{j_k} \cap Z_l^{j_l} \subset Z_k^{i_k j_k} \cap Z_l^{i_l j_l},$$

where $i = (i_k, i_l)$, $j = (j_k, j_l) \in 0(2)$.

We observe (see [12]) that the function f satisfying the condition (3.6) is continous if and only if the sets $Z_1, \ldots, Z_m$ fulfilling the condition (3.8) are such that $Z_\nu = \emptyset$ or $Z_\nu = \mathbb{R}(n)$ and the additive functions a_ν for $\nu \in \{1, \ldots, m\}$ are continous.

We notice also that the function f satisfying the condition (3.6) can be measurable without being continuous. For example, for $m = n = 2$ it is enough to consider the function f_1 given by the formula (3.5).

Let us make the following definitions.

Definition 3.1 Let $C \subset \mathbb{R}(n)$ be a cone over $\mathbb{Q}$ ($x + y \in C$ and $qx \in C$ for all $x, y \in C$, $q \in \mathbb{Q}_+$). Denote:

$< C >$ – the linear subspace of $\mathbb{R}^n$ over the field $\mathbb{R}$ generated by C;
C^* – the interior of the set C in $< C >$.

Definition 3.2 For every subset $\{l_1, \ldots, l_k\} \subset \{1, \ldots, n\}$ we define the set

$$B_{l_1,\ldots,l_k} := \{(x_1, \ldots, x_n) \in \mathbb{R}(n) : x_{l_1} = \cdots = x_{l_k} = 0\},$$

and then we define the set

$$\mathbb{B} := \{B_{l_1,\ldots,l_k} : \{l_1, \ldots, l_k\} \subset \{1, \ldots, n\}\}.$$

Theorem 3.7 leads to the following.

Corollary 3.1 *If the sets* $Z_1, \ldots, Z_m$ *are pairwise disjoint and satisfy condition* (3.8), *then condition* (3.9) *is equivalent to the following condition:* $Z_1, \ldots, Z_m$ *are cones over* $\mathbb{Q}$.

Corollary 3.2 *If a system of sets* $Z_1, \ldots, Z_m$ *satisfies conditions* (3.8) *and* (3.9), *then for every non-empty subset* $\{l_1, \ldots, l_p\}$ *of the set* $\{1, \ldots, m\}$ *and for every* $(i_{l_1}, \ldots, i_{l_p}) \in 0(p)$ *the set* $Z_{l_1}^{i_{l_1}} \cap \cdots \cap Z_{l_p}^{i_{l_p}}$ *is a cone over* $\mathbb{Q}$.
Let us observe that if the sets $Z_1, \ldots, Z_m$ satisfy conditions (3.8) and (3.9), then for $m \in \{1, 2\}$ the set Z_1^0 in the case of $m = 1$ and the sets Z_1^0, Z_2^0, $Z_1^0 \cap Z_2^0$ for $m = 2$

are also cones over $\mathbb{Q}$. If $m = 1$, then $Z_1 = \mathbb{R}(n)$, so $Z_1^0 = \emptyset$. If $m = 2$, then $Z_1^0 = Z_1^0 \cap Z_2^1$ and $Z_2^0 = Z_2^0 \cap Z_1^1$ are cones over $\mathbb{Q}$ and the set $Z_1^0 \cap Z_2^0$ is empty.

If the sets $Z_1, \ldots, Z_m$ satisfy conditions (3.8) and (3.9), then for $m > 2$ not every set $Z_{l_1}^0 \cap \cdots \cap Z_{l_p}^0$, where $\emptyset \neq \{l_1, \ldots, l_p\} \subset \{1, \ldots, m\}$, is necessarily a cone over $\mathbb{Q}$. Here is a suitable example for $n = 2$ and $m = 3$. Define

$$\begin{aligned} Z_1 &:= \{(x, y) \in \mathbb{R}(2) : y \leq \tfrac{1}{2}x\}, \\ Z_2 &:= \{(x, y) \in \mathbb{R}(2) : \tfrac{1}{2}x < y \leq 2x\}, \\ Z_3 &:= \{(x, y) \in \mathbb{R}(2) : y > 2x\}. \end{aligned}$$

The sets Z_1, Z_2, Z_3 satisfy conditions (3.8) and (3.9) but Z_2^0 is not a cone over $\mathbb{Q}$.

Corollary 3.3 *Let the sets $Z_1, \ldots, Z_m$ satisfy the conditions* (3.8) *and* (3.9). *If there exist $k, l \in \{1, \ldots, m\}$ such that $k \neq l$ and $(Z_k \cap Z_l)^* \neq \emptyset$, then*

$$Z_k \cap < Z_k \cap Z_l > = Z_l \cap < Z_k \cap Z_l >.$$

Corollary 3.4 *If the sets $Z_1, \ldots, Z_m$ satisfy the conditions* (3.8) *and* (3.9)*, then for all $k, l \in \{1, \ldots, m\}$ $Z_k = Z_l$ or $Z_k \cap Z_l$ is a set with empty interior in $\mathbb{R}^n$.*

Corollary 3.5 *If a system $Z_1, \ldots, Z_m$ satisfies the conditions* (3.8) *and* (3.9) *and if there exists such $k \in \{1, \ldots, m\}$ that $Z_k = \mathbb{R}(n)$, then $Z_i \in \mathbb{B}$ for every $i \in \{1, \ldots, m\}$.*

From the above Corollary we obtain, for example, that if the sets Z_1, Z_2 satisfy the conditions (3.8) and (3.9) with $n = m = 2$ and $Z_1 = \mathbb{R}(2)$, then Z_2 must be equal to one of the sets $B_\emptyset = \mathbb{R}(2), B_1, B_2, B_{1,2} = \emptyset$.

It may be proved (see [12] for $n = m$) that every function $f : \mathbb{R}(n) \to \mathbb{R}(m)$ satisfying (3.6) and the condition

$$\exists_{r>0} : [r \neq 1 \wedge \forall_{x \in \mathbb{R}(n)} : \ f(rx) = f(x)] \tag{3.12}$$

with some r being an algebraic number must have values in the set $0(m)$. It is known (see [3] for $n = m$) that this property holds also with a transcendental number r if $m \leq 2$ and in the case when $m > 2$ there exists a solution of Equation (3.6) satisfying (3.12) with some transcedental number r which range is not contained in $0(m)$. In a very long construction of such function the Axiom of Choice is used. Moreover in [7] was shown that one cannot give this construction without using non-measurable set.

Theorem 3.8 *If a function $f : \mathbb{R}(n) \to \mathbb{R}(m)$ fulfils the conditions* (3.6), (3.12) *and additionally for every $x \in \mathbb{R}(n)$ the set*

$$M_i(x) = \{tx \in \mathbb{R}(n) : f_i(tc) \neq 0\} \quad \textit{for} \ \ i \in \{1, \ldots, m\}$$

are Lebesgue linearly measurable, then f must have its values in the set $0(m)$.

From the description of the solution of Equation (3.6) follows that the function satisfying the conditions (3.6) and (3.12) has values in the set $0(m)$ if and only if all additive functions a_ν are identically equal to zero. The condition (3.12) imposes on the functions a_ν and the sets Z_ν ($\nu = 1, \ldots, m$) the conditions

$$rZ_\nu = Z_\nu \tag{3.13}$$

and

$$a_\nu(rx) = a_\nu(x) \quad \text{for } x \in Z_\nu, \tag{3.14}$$

and we have the following

Theorem 3.9 *The function* $f : \mathbb{R}(n) \to \mathbb{R}(m)$ *satisfying the conditions* (3.6) *and* (3.12) *has values only in the set* $0(m)$ *if and only if the sets* Z_ν *fulfilling the conditions* (3.8), (3.9), *and* (3.13) *satisfy the condition*

$$Z_\nu \subset (r-1)lin_{\mathbb{Q}} Z_\nu \quad for\ \nu = 1, \ldots, m \tag{3.15}$$

with r occuring in (3.12).
The above Theorem was proved in [7] for $n = m$, but from Lemma 1 from the same paper we can obtain this fact for the arbitrary $n, m \in \mathbb{N}$.

We notice that for $m = 1$ we have $Z_1 = \mathbb{R}(n)$ and the condition (3.15) is obviously fulfilled. This condition is also satisfied for $m = 2$, because then the sets Z_1^0 and Z_2^0 are cones over $\mathbb{Q}$. In the paper [3] such a cone is constructed for which the condition (3.15) is not satisfied.

In [9, 10] was given a description of the construction of the solutions of a system of functional equations: (3.6) (with $n = m$) and equation

$$\forall_{r>0} \forall_{x \in \mathbb{R}(m)} : \ f(rx) = f(x). \tag{3.16}$$

To each function $f : \mathbb{R}(m) \to 0(m)$ a partition of $\mathbb{R}(m)$ is associated, given by the family of the non-empty level sets of f, i.e., the family $\{A_i,\ i \in \mathscr{I} \subset 0(m)\}$ where $A_i = \{x \in \mathbb{R}(m) : f(x) = i\}$ and $i \in \mathscr{I}$ if and only if $A_i \neq \emptyset$. The following theorem characterizes the solutions of the system of the functional equations (3.6) and (3.16) through the properties of the corresponding families of the non-empty level sets.

Theorem 3.10 *Let* $\{A_i,\ i \in \mathscr{I} \subset 0(m)\}$ *be the family of the non-empty level sets of a function* $f : \mathbb{R}(m) \to 0(m)$. *Then f is a solution of Equation* (3.6) *satisfying* (3.16) *if and only if*

(i) A_i *is a cone over* $\mathbb{R}$ *for all* $i \in \mathscr{I}$;
(ii) $ij \neq \underline{0} \Rightarrow A_i + A_j \subset A_{ij}$ *for all* $i, j \in \mathscr{I}$.

In [9] were described explicitly all solutions of that system in the case of dimension less or equal to three. The authors wrote that for higher dimension the task of giving an analogous description seemed hopeless. We present these results only for $m \in$

$\{1,2\}$ because the construction of all solutions of that system for $m = 3$ in [9] occupies more than 12 pages.

For $m = 1$ the only solution is given by $f(x) = 1$ for $x \in \mathbb{R}(1)$.

In the case $m = 2$ we have the following possibilities:

(a) $f(x) = i$ for $x \in \mathbb{R}(2)$,

where $i \in \{(1,0), (0,1), (1,1)\}$;

(b) $$f(x) = \begin{cases} i & \text{for } x \in \mathbb{R}(2) \setminus U, \\ (1,1) \text{ or } (1,1) - i & \text{for } x \in U, \end{cases}$$

where U is one of the semiaxes of $\mathbb{R}(2)$ and $i \in \{(1,0), (0,1)\}$;

(c) $$f(x) = \begin{cases} i & \text{for } x \in Z, \\ (1,1) - i & \text{for } x \in Z', \\ i \text{ or } (1,1) - i \text{ or } (1,1) & \text{for } x \in L, \end{cases}$$

where L is a half-line in $\mathbb{R}(2)$ from the origin, Z, Z' are two non-empty and disjoint cones over $\mathbb{R}$ whose union is $\mathbb{R}(2) \setminus L$ and $i \in \{(1,0), (0,1)\}$.

In [10] the above problem was studied in a completely different way: first the authors have proved some lemmas of geometric-combinatorial type which highlight some properties that were the guidelines for developing the procedure for the construction of the solutions, then they have described an operative procedure to construct all solutions.

3.4 Construction of All m-Elements Consistent System

In this section motivated by problem of Aczel [1] and Roberts [19] we provide a way of construction of all families of the sets $Z_1, \ldots, Z_m$ satisfying the conditions (3.8) and (3.9) from the paper [6].

We start from the following:

Definition 3.3 A system of sets $(Z_1, \ldots, Z_m)$ is called an m-elements consistent system if it satisfies the conditions (3.8) and (3.9).

Definition 3.4 We call a system of sets $Z_1, \ldots, Z_p \subset \mathbb{R}(n)$ can be extended to an m-elements consistent system ($p < m$), if there exists a system of sets $Z_{p+1}, \ldots, Z_m \subset \mathbb{R}(n)$ such that $(Z_1, \ldots, Z_m)$ is m-elements consistent system.
From now on, we assume that $p < m$. We have the following:

Theorem 3.11 *A system of sets $Z_1, \ldots, Z_p \subset \mathbb{R}(n)$ can be extended to an m-elements consistent system if and only if the sets $Z_1, \ldots, Z_p$ are cones over $\mathbb{Q}$ satisfying the condition* (3.10) *for every $k, l \in \{1, \ldots, p\}$ such that $k \neq l$ and the set $\mathbb{R}(n) \setminus$*

$\bigcup_{i=1}^{p} Z_i$ *has a representation as a sum of system* $A_{p+1}, \ldots, A_m$, *which elements are pairwise disjoint cones over* $\mathbb{Q}$. *Then the system* $(Z_1, \ldots, Z_p, A_{p+1}, \ldots, A_m)$ *is* m*-elements consistent system extending the system* $Z_1, \ldots, Z_p$.
Now we describe a way of construction of the set of all m-elements consistent systems extending the system $Z_1, \ldots, Z_p$.

Let a system $Z_1, \ldots, Z_p \subset \mathbb{R}(n)$ be such that it can be extended to an m-elements consistent system. We denote

$$U_p := \{C_p = (Z_1, \ldots, Z_p, A_{p+1}, \ldots, A_m) : A_{p+1}, \ldots, A_m \text{ are pairwise disjoint cones over } \mathbb{Q} \text{ such that } A_{p+1} \cup \cdots \cup A_m = \mathbb{R}(n) \setminus \bigcup_{i=1}^{p} Z_i\}.$$

For every $C_p = (Z_1, \ldots, Z_p, A_{p+1}, \ldots, A_m) \in U_p$ we construct corresponding sets

$$Z_{p+1}^{C_p} := \{Z_{p+1} : \ Z_{p+1} \text{ is a cone over } \mathbb{Q} \text{ satisfying the following conditions: } A_{p+1} \subset Z_{p+1} \subset A_{p+1} \cup Z_1 \cup \ldots \cup Z_p, \ Z_k^1 + Z_k^1 \cap Z_{p+1}^0 \subset Z_k^1 \cap Z_{p+1}^0 \text{ and } Z_{p+1}^1 + Z_{p+1}^1 \cap Z_k^0 \subset Z_{p+1}^1 \cap Z_k^0 \text{ for every } k \in \{1, \ldots, p\}\},$$

$$U_{p+1}^{C_p} := \{(Z_1, \ldots, Z_p, Z_{p+1}, A_{p+2}, \ldots, A_m) : Z_{p+1} \in Z_{p+1}^{C_p}\},$$

$$U_{p+1} := \bigcup_{C_p \in U_p} U_{p+1}^{C_p}.$$

Each element of the set U_{p+1} is an m-elements consistent system and if $m-p > 1$, then satisfies the condition $A_{p+2} \cup \cdots \cup A_m = \mathbb{R}(n) \setminus \bigcup_{i=1}^{p+1} Z_i$. Proceeding in the analogous way, after $(m-p)$ steps, we can construct the set U_m, which each element is an m-elements consistent system extending the system $Z_1, \ldots, Z_p$.

From the above consideration we have the following:

Theorem 3.12 *Constructing in the above way the set* U_m *is the set of all* m*-elements consistent systems extending the system* $Z_1, \ldots, Z_p$.
Let us make the following definition:

Definition 3.5 A system $(A_1, \ldots, A_m)$ is called an m-elements basis system of $\mathbb{R}(n)$ if the sets $A_1, \ldots, A_m$ are pairwise disjoint cones over $\mathbb{Q}$, such that $A_1 \cup \cdots \cup A_m = \mathbb{R}(n)$.
We denote

$$\alpha := \{B = (A_1, \ldots, A_m) : A_1, \ldots, A_m \text{ is an } m\text{-elements basis system of } \mathbb{R}(n)\}.$$

We define an equivalence relation $\sim$ on the set α in the following way:

$$\forall_{B=(A_1,\ldots,A_m), B^*=(A_1^*,\ldots,A_m^*) \in \alpha} (B \sim B^* \Leftrightarrow A_1 = A_1^*).$$

Each equivalence class $[B]_\sim$, where $B = (A_1, \ldots, A_m) \in \alpha$ is the set of those elements belonging to the set α which first element equals A_1. By β we denote the set of all equivalence classes given an equivalence relation $\sim$ on α.

We notice that the equivalence class $[B]_\sim \in \beta$, where $B = (A_1, \dots, A_m)$, is equal to the set U_1 constructing according to above description for one-element set A_1, i.e., the set of all m-elements consistent systems extending the system A_1, such that $m-1$ remaining cones are pairwise disjoint and their sum is equal to $\mathbb{R}(n) \setminus A_1$. By Theorem 3.12, for each equivalence class $[B = (A_1, \dots, A_m)]_\sim \in \beta$ we can construct corresponding set $U_m([B]_\sim)$, which is the set of all m-elements consistent systems extending the system A_1. Hence we derive the following result:

Theorem 3.13 *A set*

$$U = \bigcup_{[B]_\sim \in \beta} U_m([B]_\sim)$$

is the set of all m-elements consistent systems of $\mathbb{R}(n)$.

The way of construction of all elements of the set α is unknown.

Denote by $\alpha^{\mathbb{R}} := \{(A_1, \dots, A_m) \in \alpha$ such that $A_1, \dots, A_m$ are cones over $\mathbb{R}\}$.

Theorem 3.14 *The set*

$$U^{\mathbb{R}} := \bigcup_{[B]_\sim \in \beta^{\mathbb{R}}} U_m([B]_\sim)$$

is the set of all m-elements consistent systems such that all their elements are cones over $\mathbb{R}$, *where* $\beta^{\mathbb{R}}$ *is the set of all equivalence classes given the equivalence relation* $\sim$ *on the set* $\alpha^{\mathbb{R}}$.

In [9, 10] was described the completely different way of construction of all m-elements consistent systems such that their elements are cones over $\mathbb{R}$ and $n = m$. Theorem 3.14 may be treated as a generalization of this construction for the case of n, m being arbitrarily chosen natural numbers, independent of each other.

3.5 On the Multifunctions Related to the Plurality Function

As a generalization, in [14] were considered the multifunctions $Z : T \to 2^G$, where T is an arbitrary non-empty set, $(G, +)$ is an arbitrary groupoid. The conditions (3.8) and (3.9) were replaced by

$$\bigcup_{t \in T} Z(t) = G, \tag{3.17}$$

and

$$(\exists_{t \in T} : i(t)j(t) \neq 0) \Rightarrow \bigcap_{t \in T} Z(t)^{i(t)} + \bigcap_{t \in T} Z(t)^{j(t)} \subset \bigcap_{t \in T} Z(t)^{i(t)j(t)}, \tag{3.18}$$

respectively, where $Z(t)^1 := Z(t), Z(t)^0 := G \setminus Z(t)$, and $i(t), j(t) : T \to \{0, 1\}$ are arbitrary functions not identically equal to zero.

It is known that the multifunction $Z(t) : T \to 2^G$ fulfilling condition (3.17), satisfies condition (3.18) if and only if $Z(t)$ satisfies condition

$$Z(t_1)^{k_1} \cap Z(t_2)^{k_2} + Z(t_1)^{l_1} \cap Z(t_2)^{l_2} \subset Z(t_1)^{k_1 l_1} \cap Z(t_2)^{k_2 l_2},$$

for all $t_1, t_2 \in T$ and for all $k_1, k_2, l_1, l_2 \in \{0, 1\}$ such that $k_1 l_1 + k_2 l_2 \neq 0$.

Moreover, we have the following theorem (see [5]):

Theorem 3.15 *Let T be an arbitrary set with at least 2 elements and let the multifunction $Z(t) : T \to 2^{\mathbb{R}(n)}$ satisfy condition*

$$\bigcup_{t \in T} Z(t) = \mathbb{R}(n). \tag{3.19}$$

If the multifunction $Z(t)$ fulfils the system of conditional equations

$$(\exists_{t \in T} : i(t) j(t) \neq 0) \Rightarrow \bigcap_{t \in T} Z(t)^{i(t)} + \bigcap_{t \in T} Z(t)^{j(t)} = \bigcap_{t \in T} Z(t)^{i(t) j(t)} \tag{3.20}$$

for the arbitrary functions $i(t), j(t) : T \to \{0, 1\}$ not identically equal to zero, then $Z(t)$ satisfies the system of equations

$$Z(t_1)^{k_1} \cap Z(t_2)^{k_2} + Z(t_1)^{l_1} \cap Z(t_2)^{l_2} = Z(t_1)^{k_1 l_1} \cap Z(t_2)^{k_2 l_2}, \tag{3.21}$$

for all $t_1, t_2 \in T$ and all $k_1, k_2, l_1, l_2 \in \{0, 1\}$ such that $k_1 l_1 + k_2 l_2 \neq 0$.

The converse of Theorem 3.15 for the set T with at least 2 elements is not true, and here is an example for $T = \{1, 2, 3\}$.

Let H be a Hamel base of the space $\mathbb{R}^n$, such that
$h_0 = (\sqrt{2}, 0, \ldots, 0) \in \mathbb{R}(n)$,
$h_i = (0, \ldots, 0, \overset{(i)}{1}, 0, \ldots, 0) \in \mathbb{R}(n)$ for $i = 1, \ldots, n$
belong to H.

Every $x \in \mathbb{R}^n$ has a representation, unique up to terms with coefficients zero

$$x = \sum_{l=0}^{k} q_l h_l,$$

where $q_l \in \mathbb{Q}$ and $h_l \in H$ for $l \in \{0, \ldots, k\}$.

We define the multifunction $Z(t) : \{1, 2, 3\} \to 2^{\mathbb{R}(n)}$ in the following way:

$$Z(t) = \begin{cases} \{x \in \mathbb{R}(n) : q_o \geq 0\} & \text{for } t = 1, \\ \{x \in \mathbb{R}(n) : q_o = 0\} & \text{for } t = 2, \\ \{x \in \mathbb{R}(n) : q_o \leq 0\} & \text{for } t = 3. \end{cases}$$

It can be easily checked that the sets $Z(1)$, $Z(2)$, $Z(3)$ satisfy the conditions (3.19) and (3.21) for all $t_1, t_2 \in \{1, 2, 3\}$ and all $k_1, k_2, l_1, l_2 \in \{0, 1\}$ such that $k_1 l_1 + k_2 l_2 \neq 0$. The condition (3.20) is not satisfied because

$$\underset{\substack{\| \\ \emptyset}}{Z(1)^1 \cap Z(2)^1 \cap Z(3)^0} + \underset{\substack{\nparallel \\ \emptyset}}{Z(1)^1 \cap Z(2)^0 \cap Z(3)^0} \subsetneq \underset{\substack{\nparallel \\ \emptyset}}{Z(1)^1 \cap Z(2)^0 \cap Z(3)^0}.$$

Acknowledgements The author would like to thank the referee for his helpful suggestions for improving the present paper.

References

1. Aczél, J.: Problem 16. In: Report of the Twenty-Ninth International Symposium on Functional Equations. Aequationes Mathematicae, vol. 43, pp. 299 (1992)
2. Bahyrycz, A.: The characterization of the indicator plurality function. Rocznik Nauk.-Dydakt. WSP w Krakowie. Prace Matematyczne **XV**, 15–35 (1998)
3. Bahyrycz, A.: On the problem concerning the indicator plurality function. Opusc. Math. **21**, 11–30 (2001)
4. Bahyrycz, A.: On sets associated to conditional equation of exponential function. Ann. Acad. Pedagog. Crac. Stud. Math. **6**, 19–33 (2007)
5. Bahyrycz, A.: On systems of equations with unknown multifunctions related to the plurality function. Ann. Univ. Paedagog. Crac. Stud. Math. **9**, 133–142 (2010)
6. Bahyrycz, A.: Construction of systems of sets related to the plurality functions. J. Math. Anal. Appl. **388**, 39–47 (2012)
7. Bahyrycz, A., Moszner, Z.: On the indicator plurality function. Publ. Math. Debr. **61**, 469–478 (2002)
8. Fishburn, P.C.: The Theory of Social Choice. Princeton University Press, Princeton (1973)
9. Forti, G.L., Paganoni, L.: Description of the solution of a system of functional equations related to plurality functions: the low-dimensional cases. Results Math. **27**, 346–361 (1995)
10. Forti, G.L., Paganoni, L.: A system of functional equations related to plurality functions. A method for the construction of the solutions. Aequationes Math. **52**, 135–156 (1996)
11. Moszner, Z.: Sur les fonctions de pluralité. Aequationes Math. **47**, 175–190 (1994)
12. Moszner, Z.: Remarques sur la fonction de pluralité. Results Math. **50**, 387–394 (1995)
13. Moszner, Z.: La fonction d'indice et la fonction exponentielle. Ann. Acad. Pedagog. Crac. Stud. Math. **2**, 23–37 (2002)
14. Moszner, Z.: Sur la fonction de choix et la fonction d'indice. Ann. Acad. Pedagog. Crac. Stud. Math. **4**, 143–169 (2004)
15. Richelson, J.T.: A characterization result for the plurality rule. J. Econ. Theory **19**, 548–550 (1978)
16. Roberts, F.S.: Characterization of the plurality function. Math. Soc. Sci. **21**, 101–127 (1991)
17. Roberts, F.S.: On the indicator function of plurality function. Math. Soc. Sci. **22**, 163–174 (1991)
18. Roberts, F.S.: Characterizations of the plurality function. In: Report of the Twenty-Eighth International Symposium on Functional Equations. Aequationes Mathematicae, vol. 41, pp. 272–273, 277–278 (1991)
19. Roberts, F.S.: P-290. In: Problems and Solutions. Aequationes Math. **44**, 124 (1992)
20. Rosenbaum, Z.S.: P290S2. Aequationes Math. **46**, 317–319 (1993)
21. Young, H.P.: An axiomatization of Borda's rule. J. Econ. Theory **9**, 43–52 (1974)
22. Young, H.P.: Social choice scoring functions. SIAM J. Appl. Math. **28**, 824–838 (1975)

Chapter 4
The Translation Equation in the Ring of Formal Power Series Over $\mathbb{C}$ and Formal Functional Equations

Harald Fripertinger and Ludwig Reich

Abstract In this survey we describe the construction of one-parameter subgroups (iteration groups) of Γ, the group of all (with respect to substitution) invertible power series in one indeterminate x over $\mathbb{C}$. In other words, we describe all solutions of the translation equation in $\mathbb{C}[\![x]\!]$, the ring of formal power series in x with complex coefficients. For doing this the method of formal functional equations will be applied. The coefficient functions of solutions of the translation equation are polynomials in additive and generalized exponential functions. Replacing these functions by indeterminates we obtain formal functional equations. Applying formal differentiation operators to these formal translation equations we obtain three types of formal differential equations. They can be solved in order to get explicit representations of the coefficient functions. For solving the formal differential equations we apply Briot–Bouquet differential equations in a systematic way.

Keywords Translation equation • Formal functional equations • Formal partial differential equations • Aczél–Jabotinsky type equations • Briot–Bouquet equations • Formal iteration groups of type I • Formal iteration groups of type (II, k) • Ring of formal power series over $\mathbb{C}$ • Lie–Gröbner series

Mathematics Subject Classification (2010) Primary 39B12; Secondary 39B52, 13F25, 30D05

H. Fripertinger (✉) • L. Reich
Institut für Mathematik und wissenschaftliches Rechnen, NAWI-Graz, Karl-Franzens-Universität Graz, Heinrichstr. 36/4, A-8010 Graz, Austria
e-mail: harald.fripertinger@uni-graz.at; ludwig.reich@uni-graz.at

J. Brzdęk et al. (eds.), *Developments in Functional Equations and Related Topics*, Springer Optimization and Its Applications 124,
DOI 10.1007/978-3-319-61732-9_4

4.1 Introduction

As a motivation we mention the embedding problem from analytic mechanics [30] or geometric complex analysis [24].

4.1.1 *The Embedding Problem*

Consider a domain $U \subseteq \mathbb{C}^n$, $n \geq 1$, $\mathbf{0} = (0,\ldots,0) \in U$, and a biholomorphic function $\tilde{F}\colon U \to U$ so that $\tilde{F}(\mathbf{0}) = \mathbf{0}$. We try to find a family $(F_t)_{t\in\mathbb{C}}$ of biholomorphic functions $F_t\colon U \to U$ so that $F_t(\mathbf{0}) = \mathbf{0}$, $t \in \mathbb{C}$, and

$$F_1 = \tilde{F}$$

$$F_s \circ F_t = F_{s+t} \qquad s,t \in \mathbb{C}. \tag{T}$$

The mapping $\mathbb{C} \times U \ni (t,x) \mapsto F_t(x) \in U$ is supposed to be holomorphic. The family $(F_t)_{t\in\mathbb{C}}$ is called a *flow*, a *one-parameter group*, an *iteration group*, or an *embedding* of $\tilde{F}$. Formula (T) is called the *translation equation*. If we represent the mappings F_t by their Taylor expansions in x and if we neglect the convergence of these series, then we obtain a solution of (T) in the ring of formal power series.

4.1.2 *The Ring of Formal Power Series with Complex Coefficients*

Now we want to study (T) in $\mathbb{C}[\![x]\!]$, the ring of all formal power series $F(x) = c_0 + c_1x + \ldots$ in the indeterminate x over $\mathbb{C}$. For a detailed introduction to formal power series we refer the reader to [1] and [13]. Together with addition $+$ and multiplication $\cdot$ the set $\mathbb{C}[\![x]\!]$ forms a commutative ring. If $F \neq 0$, then the order of $F(x) = c_0 + c_1x + \ldots$ is defined as $\operatorname{ord}(F) = \min\{n \geq 0 \mid c_n \neq 0\}$. Moreover, $\operatorname{ord}(0) = \infty$. The *composition* $\circ$ of formal series is defined as follows: Let $F, G \in \mathbb{C}[\![x]\!]$, $\operatorname{ord}(G) \geq 1$, then $(F \circ G)(x)$ is $F(G(x)) = \sum_{n\geq 0} c_\nu [G(x)]^\nu$. (This converges in the order topology.) Consider

$$\Gamma = \{F \in \mathbb{C}[\![x]\!] \mid F(x) = c_1x + \ldots,\ c_1 \neq 0\} = \{F \in \mathbb{C}[\![x]\!] \mid \operatorname{ord}(F) = 1\}$$

and

$$\Gamma_1 = \{F \in \Gamma \mid c_1 = 1\}.$$

Then $(\Gamma, \circ)$ is the group of all invertible formal power series (with respect to $\circ$), and $(\Gamma_1, \circ)$ is a subgroup of $(\Gamma, \circ)$. It will be necessary to consider rings of formal power series in more than one variable, e.g., $\mathbb{C}[\![x,y]\!] = (\mathbb{C}[\![x]\!])[\![y]\!]$, $\mathbb{C}[\![x,y,z]\!]$, etc., and also rings of the form $(\mathbb{C}[y])[\![x]\!]$, where $\mathbb{C}[y]$ is the polynomial ring in y over $\mathbb{C}$, which are subrings of $\mathbb{C}[\![x,y]\!]$.

The derivation of $F \in \mathbb{C}[\![x]\!]$, $F(x) = \sum_{n\geq 0} c_n x^n$ is

$$F'(x) = \frac{dF}{dx}(x) = \sum_{n\geq 0}(n+1)c_{n+1}x^n.$$

In $\mathbb{C}[\![x,y]\!]$ or $\mathbb{C}[\![x,y,z]\!]$ we have derivations with respect to x, y, z. The chain rule is valid which means that for $F, G \in \mathbb{C}[\![x]\!]$, $\mathrm{ord}(G) \geq 1$, the derivation of $F \circ G \in \mathbb{C}[\![x]\!]$ is of the form $(F \circ G)'(x) = F'(G(x))G'(x)$. In rings of the form $\mathbb{C}[\![x,y]\!]$ or $\mathbb{C}[\![x,y,z]\!]$ the mixed chain rule holds true.

4.1.3 Iteration Groups

Iteration groups or *one-parameter groups* in $\mathbb{C}[\![x]\!]$ are families $(F_t)_{t\in\mathbb{C}}$, $F_t \in \Gamma$, $t \in \mathbb{C}$, satisfying (T). If we write $F_t(x)$ as

$$F(t,x) = \sum_{n\geq 1} c_n(t)x^n, \qquad t \in \mathbb{C},$$

then (T) is equivalent to

$$F(s+t,x) = F(s,F(t,x)), \qquad s,t \in \mathbb{C}.$$

Therefore $F_0(x) = x$ and $F_{-t}(x) = F_t^{-1}(x)$.

An iteration group in Γ can be seen as a homomorphism

$$\theta\colon (\mathbb{C},+) \to (\Gamma,\circ), \qquad \theta(t) = F_t.$$

Moreover, in [17–19] and [16], Jabłoński and Reich were studying homomorphisms $\theta\colon (G,+) \to (\Gamma,\circ)$, where $(G,+)$ is a commutative group. In general the situation $G \neq \mathbb{C}$ is even more involved. In the present paper we will only deal with $G = \mathbb{C}$.

The problem to describe the one-parameter groups in the group of invertible formal power series in one indeterminate with complex coefficients and, more generally, to describe one-parameter groups of invertible formal power series transformations ("formally biholomorphic mappings") was studied by several authors, mainly in connection with the embedding problem, that is, whether a given formal power series (a formally biholomorphic mapping) can be embedded in such an iteration group. We mention Lewis [21], Sternberg [30], Chen [2], Peschl and Reich [24], Reich and Schwaiger [28], Mehring [23], and Praagman [25].

If $(F_t)_{t\in\mathbb{C}}$ is an iteration group in Γ and $S \in \Gamma$, then $(S^{-1}\circ F_t \circ S)_{t\in\mathbb{C}}$ is an iteration group as well. Two iteration groups $(F_t)_{t\in\mathbb{C}}$ and $(G_t)_{t\in\mathbb{C}}$ are called *conjugate* if there is some $S \in \Gamma$ so that $G_t = S^{-1} \circ F_t \circ S$ for all $t \in \mathbb{C}$.

4.1.4 *The Main Problems*

Motivated by the question of embeddability the problem arises to find the structure and the explicit form of iteration groups in detail, not necessarily as a part of the embedding problem. In the sequel we will study the following topics:

1. Construction of all iteration groups in Γ.
2. Find the detailed structure and explicit form of the coefficient functions $c_n\colon \mathbb{C} \to \mathbb{C}$ ($n \geq 1$) of the solutions $F_t(x) = \sum_{n\geq 1} c_n(t)x^n$, $t \in \mathbb{C}$, of (T).
3. Describe the structure of all iteration groups and their normal forms with respect to conjugation.

The construction of iteration groups is strongly connected with the *maximal abelian subgroups of* $(\Gamma, \circ)$ (cf. [26]).

In the present paper we apply the method of formal functional equations which differs in many aspects from the approach by Jabłoński and Reich [17, 18]. This approach combines a detailed investigation of the systems (FE,I) and (FE, (II, k)) (see Section 4.2) for the coefficient functions of iteration groups with the a priori construction of the so-called *analytic* iteration groups, which have by definition entire coefficient functions, and with the application of certain polynomial relations associated with the coefficient functions. In our paper, however, we do not use any knowledge in analytic iteration groups.

We hardly ever present complete proofs, in some places we indicate some sketch of the proof. For details the reader is referred to the publications [4] in connection with iteration groups of type I and [5] for iteration groups of type (II, k).

We finish the introduction by giving an outline of the results and adding several comments. In Section 4.2 we describe the basic distinction between iteration groups of type I and iteration groups of type (II, k), $k \geq 2$. After studying the infinite systems of functional equations characterizing the coefficient functions of iteration groups, namely (FE,I) for iteration groups of type I and (FE, (II, k)) for iteration groups of type (II, k) (see Lemmas 4.1 and 4.2), we reduce the construction to the investigation of the so-called formal iteration groups of type I and formal iteration groups of type (II, k) (Theorem 4.1). These objects are elements in $(\mathbb{C}[y])[\![x]\!]$ which are solutions of certain relations in $(\mathbb{C}[y,z])[\![x]\!]$, namely the formal translation equations (Tform, I) and (Tform, (II, k)), together with appropriate boundary conditions. The basic idea of this reduction is the possibility to replace in the case of iteration groups of type I, say $F_t(x) = c_1(t)x + \ldots$, $t \in \mathbb{C}$, the generalized exponential function $c_1 \neq 1$ by an indeterminate y and similarly in the case of iteration groups of type (II, k), $F_t(x) = x + c_k(t)x^k + \ldots$, $t \in \mathbb{C}$, the additive function $c_k \neq 0$ by an indeterminate y, in the systems (FE,I) and (FE, (II, k)), respectively. Furthermore, we deduce from (Tform, I) and (Tform, (II, k)) by formal differentiation two formal differential equations, namely (Dform, I), (PDform, I) and by combining these two (AJform, I) for formal iteration groups of type I, and (Dform, (II, k)), (PDform, (II, k)), and (AJform, (II, k)) for formal iteration groups of type (II, k). The partial differential equations (PDform, I) and (PDform, (II, k)) may be considered as the simplest since they do not require a substitution of the unknown series $G(y, x)$. The Aczél–Jabotinsky equations (AJform, I) and (AJform, (II, k)) are weaker than the other differential equations just mentioned, and we add a remark (Theorem 4.2) how these Aczél–Jabotinsky differential equations can be used to construct and describe maximal abelian subgroups of Γ. All these differential equations contain the generator $H(x)$ where $H(x) = \frac{\partial}{\partial y} G(y,x)|_{y=1} = x + h_2x^2 + \dots$ for formal iteration groups of type I and $H(x) = \frac{\partial}{\partial y} G(y,x)|_{y=1} = x^k + h_{k+1}x^{k+1} + \dots$, $k \geq 2$, for formal iteration groups of type (II, k). The coefficients h_ν of the generators play an important role as natural parameters in the representations we are going to obtain in the following sections. In Section 4.2.5 we draw attention to the reordering of a formal iteration group $G(y,x) \in (\mathbb{C}[y])[\![x]\!]$ as $G(y,x) = \sum_{n\geq 1} \phi_n(x)y^n$ (for type I) or $G(y,x) = \sum_{n\geq 0} \phi_n(x)y^n$ (for type (II, k)) which allows in several situations a simpler and more elegant integration of the differential system.

In Section 4.3 we present the main results about the explicit form of formal iteration groups. Theorems 4.3 and 4.4 give the form of the coefficient functions P_n as derived from (PDform, I) for formal iteration groups $G(y,x) = yx + \sum_{n\geq 2} P_n(y)x^n$ of type I. The coefficient functions $P_n(y)$ are not only polynomials in y, but also universal polynomials in y and the coefficients $h_2, \dots, h_n$ of the generator H, where $H(x) = x + \dots$ can be chosen arbitrarily. We obtain rather explicit formulas for the P_n, including recursive relations describing the dependence on the parameters $(h_n)_{n\geq 2}$, as well as estimates of the degree of P_n. Using the reordering $G(y,x) = \sum_{n\geq 1} \phi_n(x)y^n$ of the formal iteration group of type I in (PDform, I) leads to Briot–Bouquet differential equations for the coefficients ϕ_n. The result is Theorem 4.5 which gives the unique representation $G(y,x) = S^{-1}(yS(x))$ with $S \in \Gamma_1$, sometimes called standard form. This means that each formal iteration group of type I is conjugate to yx which has generator x.

Theorems 4.6 and 4.7 show another representation of the coefficient functions $P_n(y)$ of formal iteration groups of type I, as deduced from (Dform, I). Theorem 4.8 contains one more description of $G(y,x) = \sum_{n\geq 1} \phi_n(x)y^n$, a formal iteration group of type I, which follows from (Dform, I), where $\phi_n(x)$ is expressed as $\varphi_n(\phi_1(x))$, $n \geq 1$, and a recurrence for $(\varphi_n)_{n\geq 1}$ without differentiation is deduced.

Theorem 4.9 refers to the solutions of (AJform, I). Here again Briot–Bouquet differential equations may be applied. The condition $G(y,x) = yx + \dots$ leads exactly to the solutions of (Tform, I) (see Theorems 4.9 and 4.10). Theorem 4.11 is also based on (AJform, I), reordering of $G(y,x)$, and using Briot–Bouquet differential equations. It gives again the standard form and the recurrence of Theorem 4.8.

In Section 4.3.4 we sketch two further approaches to obtain the standard form, here directly without formal functional equations. In connection with the first approach we discuss the important connection (4.1) of the generators of two conjugate formal iteration groups of type I. We formulate this connection as a differential equation for the conjugating series $S \in \Gamma$, involving the generators H and $\tilde{H}$ of the conjugate formal iteration groups. Formula (4.1), also valid for formal iteration groups of type (II, k), will also appear later in the paper. The second approach to the standard form is a calculation in the field $\mathbb{C}\langle\langle x\rangle\rangle$ of formal Laurent series with finite principal part.

The results for formal iteration groups of type (II, k) follow in the next section. The situation is not only much more complicated from a technical point of view, but also offers new "aspects." Theorems 4.12 and 4.13 refer to explicit formulas for the coefficient functions of the formal iteration groups of type (II, k), $G(y, x) = x + yx^k + \sum_{n>k} P_n(y)x^n$, derived from (PDform, (II, k)) or (Dform, (II, k)). Here P_n is a universal polynomial in y and the coefficients $h_{k+1}, \ldots, h_{n-k+1}, h_n$ of the generator $H(x) = x^k + h_{k+1}x^{k+1} + \ldots$. As a matter of fact, P_n does not depend on $h_{n-k+2}, \ldots, h_{n-1}$. Estimates of the degree of the P_n are given.

A similar result follows from (AJform, (II, k)) (see Theorems 4.14 and 4.15).

Writing $G(y, x) = \sum_{n\geq 0} \phi_n(x)y^n$ and substituting it into (PDform, (II, k)) we find a simple recurrence formula (PDR$_n$, (II, k)) for ϕ_n, this time with differentiation. Its solution under the boundary condition (BR, (II, k)) is contained in Theorems 4.16–4.18. The explicit formula for ϕ_{n+1} in Theorem 4.17 has as parameters certain coefficients h_ν of the generator H and certain coefficients of the P_n which are the coefficients of $G(y, x) = x + yx^k + \sum_{n>k} P_n(y)x^n$, whereas the explicit formula for ϕ_n in Theorem 4.18 has as parameters the coefficients of the generator only.

Formal iteration groups of type I and those of type (II, k) have very different properties with respect to conjugation. We start Section 4.3.6 by claiming that each formal iteration group of type (II, k) is conjugate to a formal iteration group of type (II, k) with generator $\tilde{H}(x) = x^k + hx^{2k-1}$, a so-called normal form. This is unique if we restrict the conjugating series S to be an element of Γ_1. To see this, we have to solve (4.1) for $S \in \Gamma_1$. Theorem 4.19 describes in detail, using (PDform, (II, k)) and (B, (II, k)), the explicit form of formal iteration groups of type (II, k) with generators $x^k + hx^{2k-1}$. These normal forms have the simplified structure $G(y, x) = \sum_{n\geq 0} P_{n(k-1)+1}(y)x^{n(k-1)+1}$ which is, however, much more complicated than the standard form $S^{-1}(yS(x))$ of formal iteration groups of type I.

It follows that the normal form $G(y, x)$ determined by the generator $x^k + hx^{2k-1}$ has an expansion $G(y, x) = \sum_{r\geq 0} G_r(y, x)h^r$ as a power series in h with coefficients $G_r(y, x) \in \mathbb{C}[\![y, x]\!]$, since h can be considered as a new indeterminate. The series $G_r(y, x)$ are determined from the recursive system (4.4) and (4.5). Their form is presented in Theorem 4.20. The differential equation (Dform, (II, k)) leads to a more compact description of $G_r(y, x)$, given in Theorem 4.21, involving a series of binomial type and a polynomial in $\ln(1 - (k - 1)yx^{k-1})$. The series $G_0(y, x) = x(1 - (k - 1)yx^{k-1})^{-1/(k-1)}$ plays a role in the theory of reversible power series (cf. [12]). Eventually Theorem 4.22 builds a bridge to Lie–Gröbner series.

We finish the paper by collecting some open problems. The most interesting one is the construction of iteration groups in higher dimensions by means of formal functional equations. So far only partial results are known.

4.2 First Classification of Iteration Groups

Let $(F_t)_{t\in\mathbb{C}}$ be an iteration group, $F_t(x) = \sum_{n\geq 1} c_n(t)x^n$, $t \in \mathbb{C}$. We consider three different types of iteration groups:

1. $F_t(x) = x$ for all $t \in \mathbb{C}$ is the *trivial iteration group.*
2. If $c_1 \neq 1$, then

$$c_1(s+t) = c_1(s)c_1(t), \qquad s,t \in \mathbb{C},$$

   thus c_1 is a non-trivial generalized exponential function. We call $(F_t)_{t\in\mathbb{C}}$ an iteration group of *type* I.
3. If $c_1 = 1$, then there exists some $k \geq 2$, so that $c_2 = \cdots = c_{k-1} = 0$, $c_k \neq 0$, and

$$c_k(s+t) = c_k(s) + c_k(t), \qquad s,t \in \mathbb{C},$$

   thus c_k is a non-trivial additive function. We say that $(F_t)_{t\in\mathbb{C}}$ is an iteration group of *type* (II, k).

This classification is compatible with the conjugation of iteration groups, i.e., if $(F_t)_{t\in\mathbb{C}}$ and $(G_t)_{t\in\mathbb{C}}$ are conjugate, then they have the same type.

4.2.1 Systems of Functional Equations for the Coefficient Functions

Consider a family $(F_t)_{t\in\mathbb{C}}$, $F_t(x) = \sum_{n\geq 1} c_n(t)x^n$, $t \in \mathbb{C}$, where $c_1 \neq 1$. Then $(F_t)_{t\in\mathbb{C}}$ is an iteration group of type I, if and only if the system

$$\begin{aligned}
c_1(s+t) &= c_1(s)c_1(t)\\
c_2(s+t) &= c_1(s)c_2(t) + c_2(s)c_1(t)^2 &&\text{(FE,I)}\\
c_n(s+t) &= c_1(s)c_n(t) + c_n(s)c_1(t)^n + \tilde{P}_n\big(c_2(s),\dots,c_{n-1}(s),c_2(t),\dots,c_{n-1}(t)\big),\\
&\qquad n \geq 2
\end{aligned}$$

is satisfied for all $s,t \in \mathbb{C}$. The $\tilde{P}_n$ are universal polynomials which are linear in $c_2(s),\dots,c_{n-1}(s)$.

Lemma 4.1 ([4, Lemma 2]) *If $(F_t)_{t\in\mathbb{C}}$ is an iteration group of type* I *of the form $F_t(x) = \sum_{n\geq 1} c_n(t)x^n$, $t \in \mathbb{C}$, then c_1 is a non-trivial generalized exponential function and there exists a sequence of polynomials $(P_n)_{n\geq 2}$ so that*

$$c_n(s) = P_n(c_1(s)) \quad \forall s \in \mathbb{C},\ \textit{and}\ P_n(0) = 0, \qquad n \geq 2.$$

Since $c_1 \neq 1$, for $n \geq 2$ there exists some $t_n \in \mathbb{C}$ so that $c_1(t_n)^n - c_1(t_n) \neq 0$. From $c_2(s+t) = c_2(t+s)$, for all $s, t \in \mathbb{C}$, we obtain

$$c_2(s) = \frac{c_2(t_2)(c_1(s)^2 - c_1(s))}{c_1(t_2) - c_1(t_2)^2} = P_2(c_1(s)), \qquad s \in \mathbb{C}.$$

Using induction on n and $c_n(s+t) = c_n(t+s)$, $\forall s, t \in \mathbb{C}$, we obtain the assertion from (FE,I).

Hence we obtain from (FE,I)

$$\begin{aligned} P_n(c_1(s)c_1(t)) &= P_n(c_1(s+t)) = c_n(s+t) \\ &= c_1(s)P_n(c_1(t)) + P_n(c_1(s))c_1(t)^n \qquad (\hat{\mathrm{P}}, \mathrm{I}) \\ &\quad + \tilde{P}_n\big(P_2(c_1(s)), \ldots, P_{n-1}(c_1(s)), P_2(c_1(t)), \ldots, P_{n-1}(c_1(t))\big) \end{aligned}$$

for all $s, t \in \mathbb{C}$ and all $n \geq 2$.

Consider a family $(F_t)_{t\in\mathbb{C}}$, $F_t(x) = x + \sum_{n\geq k} c_n(t)x^n$, $t \in \mathbb{C}$, where $c_k \neq 0$. Then $(F_t)_{t\in\mathbb{C}}$ is an iteration group of type (II, k), if and only if the system

$$\begin{aligned} c_n(s+t) &= c_n(s) + c_n(s), \qquad k \leq n \leq 2k-2, \\ c_{2k-1}(s+t) &= c_{2k-1}(s) + c_{2k-1}(t) + kc_k(s)c_k(t) \\ c_{2k}(s+t) &= c_{2k}(s) + c_{2k}(t) + kc_k(s)c_{k+1}(t) + (k+1)c_{k+1}(s)c_k(t) \\ c_n(s+t) &= c_n(s) + c_n(t) + kc_k(s)c_{n-(k-1)}(t) \qquad (\mathrm{FE}, (\mathrm{II}, k)) \\ &\quad + \big(n-(k-1)\big)c_{n-(k-1)}(s)c_k(t) \\ &\quad + \tilde{P}_n\big(c_k(s), \ldots, c_{n-k}(s), c_k(t), \ldots, c_{n-k}(t)\big), \qquad n > 2k, \end{aligned}$$

for all $s, t \in \mathbb{C}$, where $\tilde{P}_n$ are universal polynomials which are linear in $c_k(s), \ldots, c_{n-k}(s)$.

Lemma 4.2 ([5, Lemma 1]) *Consider some integer $k \geq 2$. If $(F_t)_{t\in\mathbb{C}}$ is an iteration group of type (II, k), $F_t(x) = x + \sum_{n\geq k} c_n(t)x^n$, $t \in \mathbb{C}$, then c_k is a non-trivial additive function and there exists a sequence of polynomials $(P_n)_{n\geq k}$ so that*

$$c_n(s) = P_n(c_k(s)), \qquad s \in \mathbb{C},\ n \geq k.$$

The reader should remember that these polynomials P_n differ from the polynomials P_n of Lemma 4.1. Since $c_k \neq 0$ there exists some $t_0 \in \mathbb{C}$ so that $c_k(t_0) \neq 0$. Using (FE, (II, k)) for $n = 2k$ we obtain from $c_{2k}(s+t) = c_{2k}(t+s)$,

$$c_{k+1}(s) = \frac{c_{k+1}(t_0)}{c_k(t_0)} c_k(s) = P_{k+1}(c_k(s)), \qquad s \in \mathbb{C}.$$

By induction on n and $c_{n+k-1}(s+t) = c_{n+k-1}(t+s)$, for all $s, t \in \mathbb{C}$, we obtain the assertion from (FE, (II, k)).

Hence we obtain from (FE, (II, k))

$$\begin{aligned} P_n(c_k(s) + c_k(t)) &= P_n(c_k(s+t)) = c_n(s+t) \\ &= P_n(c_k(s)) + P_n(c_k(t)) + kc_k(s)P_{n-(k-1)}(c_k(t)) \qquad (\hat{\mathrm{P}}, (\mathrm{II}, k)) \\ &\quad + \big(n-(k-1)\big)P_{n-(k-1)}(c_k(s))c_k(t) \\ &\quad + \tilde{P}_n\big(c_k(s), \ldots, P_{n-k}(c_k(s)), c_k(t), \ldots, P_{n-k}(c_k(t))\big), \end{aligned}$$

for all $s, t \in \mathbb{C}$ and $n \geq k$, where $P_j = 0$ for $j < k$ and $\tilde{P}_j = 0$ for $j \leq 2k$.

4.2.2 Formal Functional Equations

Formal functional equations in connection with the translation equation were studied by Gronau [10, 11], and the present authors [4, 5]. Similar methods were also applied for the study of cocycle equations which occur in connection with covariant embeddings of the linear functional equation (cf. [3, 6, 7]). Assume that $(F_t)_{t\in\mathbb{C}}$ is an iteration group of type I, $F_t(X) = \sum_{n\geq 1} c_n(t)x^n$, $t \in \mathbb{C}$, where $c_1(s+t) = c_1(s)c_1(t)$, $s, t \in \mathbb{C}$, $c_1 \neq 1$ and $c_1 \neq 0$. Since the image of c_1 contains infinitely many elements we can prove for any polynomial $Q(x, y) \in \mathbb{C}[x, y]$ that $Q(c_1(s), c_1(t)) = 0$ for all $s, t \in \mathbb{C}$ implies $Q = 0$. From ($\hat{\mathrm{P}}$, I) we obtain by replacing $c_1(s)$ and $c_1(t)$ by independent variables y, z, that

$$P_n(yz) = yP_n(z) + P_n(y)z^n + \tilde{P}_n\big(P_2(y), \ldots, P_{n-1}(y), P_2(z), \ldots, P_{n-1}(z)\big) \qquad (\mathrm{P}, \mathrm{I})$$

in $\mathbb{C}[y, z]$ for $n \geq 2$. Writing $G(y, x) = yx + \sum_{n\geq 2} P_n(y)x^n \in (\mathbb{C}[y])[\![x]\!]$ we deduce from (P, I) that G satisfies the formal translation equation of type I

$$G(yz, x) = G(y, G(z, x)) \qquad (\mathrm{Tform}, \mathrm{I})$$

in $(\mathbb{C}[y, z])[\![x]\!]$. We call $G(y, x)$ a *formal iteration group of type* I. It also satisfies the condition

$$G(1, x) = x. \qquad (\mathrm{B}, \mathrm{I})$$

Assume that $(F_t)_{t\in\mathbb{C}}$ is an iteration group of type (II, k) for some $k \geq 2$, $F_t(x) = x + \sum_{n\geq k} c_n(t)x^n$, $t \in \mathbb{C}$, where $c_k(s+t) = c_k(s) + c_k(t)$, $s, t \in \mathbb{C}$, $c_k \neq 0$. Since the image of c_k contains infinitely many elements we can prove for any polynomial $Q(x, y) \in \mathbb{C}[x, y]$ that $Q(c_k(s), c_k(t)) = 0$ for all $s, t \in \mathbb{C}$ implies $Q = 0$.

From $(\hat{\mathrm{P}}, (\mathrm{II}, k))$ we obtain by replacing $c_k(s)$ and $c_k(t)$ by independent variables y, z, that

$$P_n(y+z) = P_n(y) + P_n(z) + kyP_{n-(k-1)}(z) + \big(n-(k-1)\big)P_{n-(k-1)}(y)z + \tilde{P}_n\big(y, \dots, P_{n-k}(y), z, \dots, P_{n-k}(z)\big) \qquad (\mathrm{P}, (\mathrm{II}, k))$$

for all $n \geq k$.

Writing $G(y,x) = x + yx^k + \sum_{n\geq k+1} P_n(y)x^n \in (\mathbb{C}[y])[\![y]\!]$ we deduce from $(\mathrm{P}, (\mathrm{II}, k))$ that G satisfies the formal translation equation of type (II, k)

$$G(y+z, x) = G(y, G(z,x)) \qquad (\mathrm{Tform}, (\mathrm{II}, k))$$

in $(\mathbb{C}[y,z])[\![x]\!]$. We call $G(y,x)$ a *formal iteration group of type* (II, k). It also satisfies the condition

$$G(0,x) = x. \qquad (\mathrm{B}, (\mathrm{II}, k))$$

Conversely, from each formal iteration group we can construct iteration groups in the following way (cf. [4, Theorem 3] and [5, Theorem 3]):

Theorem 4.1 1. *If $G(y,x)$ is a formal iteration group of type* I, *c_1 a generalized exponential function, $c_1 \neq 1$, then $(G(c_1(t), x))_{t\in\mathbb{C}}$ is an iteration group of type* I.
2. *If $G(y,x)$ is a formal iteration group of type (II, k), $k \geq 2$, c_k an additive function, $c_k \neq 0$, then $(G(c_k(t), x))_{t\in\mathbb{C}}$ is an iteration group of type (II, k).*

4.2.3 Differential Equations Obtained from the Translation Equation

Let $G(y,x) \in (\mathbb{C}[y])[\![x]\!]$ be a formal iteration group of type I. Then the infinitesimal generator of G is defined as

$$H(x) = \frac{\partial}{\partial y} G(y,x)\Big|_{y=1}.$$

It is of the form $H(x) = x + \sum_{n\geq 2} h_n x^n$. Differentiation of (Tform, I) with respect to y yields

$$z\frac{\partial}{\partial t} G(t,x)|_{t=yz} = \frac{\partial}{\partial y} G\big(y, G(z,x)\big).$$

For $y = 1$ we get

$$z\frac{\partial}{\partial z} G(z,x) = H(G(z,x)). \qquad (\mathrm{Dform, I})$$

Differentiation of (Tform, I) with respect to z and application of the mixed chain rule yields

$$y\frac{\partial}{\partial t}G(t,x)|_{t=yz} = \frac{\partial}{\partial t}G(y,t)|_{t=G(z,x)}\frac{\partial}{\partial z}G(z,x).$$

For $z = 1$ we get

$$y\frac{\partial}{\partial y}G(y,x) = H(x)\frac{\partial}{\partial x}G(y,x). \qquad \text{(PDform, I)}$$

The advantage of this equation lies in the circumstance that no substitution of the unknown series $G(y,x)$ is needed and that (PDform, I) is a linear equation.

Combining (Dform, I) and (PDform, I), we obtain an Aczél–Jabotinsky differential equation of the form

$$H(x)\frac{\partial}{\partial x}G(y,x) = H(G(y,x)). \qquad \text{(AJform, I)}$$

In this equation the variable y is an internal parameter since it does not appear explicitly in (AJform, I).

Let $G(y,x) \in (\mathbb{C}[y])[\![x]\!]$ be a formal iteration group of type (II, k) for some $k \geq 2$. Then the infinitesimal generator of G is defined as

$$H(x) = \frac{\partial}{\partial y}G(y,x)\Big|_{y=0}.$$

It is of the form $H(x) = x^k + \sum_{n\geq k+1} h_n x^n$. Differentiation of (Tform, (II, k)) with respect to y yields

$$\frac{\partial}{\partial t}G(t,x)|_{t=y+z} = \frac{\partial}{\partial y}G\big(y, G(z,x)\big).$$

For $y = 0$ we get

$$\frac{\partial}{\partial z}G(z,x) = H(G(z,x)). \qquad \text{(Dform, (II, }k\text{))}$$

Differentiation of (Tform, (II, k)) with respect to z and application of the mixed chain rule yields

$$\frac{\partial}{\partial t}G(t,x)|_{t=y+z} = \frac{\partial}{\partial t}G(y,t)|_{t=G(z,x)}\frac{\partial}{\partial z}G(z,x).$$

For $z = 0$ we get

$$\frac{\partial}{\partial y}G(y,x) = H(x)\frac{\partial}{\partial x}G(y,x). \qquad \text{(PDform, (II, }k\text{))}$$

The advantage of this equation lies in the circumstance that no substitution of the unknown series $G(y,x)$ is needed and that (PDform, (II, k)) is a linear equation.

Combining (Dform, (II, k)) and (PDform, (II, k)), we obtain an Aczél–Jabotinsky differential equation of the form

$$H(x)\frac{\partial}{\partial x}G(y,x) = H(G(y,x)). \qquad \text{(AJform, (II, } k\text{))}$$

In this equation the variable y is an internal parameter since it does not appear explicitly in (AJform, (II, k)).

4.2.4 The Relevance of Aczél–Jabotinsky Differential Equations

The Aczél–Jabotinsky differential equations can be used to characterize maximal abelian subgroups of Γ (cf. [26]). The main result reads as follows:

Theorem 4.2 *A set $\mathscr{F} \subset \Gamma$ is a maximal abelian subgroup of Γ if and only if there exists some $H \in \mathbb{C}[\![x]\!]$, $H \neq 0$, $\operatorname{ord}(H) \geq 1$, so that*

$$\phi \in \mathscr{F} \iff H(x)\phi'(x) = H(\phi(x)).$$

It can be shown that either $\mathscr{F}$ is isomorphic to $\mathbb{C}^*$, or $\mathscr{F}$ is isomorphic to

$$\left\{ \begin{pmatrix} \rho & t \\ 0 & \rho \end{pmatrix} \,\middle|\, \rho^m = 1,\ t \in \mathbb{C} \right\},$$

where m is uniquely determined by $\mathscr{F}$.

4.2.5 Reordering the Summands

Let $G(y,x) = \sum_{n\geq 1} P_n(y)x^n \in (\mathbb{C}[y])[\![x]\!] \subset \mathbb{C}[\![y,x]\!]$ be a formal iteration group of type I, then it is possible to write $G(y,x)$ in the form

$$G(y,x) = \sum_{n\geq 1} \phi_n(x)y^n \in (\mathbb{C}[\![x]\!])[\![y]\!] \qquad \text{(R, I)}$$

where $\phi_1 \in \Gamma_1$, and $(\phi_n(x))_{n\geq 1}$ is a summable family in $\mathbb{C}[\![x]\!]$. Therefore the boundary condition (B, I)

$$G(1,x) = \sum_{n\geq 1} \phi_n(x) = x \qquad \text{(BR, I)}$$

makes sense. It is possible to use this representation of G in the differential equations (Dform, I), (PDform, I), and (AJform, I).

Let

$$G(y,x) = x + yx^k + \sum_{n\geq k+1} P_n(y)x^n \in (\mathbb{C}[y])[\![x]\!] \subset \mathbb{C}[\![y,x]\!],$$

$k \geq 2$, be a formal iteration group of type (II, k), then it is possible to write $G(y,x)$ in the form

$$G(y,x) = \sum_{n\geq 0} \phi_n(x)y^n \in (\mathbb{C}[\![x]\!])[\![y]\!] \qquad \text{(R, (II, }k\text{))}$$

where $(\phi_n(x))_{n\geq 0}$ is a summable family in $\mathbb{C}[\![x]\!]$. The boundary condition (B, (II, k)) reads as

$$G(0,x) = \phi_0(x) = x. \qquad \text{(BR, (II, }k\text{))}$$

It is possible to use this representation of G in the differential equations (Dform, (II, k)), (PDform, (II, k)), and (AJform, (II, k)).

4.3 Solving the Translation Equation by a Purely Algebraic Differentiation Process

Here we present the construction of formal iteration groups by solving the differential equations (Dform, I), (PDform, I), or (AJform, I) for formal iteration groups of type I and (Dform, (II, k)), (PDform, (II, k)), or (AJform, (II, k)) for formal iteration groups of type (II, k) under the appropriate boundary conditions.

4.3.1 *Formal Iteration Groups of Type* I *Obtained from* (PDform, I) *and* (B, I)

Using the partial differential equation (PDform, I) we describe how the polynomials P_n, $n \geq 2$, depend on the coefficients h_j, $j \geq 2$, of the infinitesimal generator H of the formal iteration group G of type I. We determine all solutions of (PDform, I) and (B, I) and we show that each of them is a solution of (Tform, I).

Theorem 4.3 ([4, Theorem 4]) *For each generator $H(x) = x + h_2x^2 + \ldots$ the partial differential equation* (PDform, I) *together with* (B, I) *has exactly one solution. It is given by*

$$G(y,x) = yx + \sum_{n\geq 2} P_n(y)x^n \in (\mathbb{C}[y])[\![x]\!].$$

The polynomials P_n, $n \geq 2$, are of formal degree n (that is an upper bound for the degree), they satisfy $P_n(0) = 0$, and they are of the form

$$P_n(y) = \frac{h_n}{n-1}(y^n - y) + \sum_{j=1}^{n-1} \frac{\Phi_j^{(n)}(h_2, \ldots, h_{n-1})}{n-j}(y^n - y^j)$$

where the polynomials $\Phi_j^{(n)}$, $1 \leq j \leq n-1$, are (recursively) determined by

$$\sum_{r=2}^{n-1} h_r(n-r+1)P_{n-r+1}(y) = \sum_{j=1}^{n-1} \Phi_j^{(n)}(h_2, \ldots, h_{n-1})y^j.$$

Theorem 4.4 ([4, Theorem 5]) *For each generator $H(x) = x + h_2x^2 + \ldots$ the solution $G(y, x)$ of* (PDform, I) *and* (B, I) *is a solution of the formal translation equation* (Tform, I).
Let G be the solution of (PDform, I) and (B, I) for the generator H. In order to prove this theorem we show that both series

$$U(y, z, x) := G(yz, x)$$
$$V(y, z, x) := G(z, G(y, x))$$

satisfy the system

$$y\frac{\partial}{\partial y}f(y, z, x) = H(x)\frac{\partial}{\partial x}f(y, z, x)$$
$$f(1, z, x) = G(z, x)$$

which has a unique solution in $(\mathbb{C}[y, z])[\![x]\!]$.

Let G be the solution of (PDform, I) and (B, I) for the generator H. Reordering the summands of G we write $G(y, x)$ as $\sum_{n\geq 1} \phi_n(x)y^n$. Then from (PDform, I) and (B, I) we obtain

$$\sum_{n\geq 1} n\phi_n(x)y^n = H(x)\sum_{n\geq 1} \phi_n'(x)y^n \qquad \text{(PDR, I)}$$

and (BR, I). Equation (PDR, I) is equivalent to

$$n\phi_n(x) = H(x)\phi_n'(x) \qquad (\text{PDR}_n, \text{I})$$

for all $n \geq 1$. Each of these equations is equivalent to a Briot–Bouquet differential equation (in the non-generic case), thus it has solutions.

A *Briot–Bouquet differential equation* (cf. [20, Section 5.2], [14, Section 11.1], [15, Section 12.6]) is a complex differential equation

$$zw'(z) = az + bw(z) + \sum_{\alpha+\beta\geq 2} a_{\alpha,\beta} z^{\alpha} [w(z)]^{\beta},$$

where $w(z)$ is a power series in z with $w(0) = 0$, and the power series on the right-hand side is given. Cauchy's theorem on existence and uniqueness cannot be applied directly. In the case $b = n$, a positive integer, a formal solution $w(z)$ exists if, and only if, a certain polynomial $P(a, b, a_{\alpha,\beta} : \alpha + \beta \leq n)$ vanishes. If so, then the equation is called *solvable* or *non-generic of type n*, and all solutions take the shape

$$w_t(z) = c_1 z + \ldots + c_{n-1} z^{n-1} + tz^n + \sum_{\nu\geq n+1} Q_\nu(t) z^\nu, \qquad t \in \mathbb{C},$$

for polynomials $Q_\nu(t)$. The coefficients c_i, $1 \leq i \leq n-1$, are uniquely determined. The series $w_t(z)$ is convergent if the given right-hand side is convergent.

Let

$$H(x) = x(1 + \sum_{n\geq 1} h_n^* x^n) = xH^*(x),$$

then $h_n^* = h_{n+1}$, $n \geq 1$, and (PDR$_n$, I) is equivalent to

$$n\phi_n(x) = xH^*(x)\phi_n'(x)$$

or

$$x\phi_n'(x) = n\phi_n(x)[1 + h_1^* x + \ldots]^{-1}.$$

Finally for each $n \geq 1$ we end up with the system

$$x\phi_n'(x) = n\phi_n(x) + n \sum_{\alpha+\beta\geq 2} d_{\alpha,\beta} x^\alpha [\phi_n(x)]^\beta$$

$$\phi_n(0) = 0.$$

The set of solutions of (PDR$_n$, I) is then given by $\{\varphi_n^{(n)} [\phi_{1,0}(x)]^n \mid \varphi_n^{(n)} \in \mathbb{C}\}$, where $\phi_{1,0}(x)$ is the unique solution of (PDR$_1$, I) which belongs to Γ_1, i.e., which is of the form $\phi_{1,0}(x) = x + \ldots$. Denote this series by $S(x) = \phi_{1,0}(x)$ and let $\sum_{n\geq 1} \phi_n(x)$ be a solution of (PDR, I). From the boundary condition (BR, I) we obtain

$$x = \sum_{n\geq 1} \phi_n(x) = \sum_{n\geq 1} \varphi_n^{(n)} [S(x)]^n,$$

whence,

$$S^{-1}(x) = \sum_{n\geq 1} \varphi_n^{(n)} x^n,$$

from which it is possible to determine the values $\varphi_n^{(n)}$, $n \geq 1$.

The main result of this section is

Theorem 4.5 ([4, Theorem 7]) *If* $G(y,x) = \sum_{n\geq 1} \phi_n(x) y^n$ *is a solution of* (Tform, I) *and* (B, I), *then there exists exactly one* $S \in \Gamma_1$ *so that*

$$G(y,x) = S^{-1}\big(yS(x)\big).$$

Using the representation (R, I) *we have* $\phi_n(x) = \varphi_n^{(n)}[S(x)]^n$, *where* $\varphi_n^{(n)} \in \mathbb{C}$, $n \geq 1$.

Conversely, for every $S \in \Gamma_1$ *the series*

$$G(y,x) = S^{-1}\big(yS(x)\big)$$

is a solution of (Tform, I) *and* (B, I).

4.3.2 *Formal Iteration Groups of Type* I *Obtained from* (Dform, I) *and* (B, I)

For the differential equation (Dform, I) we obtain similar results as in the previous section (see also [4, Theorems 9, 10, 11]).

Theorem 4.6 *For each generator*

$$H(x) = x + h_2 x^2 + \dots$$

the differential equation (Dform, I) *together with* (B, I) *has exactly one solution. It is given by*

$$G(z,x) = zx + \sum_{n\geq 2} P_n(z) x^n \in (\mathbb{C}[z])[\![x]\!].$$

The polynomials P_n, $n \geq 2$, *are of formal degree n, they satisfy* $P_n(0) = 0$, *and they are of the form*

$$P_n(z) = \frac{h_n}{n-1}(z^n - z) + \sum_{j=2}^{n} \frac{\Psi_j^{(n)}(h_2, \dots, h_{n-1})}{j-1}(z^j - z)$$

where the polynomials $\Psi_j^{(n)}$, $2 \le j \le n$, *are (recursively) determined by*

$$\sum_{\nu=2}^{n-1} h_\nu \sum_{\substack{r_1+\ldots+r_\nu=n \\ r_j \ge 1}} \left(\prod_{j=1}^{\nu} P_{r_j}(z) \right) = \sum_{j=2}^{n} \Psi_j^{(n)}(h_2, \ldots, h_{n-1}) z^j.$$

Theorem 4.7 *For each generator*

$$H(x) = x + h_2 x^2 + \ldots$$

the solution $G(z, x)$ *of* (Dform, I) *and* (B, I) *is a solution of the formal translation equation* (Tform, I).

Using the representation (R, I) we obtain from (Dform, I)

$$\sum_{n \ge 1} n\phi_n(x) z^n = \sum_{\nu \ge 1} h_\nu \left[\sum_{n \ge 1} \phi_n(x) z^n \right]^\nu \tag{DR, I}$$

which is equivalent to

$$n\phi_n(x) = \sum_{\nu=1}^{n} h_\nu \sum_{\substack{r_1+\ldots+r_\nu=n \\ r_j \ge 1}} \left(\prod_{j=1}^{\nu} \phi_{r_j}(x) \right) \tag{DR$_n$, I}$$

for all $n \ge 1$. This is a recursive formula for the ϕ_n without any differentiation process. The solutions of (DR, I) are given in

Theorem 4.8 *Consider* $H(x) = x + h_2 x^2 + \ldots$.

1. *Every* $\phi_1(x) \in \mathbb{C}[\![x]\!]$ *satisfies* (DR$_1$, I).
2. *Let* $\phi_1 \in \mathbb{C}[\![x]\!] \setminus \{0\}$. *For each* $n \ge 2$ *there exists exactly one solution* ϕ_n *of* (DR$_n$, I), *depending on* ϕ_1. *It is given by* $\phi_n(x) := \varphi_n[\phi_1(x)]^n$, *where* $\varphi_1 = 1$ *and*

$$\varphi_n = \frac{1}{n-1} \sum_{\nu=2}^{n} h_\nu \sum_{r_1+\ldots+r_\nu=n} \prod_{j=1}^{\nu} \varphi_{r_j}, \qquad n \ge 2.$$

   Consequently, φ_n *does not depend on the choice of* ϕ_1.
3. *The system* (DR, I) *and* (BR, I) *has a unique solution. It is given by*

$$\sum_{n \ge 1} \varphi_n [\phi_1(x)]^n z^n$$

for $\varphi_1 = 1$, φ_n *for* $n \geq 2$ *given as above, and*

$$\phi_1(x) = (x + \sum_{n\geq 2} \varphi_n x^n)^{-1},$$

which is an element of Γ_1.

4.3.3 *Formal Iteration Groups of Type* I *Obtained from* (AJform, I) *and* (B, I)

Here we present some facts from [4, Section 2.3]. Writing the series H as $x(1 + h_2x + \ldots)$ and $\phi(x) := G(y, x)$ motivates that (AJform, I) is equivalent to

$$x\phi'(x) = [1 + h_2x + \ldots]^{-1} H(\phi(x))$$

or

$$x\phi'(x) = \phi(x) + \sum_{\substack{\alpha+\beta\geq 2\\ \beta\geq 1}} d_{\alpha,\beta}(h) x^\alpha [\phi(x)]^\beta$$

which is a Briot–Bouquet differential equation. It is well known that for each $\tilde{P}_1(y) \in \mathbb{C}[y]$ there exists exactly one solution

$$\tilde{G}(y, x) = \tilde{P}_1(y)x + \sum_{n\geq 2} \tilde{P}_n(y)x^n$$

of this Briot–Bouquet equation with coefficients $\tilde{P}_n(y)$ which are polynomials, $n \geq 2$.

The solutions of (AJform, I) with $\tilde{P}_1(y) = y$ are determined in the next theorem.

Theorem 4.9 *1. For each generator* $H(x) = x + h_2x^2 + \ldots$ *the differential equation* (AJform, I) *has exactly one solution of the form*

$$G(y, x) = yx + \sum_{n\geq 2} P_n(y)x^n \in (\mathbb{C}[y])[\![x]\!].$$

2. *The polynomials* P_n, $n \geq 2$, *(from the unique solution* $G(y, x) \equiv yx \bmod x^2$*) are of formal degree n, they satisfy* $P_n(0) = 0$*, and they are of the form*

$$P_n(y) = \frac{h_n}{n-1}(y^n - y) + \sum_{j=2}^{n} \frac{\Theta_j^{(n)}(h_2, \ldots, h_{n-1})}{n-1}(y^j - y)$$

where the polynomials $\Theta_j^{(n)}$, $2 \le j \le n$, *are (recursively) determined by*

$$\sum_{\nu=2}^{n-1} h_\nu \left(\sum_{\substack{r_1+\ldots+r_\nu=n \\ r_j \ge 1}} \left(\prod_{j=1}^{\nu} P_{r_j}(y) \right) - (n-\nu+1) P_{n-\nu+1}(y) \right) =$$

$$\sum_{j=2}^{n} \Theta_j^{(n)}(h_2, \ldots, h_{n-1})(y^j - y).$$

Applying the same method as in the proof of Theorem 4.4 we obtain

Theorem 4.10 *For each generator* $H(x) = x + h_2x^2 + \ldots$ *the solution* $G(y, x)$ *of the differential equation* (AJform, I) *with* $G(y, x) \equiv yx \bmod x^2$ *is a solution of* (Tform, I).

Using the representation (R, I) we obtain from (AJform, I)

$$H(x) \sum_{n\ge1} \phi_n'(x) y^n = \sum_{\nu\ge1} h_\nu \left[\sum_{n\ge1} \phi_n(x) y^n \right]^\nu \qquad \text{(AJR, I)}$$

which is equivalent to

$$H(x)\phi_n'(x) = \phi_n(x) + \sum_{\nu=2}^{n} h_\nu \sum_{\substack{r_1+\ldots+r_\nu=n \\ r_j \ge 1}} \left(\prod_{j=1}^{\nu} \phi_{r_j}(x) \right) \qquad (\text{AJR}_n, \text{I})$$

for all $n \ge 1$. Again these equations are Briot–Bouquet differential equations since, for all $n \ge 1$,

$$x\phi_n'(x) = [1 + h_2x + \ldots]^{-1} \left(\phi_n(x) + \sum_{\nu=2}^{n} h_\nu \sum_{\substack{r_1+\ldots+r_\nu=n \\ r_j \ge 1}} \left(\prod_{j=1}^{\nu} \phi_{r_j}(x) \right) \right).$$

We are mainly interested in solutions where $\phi_1(x) = x + \ldots$ since they lead to iteration groups. The set of all solutions of (AJR, I) is described in

Theorem 4.11 *Consider* $H(x) = x + h_2x^2 + \ldots$.

1. *For every* $c \in \mathbb{C}$, *there is exactly one solution*

$$\phi_1(x) \equiv cx \bmod x^2$$

of (AJR_1, I).

2. *Assume that* $\phi_1 = cx + \ldots$, $c \neq 0$, *is a solution of* (AJR_1, I). *Then for each* $n \geq 2$ *there exists exactly one solution* $\phi_n(x)$ *of* (AJR_n, I). *It is given by* $\phi_n(x) = \varphi_n[\phi_1(x)]^n$, *where* $\varphi_1 = 1$ *and*

$$\varphi_n = \sum_{\nu=2}^{n} \frac{h_\nu}{n-1} \sum_{r_1+\ldots+r_\nu=n} \prod_{j=1}^{\nu} \varphi_{r_j}, \qquad n \geq 2.$$

Consequently, φ_n *does not depend on the choice of* ϕ_1.
3. *The unique solution* ϕ_1 *of system* (AJR_1, I) *which belongs to* Γ_1, *(i.e.,* $c = 1$*) leads to the solution*

$$\sum_{n\geq 1} \varphi_n[\phi_1(x)]^n y^n$$

of (AJR, I), *where* $\varphi_1 = 1$ *and* φ_n *for* $n \geq 2$ *given as above. Moreover* $\phi_1(x) = (x + \sum_{n\geq 2} \varphi_n x^n)^{-1}$.

Based on these results it is possible to give another simple proof of Theorem 4.5.

4.3.4 Normal Forms of Iteration Groups of Type I

From Theorem 4.5 we know that each formal iteration group $G(y, x)$ of type I is conjugate to yx. We call it the *normal form of formal iteration groups of type* I. Let $(F_t)_{t\in\mathbb{C}}$ be an iteration group of type I, $F_t(x) = \sum_{n\geq 1} c_n(t)x^n$ for all $t \in \mathbb{C}$. Then there exists some $S \in \Gamma_1$ so that $F_t(x) = S^{-1}(c_1(t)S(x))$, $t \in \mathbb{C}$.

We want to present two further methods for finding this normal form.

1. Consider the generator $H(x) = \frac{\partial}{\partial y} G(y, x)|_{y=1} = x + h_2x^2 + \ldots$ of a formal iteration group of type I, and some $S \in \Gamma$. Then $\tilde{G}(y, x) = S^{-1}(G(y, S(x)))$ is a solution of (Tform, I). We calculate its generator

$$\tilde{H}(x) = \frac{\partial}{\partial y} S^{-1}(G(y, x))|_{y=1}$$

by an application of the chain rule:

$$\frac{\partial}{\partial y} S^{-1}(G(y, Sx)) = (S^{-1})'(G(y, Sx)) \frac{\partial}{\partial y} G(y, Sx).$$

Putting $y = 1$ we obtain $\tilde{H}(x) = (S^{-1})'(Sx)H(Sx)$. Since $(S^{-1})'(Sx)S'(x) = 1$ we get

$$\left(\frac{\partial}{\partial x} S(x)\right) \tilde{H}(x) = H(S(x)). \tag{4.1}$$

If we choose $\tilde{H}(x) = x$, then (4.1) yields the Briot–Bouquet differential equation

$$x\frac{\partial}{\partial x}S(x) = S(x) + h_2[S(x)]^2 + \dots \tag{4.2}$$

(see [20, Section 5.2], [14, Section 11.1], [15, Section 12.6]). It is known that (4.2) has exactly one solution in $S \in \Gamma_1$. Using this S it follows that $\tilde{G}(y,x)$ has the generator $\tilde{H}(x) = x$, hence from (B, I) we get $yx = \tilde{G}(y,x) = S^{-1}(G(y,S(x)))$, or equivalently

$$G(y,x) = S(yS^{-1}(x)). \tag{4.3}$$

2. Consider for some $H(x) = x + h_2x^2 + \ldots \in \mathbb{C}[\![x]\!]$ the Aczél–Jabotinsky equation

$$H(x)\Phi'(x) = H(\Phi(x)), \text{ for } \Phi(x) = \rho x + \dots, \ \rho \neq 0. \tag{AJ}$$

We compute the standard form of its set of solutions by computation in $\mathbb{C}\langle\langle x\rangle\rangle$, the ring of formal Laurent series with finite principal part. Again we write $H(x) = xH^*(x)$ and assume that $[H^*(x)]^{-1} = 1 + h_1^*x + h_2^*x^2 + \ldots$. Then from (AJ) we get $xH^*(x)\Phi'(x) = \Phi(x)H^*(\Phi(x))$ thus

$$\frac{\Phi'(x)}{\Phi(x)}\left(1 + \sum_{n\geq 1} h_n^*[\Phi(x)]^n\right) = \frac{1}{x}\left(1 + \sum_{n\geq 1} h_n^* x^n\right)$$

and

$$\frac{\Phi'(x)}{\Phi(x)} - \frac{1}{x} = -\sum_{n\geq 1} h_n^*\Phi'(x)[\Phi(x)]^{n-1} + \sum_{n\geq 1} h_n^* x^{n-1}.$$

Using the differentiation operator this can be written as

$$\frac{\partial}{\partial x}\left(\ln\frac{\Phi(x)}{\rho x}\right) = -\frac{\partial}{\partial x}\left(\sum_{n\geq 1}\frac{h_n^*}{n}[\Phi(x)]^n\right) + \frac{\partial}{\partial x}\left(\sum_{n\geq 1}\frac{h_n^*}{n}x^n\right),$$

therefore

$$\ln\frac{\Phi(x)}{\rho x} = -T(\Phi(x)) + T(x) \text{ for } T(x) = \sum_{n\geq 1}\frac{h_n^*}{n}x^n.$$

Applying the exponential series we deduce

$$\frac{\Phi(x)}{\rho x} = \frac{\exp(T(x))}{\exp(T(\Phi(x)))}$$

or equivalently $\Phi(x)\exp(T(\Phi(x))) = \rho x \exp(T(x))$. The series S given by $S(x) = x\exp(T(x))$ is in Γ_1 and satisfies $S(\Phi(x)) = \rho S(x)$, whence $\Phi(x) = S^{-1}(\rho S(x))$. The coefficients of S are polynomials in the coefficients h_n.

4.3.5 Formal Iteration Groups of Type (II, k) Obtained from the Three Differential Equations

The solutions of (PDform, (II, k)) [or (Dform, (II, k))] together with (B, (II, k)) and the polynomials $P_n(y)$, $n > k$, occurring as their coefficient functions are completely described in

Theorem 4.12 ([5, Theorems 4, 9]) *Consider some* $k \geq 2$.

1. *For each generator* $H(x) = x^k + \sum_{n>k} h_n x^n$ *the system of* (PDform, (II, k)) *and* (B, (II, k)) *(or* (Dform, (II, k)) *and* (B, (II, k))*] has exactly one solution. It is given by*
$$G(y,x) = x + yx^k + \sum_{n>k} P_n(y)x^n \in (\mathbb{C}[y])[\![x]\!].$$
2. *The polynomials* P_n, $n \geq k$, *have a formal degree* $\lfloor (n-1)/(k-1) \rfloor$ *and they are of the form*
$$P_n(y) = \begin{cases} h_n y & k \leq n < 2k-1 \\ h_{2k-1}y + \frac{k}{2}y^2 & n = 2k-1 \\ h_n y + \frac{n+1}{2} h_{n-k+1} y^2 + \Phi_n(y, h_{k+1}, \dots, h_{n-k}) & n \geq 2k, \end{cases}$$
where Φ_n *are polynomials in* y *and in the coefficients* $h_{k+1}, \dots, h_{n-k}$. *They satisfy* $\Phi_n(0, h_{k+1}, \dots, h_{n-k}) = 0$. *For* $n > 2k$ *a formal degree of* Φ_n *as a polynomial in* y *is* $\lfloor (n-1)/(k-1) \rfloor$.

Theorem 4.13 ([5, Theorems 5, 10]) *For each generator* $H(x) = x^k + \sum_{n>k} h_n x^n$ *the solution* $G(y,x)$ *of the system* (PDform, (II, k)) *and* (B, (II, k)) *[or* (Dform, (II, k)) *and* (B, (II, k))*] is a solution of* (Tform, (II, k)).

For the Aczél–Jabotinsky equation we obtain

Theorem 4.14 ([5, Theorem 13]) *Consider some* $k \geq 2$.

1. *For each generator* $H(x) = x^k + \sum_{n>k} h_n x^n$ *and for any polynomial* $P_k(y) \in \mathbb{C}[y]$ *with* $P_k(0) = 0$ *the differential equation* (AJform, (II, k)) *together with* (B, (II, k)) *has exactly one solution of the form*
$$G(y,x) = x + P_k(y)x^k + \sum_{n>k} P_n(y)x^n \in (\mathbb{C}[y])[\![x]\!].$$

The polynomials $P_n(y)$ for $n > k$ are given by

$$P_n(y) = \begin{cases} h_n P_k(y) & \text{if } n < 2k-1 \\ h_{2k-1}P_k(y) + \frac{k}{2}P_k(y)^2 & \text{if } n = 2k-1 \\ h_n P_k(y) + \frac{n+1}{2} h_{n-k+1} P_k(y)^2 & \\ \qquad + \Phi_n(P_k(y), h_{k+1}, \dots, h_{n-k}) & \text{if } n \geq 2k, \end{cases}$$

with polynomials Φ_n, $n \geq 2k$, in $P_k(y)$ and $h_{k+1}, \dots, h_{n-k}$.

2. *Assume that $P_k(y) = y$. The polynomials P_n, $n \geq k$, have a formal degree $\lfloor (n-1)/(k-1) \rfloor$ and their coefficients are given in Theorem 4.12.*

Applying the same method as in the proof of Theorem 4.4 we obtain

Theorem 4.15 ([5, Theorem 14]) *For each generator $H(x) = x^k + \sum_{n>k} h_n x^n$ the solution $G(y,x)$ of* (AJform, (II, k)) *with $G(y,x) \equiv x + yx^k \bmod x^{k+1}$ is a solution of* (Tform, (II, k)).

Let G be the solution of (PDform, (II, k)) and (B, (II, k)) for the generator $H(x) = x^k + \sum_{n>k} h_n x^n$. Reordering the summands of G we write $G(y,x)$ as $\sum_{n\geq 0} \phi_n(x) y^n$. Then (PDform, (II, k)) yields

$$\sum_{n\geq 1} n\phi_n(x) y^{n-1} = H(x) \sum_{n\geq 0} \phi_n'(x) y^n, \qquad \text{(PDR, (II, } k))$$

where $(\phi_n'(x)y^n)_{n\geq 0}$ is a summable family. We note that (PDR, (II, k)) is satisfied if and only if

$$\phi_{n+1}(x) = \frac{1}{n+1} H(x) \phi_n'(x) \qquad \text{(PDR}_n\text{, (II, } k))$$

holds true for all $n \geq 0$.

The solutions of (PDR, (II, k)) and (BR, (II, k)) are thoroughly analyzed in the following theorems.

Theorem 4.16 ([5, Theorem 15]) *For each generator $H(x) = \sum_{n\geq k} h_n x^n$, $k \geq 2$, $h_k = 1$, the system* (PDR$_n$, (II, k)) *and* (BR, (II, k)) *has a unique solution. For $n \geq 0$ the order of $\phi_n(x)$ is equal to $n(k-1)+1$ and $\phi_n(0) = 0$.*

Theorem 4.17 ([5, Corollary 16, Theorem 18]) *Consider some $k \geq 2$ and assume that $\sum_{n\geq 0} \phi_n(x) y^n = \sum_{r\geq 1} P_r(y) x^r$ is the solution of* (PDR$_n$, (II, k)) *and* (BR, (II, k)) *for a given generator $H(x)$. Writing*

$$P_r(y) = \sum_{j\geq 0} P_{r,j} y^j, \quad r \geq 1, \text{ and } \phi_n(x) = \sum_{r\geq 1} P_{r,n} x^r, \quad n \geq 0,$$

we deduce that $P_r = 0$ for $2 \leq r < k$. Moreover for $r \geq k$ the series $P_r(y)$ is a polynomial which has a formal degree $\lfloor (r-1)/(k-1) \rfloor$ and which satisfies $P_r(0) = 0$. Consequently

$$\sum_{n\geq 0}\phi_n(x)y^n = x+\sum_{r\geq k}P_r(y)x^r \in (\mathbb{C}[y])[\![x]\!].$$

If $\phi_n(x) = \sum_{r\geq n(k-1)+1} P_{r,n}x^r$ *and* $H(x) = \sum_{r\geq k} h_r x^r$, *then*

$$\phi_{n+1}(x) = \frac{1}{n+1}\sum_{r\geq(n+1)(k-1)+1}\left(\sum_{\nu=n(k-1)+1}^{r+1-k} \nu h_{r+1-\nu}P_{\nu,n}\right)x^r, \qquad n\geq 0.$$

Theorem 4.18 ([5, Theorem 19]) *Let* $H(x) = \sum_{n\geq k} h_n x^n$, $k\geq 2$, $h_k = 1$, *be a generator and assume that* $\sum_{n\geq 0}\phi_n(x)y^n$ *is the solution of* (PDR$_n$, (II, k)) *and* (BR, (II, k)). *Then*

$$\phi_n(x) = \frac{1}{n!}\sum_{r\geq n(k-1)+1}\left(\sum_{(\nu_1,\dots,\nu_{n-1})}^{*r}\prod_{s=1}^{n-1} h_{\nu_s}\left(r+s-\sum_{t=1}^{s}\nu_t\right)h_{r+(n-1)-\sum_{t=1}^{n-1}\nu_t}\right)x^r$$

for $n\geq 1$. *In* $\sum_{(\nu_1,\dots,\nu_{n-1})}^{*r}$ *we are taking the sum over all* $(n-1)$*-tuples* $(\nu_1,\dots,\nu_{n-1})$ *of integers, such that* $k\leq \nu_s\leq r-(n-s)k+(n-1)-\sum_{t=1}^{s-1}\nu_t$.
This theorem shows that the coefficient $P_{r,n}$ of x^r in $\phi_n(x)$ depends only on the elements $h_k,\dots,h_{r-(n-1)(k-1)}$.

4.3.6 Normal Forms of Iteration Groups of Type (II, k)

Assume that $G(y,x)$ is a formal iteration group of type (II, k) for some $k\geq 2$, i.e., G is a solution of (Tform, (II, k)) and (B, (II, k)). For all $S\in\Gamma_1$ the series

$$\tilde{G}(y,x) := S^{-1}\big(G(y,S(x))\big)$$

is also a solution of (Tform, (II, k)) and (B, (II, k)). Assume that H is the infinitesimal generator of G, then according to (4.1) the infinitesimal generator of $\tilde{G}$ is

$$\tilde{H}(x) = \left[S'(x)\right]^{-1}H\big(S(x)\big).$$

This differential equation for S is not a Briot–Bouquet equation. However, it can be reduced to such an equation by putting $S(x) = x\exp(\theta(x))$, where $\theta(x)\in\mathbb{C}[\![x]\!]$, $\theta(0) = 0$. For each $H(x) = \sum_{n\geq k} h_n x^n$, $k\geq 2$, $h_k = 1$, there exist some $S(x)\in\Gamma_1$ and exactly one $h\in\mathbb{C}$, so that

$$\tilde{H}(x) = x^k + hx^{2k-1}.$$

This is the *normal form* of the generator of a formal iteration group of type (II, k). (A direct proof not using the theory of Briot–Bouquet equations can be found in [5, Theorem 28].) We say that a (formal) iteration group of type (II, k) with generator

$$H(x) = x^k + hx^{2k-1}, \qquad h \in \mathbb{C},$$

is a *normal form* and we describe these normal forms in the next theorems.

Theorem 4.19 ([5, Theorem 29]) *Consider some $k \geq 2$. The solution of* (PDform, (II, k)) *and* (B, (II, k)) *for $H(x) = x^k + hx^{2k-1}$ is given by*

$$G(y, x) = \sum_{n \geq 0} P_{n(k-1)+1}(y) x^{n(k-1)+1}$$

where

$$P_{n(k-1)+1}(y) = \begin{cases} 1 & \text{if } n = 0 \\ y & \text{if } n = 1 \\ \prod_{i=1}^{n-1} \big(i(k-1)+1\big) \frac{y^n}{n!} + hQ_n(y, h) & \text{if } n \geq 2, \end{cases}$$

and where $Q_n(y, h)$, $n \geq 2$, is a polynomial in y of degree $n - 1$ and a polynomial in h of degree $\lfloor n/2 \rfloor - 1$.

Now we assume that h is an indeterminate over $(\mathbb{C}[y])[\![x]\!]$. It is interesting to note that the normal forms of iteration groups of type (II, k) have expansions in powers of the parameter h. Since for $n \geq 2$ the degree of $P_{n(k-1)+1}(y)$ as a polynomial in h is $\lfloor n/2 \rfloor$, we can write $G(y, x)$ as

$$G(y, x) = \sum_{r \geq 0} G_r(y, x) h^r \in (\mathbb{C}[\![x, y]\!])[\![h]\!].$$

From (B, (II, k)) we deduce that $G_0(0, x) = x$ and $G_r(0, x) = 0$ for $r \geq 1$. Instead of (PDform, (II, k)) we obtain

$$\begin{aligned} \sum_{r \geq 0} \frac{\partial}{\partial y} G_r(y, x) h^r &= \big(x^k + hx^{2k-1}\big) \left(\sum_{r \geq 0} \frac{\partial}{\partial x} G_r(y, x) h^r \right) \\ &= \sum_{r \geq 0} x^k \frac{\partial}{\partial x} G_r(y, x) h^r + \sum_{r \geq 0} x^{2k-1} \frac{\partial}{\partial x} G_r(y, x) h^{r+1} \end{aligned}$$

This is a system of equations for $G_r(y, x)$, $r \geq 0$, given by

$$\frac{\partial}{\partial y} G_0(y, x) = x^k \frac{\partial}{\partial x} G_0(y, x) \tag{4.4}$$

and

$$\frac{\partial}{\partial y} G_r(y,x) = x^k \frac{\partial}{\partial x} G_r(y,x) + x^{2k-1} \frac{\partial}{\partial x} G_{r-1}(y,x), \qquad r \geq 1. \tag{4.5}$$

Theorem 4.20 ([5, Theorem 30]) *Consider* $H(x) = x^k + hx^{2k-1}$ *where* h *is an indeterminate over* $\mathbb{C}[\![x,y]\!]$. *The solution of* (4.4), (4.5), *and* (B, (II, k)) *is given by*

$$\sum_{r\geq 0} G_r(y,x)h^r$$

where

$$G_r(y,x) = \sum_{n\geq r} \sum_{\substack{(j_1,\dots,j_r)\\ 1\leq j_1\\ j_s\geq j_{s-1}+2,\ s\geq 2\\ j_r\leq n+r-1}} \frac{\prod_{i=1}^{n+r-1}[i]}{\prod_{s=1}^{r}[j_s]} \frac{x^{[n+r]}}{n!} y^n, \qquad r\geq 0,$$

where $[r] = r(k-1)+1$.

Concerning the differential equation (Dform, (II, k)) we have

Theorem 4.21 ([5, Theorem 33]) *Consider some* $k \geq 2$. *The solution of* (Dform, (II, k)) *and* (B, (II, k)) *for* $H(x) = x^k + hx^{2k-1}$ *is given by*

$$G(y,x) = \sum_{r\geq 0} G_r(y,x)h^r$$

with

$$G_r(y,x) = x^{[r]}(1-(k-1)yx^{k-1})^{-[r]/(k-1)} P_r(\ln(1-(k-1)yx^{k-1})), \qquad r\geq 0,$$

where $[r] = r(k-1)+1$ *and* P_r *are polynomials of degree* r. *Moreover* $P_0 = 1$ *and*

$$P_1(z) = -z/(k-1).$$

The binomial series is used in order to compute

$$(1-(k-1)yx^{k-1})^{-[r]/(k-1)}.$$

The particular situation $r = 0$ yields

$$G_0(y,x) = x(1-(k-1)yx^{k-1})^{-1/(k-1)}.$$

$G_0(y,x)$ together with its conjugates occur in the problem of reversible power series (c.f. [12, Section 0.3]).

There exists also an approach with *Lie–Gröbner-series* (cf. [8] or [9, Chapter 1]) to solve (PDR, (II, k)) and (BR, (II, k)). We note that Lie–Gröbner-series in the context of iteration groups have already been used by St. Scheinberg [29] and also by Reich and Schwaiger in [27]. Define an operator

$$D: \mathbb{C}[\![x]\!] \to \mathbb{C}[\![x]\!], \qquad D(f(x)) := H(x)f'(x).$$

Lemma 4.3 ([5, Lemma 23]) *Let H be a generator of order $k \geq 2$. If $(\phi_n)_{n\geq 0}$ satisfies the system* (PDR, (II, k)) *and* (BR, (II, k))*, then*

$$\phi_n(x) = \frac{1}{n!} D^n(x), \qquad n \geq 0.$$

Theorem 4.22 ([5, Theorem 24]) *The series*

$$G(y, x) := \sum_{n\geq 0} \frac{1}{n!} D^n(x)y^n,$$

is a Lie–Gröbner-series. It satisfies (Tform, (II, k)) *and* (B, (II, k)).

4.4 Concluding Remarks and Open Problems

At the end of this paper we present some open problems concerning the construction of iteration groups.

1. It is an important problem to study iteration groups in higher dimension. This means in our situation to change to the ring $\mathbb{C}[\![x_1, \ldots, x_n]\!]$ of formal power series in $n \geq 2$ indeterminates $\mathbf{x} = (x_1, \ldots, x_n)^T$ over $\mathbb{C}$ and to consider n-tuples

$$F(\mathbf{x}) = F\begin{pmatrix} x_1 \\ \vdots \\ x_n \end{pmatrix} = \begin{pmatrix} F_1((x_1, \ldots, x_n)^T) \\ \vdots \\ F_n((x_1, \ldots, x_n)^T) \end{pmatrix} = \begin{pmatrix} F_1(\mathbf{x}) \\ \vdots \\ F_n(\mathbf{x}) \end{pmatrix},$$

i.e., elements of $(\mathbb{C}[\![\mathbf{x}]\!])^n$. By $\operatorname{ord}(F(\mathbf{x}))$ we understand $\min\{\operatorname{ord}(F_1), \ldots, \operatorname{ord}(F_n)\}$. We consider the substitution of $G(\mathbf{x}) \in (\mathbb{C}[\![\mathbf{x}]\!])^n$ into $F(\mathbf{x}) \in (\mathbb{C}[\![\mathbf{x}]\!])^n$ provided that $\operatorname{ord}(G) \geq 1$.

Each $F(\mathbf{x}) \in (\mathbb{C}[\![\mathbf{x}]\!])^n$ can be written as $F(\mathbf{x}) = A \cdot \mathbf{x} + R(\mathbf{x})$, where A is a complex $n \times n$-matrix and $R(\mathbf{x}) \in (\mathbb{C}[\![\mathbf{x}]\!])^n$ with $\operatorname{ord}(R) \geq 2$. If $\det(A) \neq 0$ we call F a *formally biholomorphic* mapping. The set of all formally biholomorphic mappings forms a group $\boldsymbol{\Gamma}$ with respect to substitution $\circ$, and a family $(F_t(\mathbf{x}))_{t\in\mathbb{C}}$, $F_t(\mathbf{x}) \in \boldsymbol{\Gamma}$, satisfying the translation equation

$$F_{s+t} = F_s \circ F_t, \qquad s, t \in \mathbb{C}, \tag{T}$$

is called an iteration group in n dimensions.

The construction of all iteration groups of dimension $n \geq 2$ is an open problem and very likely the method of formal functional equations and differential equations will lead to a solution.

Mehring has shown in [22, 23] that the coefficient functions of an iteration group are polynomials in a finite number of additive or generalized exponential functions, however, the detailed structure is not known.

2. Jabłoński and Reich studied in [19] the iteration groups of truncated formal power series. It is an open question how to construct these groups using the method of formal functional equations.
3. The method of formal functional equations should also be applied in the problem of constructing maximal abelian subgroups of Γ or $\boldsymbol{\Gamma}$, in particular in higher dimension.
4. The various representations of the coefficient functions of iteration groups presented in this paper and the representations obtained by Jabłoński and Reich have so far not been compared by direct computation. This could yield interesting polynomial identities.
5. We notice that from the representation $G(y,x) = S(yS^{-1}(x))$ given in (4.3) we can derive a representation

$$G(y,x) = yx + \sum_{\nu \geq 2} Q_\nu(y, s_2, \ldots, s_\nu)x^\nu$$

where each Q_ν is a polynomial in y and in the coefficients $s_2, \ldots, s_\nu$ of $S(x) = x + s_2x^2 + \ldots$. Formula (4.2) describes a connection between the generator $H(x) = x + h_2x^2 + \ldots$ and the conjugating series $S(x)$. This gives eventually another (maybe new) representation of the coefficients P_n of G from Theorem 4.3 as polynomials in y and $h_2, \ldots, h_n$.

Acknowledgements We would like to thank the anonymous referee for helpful comments.

References

1. Cartan, H.: Elementary Theory of Analytic Functions of One or Several Complex Variables. Addison-Wesley, Reading, Palo Alto, London (1963)
2. Chen, K.T.: Local diffeomorphisms – C^∞-realization of formal properties. Am. J. Math. **87**, 140–157 (1965)
3. Fripertinger, H., Reich, L.: Covariant embeddings of the linear functional equation with respect to an iteration group in the ring of complex formal power series. Grazer Math. Ber. **350**, 96–121 (2006)
4. Fripertinger, H., Reich, L.: The formal translation equation and formal cocycle equations for iteration groups of type I. Aequationes Math. **76**, 54–91 (2008)
5. Fripertinger, H., Reich, L.: The formal translation equation for iteration groups of type II. Aequationes Math. **79**, 111–156 (2010)
6. Fripertinger, H., Reich, L.: On the formal first cocycle equation for iteration groups of type II. In: Fournier-Prunaret, D., Gardini, L., Reich, L. (eds.) European Conference on Iteration Theory 2010, pp. 32–47. EDP Science, Les Ulis (2012)

7. Fripertinger, H., Reich, L.: On the formal second cocycle equation for iteration groups of type II. J. Differ. Equ. Appl. **21**, 564–578 (2015)
8. Gröbner, W.: Die Lie-Reihen und ihre Anwendungen. 2. VEB Deutscher Verlag der Wissenschaften, Berlin (1967)
9. Gröbner, W., Knapp, H.G.: Contributions to the Method of Lie Series. Bibliographisches Institut, Mannheim (1967)
10. Gronau, D.: Two iterative functional equations for power series. Aequationes Math. **25**, 233–246 (1982)
11. Gronau, D.: Über die multiplikative Translationsgleichung und idempotente Potenzreihenvektoren. Aequationes Math. **28**, 312–320 (1985)
12. Haneczok, J.: Conjugacy type problems in the ring of formal power series. Grazer Math. Ber. **353**, 96 pp. (2009)
13. Henrici, P.: Applied and Computational Complex Analysis. Vol. I: Power Series, Integration, Conformal Mapping, Location of Zeros. Wiley, New York, London, Sydney (1974)
14. Hille, E.: Ordinary Differential Equations in the Complex Domain. Wiley, New York, London, Sydney (1976)
15. Ince, E.: Ordinary Differential Equations. Dover, New York (1956)
16. Jabłoński, W.: One-parameter groups of formal power series of one indeterminate. In: Rassias, Th.M., Brzdek, J. (eds.) Functional Equations in Mathematical Analysis, pp. 523–545. Springer, Berlin (2011)
17. Jabłoński, W., Reich, L.: On the form of homomorphisms into the differential group L_s^1 and their extensibility. Results Math. **47**, 61–68 (2005)
18. Jabłoński, W., Reich, L.: On the solutions of the translation equation in rings of formal power series. Abh. Math. Sem. Univ. Hamburg **75**, 179–201 (2005)
19. Jabłoński, W., Reich, L.: A new approach to the description of one-parameter groups of formal power series in one indeterminate. Aequationes Math. **87**, 247–284 (2014)
20. Laine, I.: Introduction to local theory of complex differential equations. In: Laine, I. (ed.) Complex Differential and Functional Equations, pp. 81–106. University of Joensuu, Joensuu (2003)
21. Lewis, D.C. Jr.: Formal power series transformations. Duke Math. J. **5**, 794–805 (1939)
22. Mehring, G.H.: Iteration im Ring der formalen Potenzreihen ohne Regularitätsvoraussetzungen. Ber. Math.-Statist. Sekt. Forsch. Graz **265**, 45 pp. (1986)
23. Mehring, G.H.: Der Hauptsatz über Iteration im Ring der formalen Potenzreihen. Aequationes Math. **32**, 274–296 (1987)
24. Peschl, E., Reich, L.: Beispiel einer kontrahierenden biholomorphen Abbildung, die in keine Liesche Gruppe biholomorpher Abbildungen einbettbar ist. Bayer. Akad. Wiss. Math.-Natur. Kl. S.-B. **1971**, 81–92 (1972)
25. Praagman, C.: Roots, iterations and logarithms of formal automorphisms. Aequationes Math. **33**, 251–259 (1987)
26. Reich, L.: On families of commuting formal power series. Berichte der Mathematisch-statistischen Sektion der Forschungsgesellschaft Joanneum Graz **294**, 285–296 (1988)
27. Reich, L., Schwaiger, J.: Über die analytische Iterierbarkeit formaler Potenzreihenvektoren. Sitzungsberichte der Österr. Akademie der Wissenschaften, Abt. II **184**, 599–617 (1975)
28. Reich, L., Schwaiger, J.: Über einen Satz von Shl. Sternberg in der Theorie der analytischen Iterationen. Monatsh. Math. **83**, 207–221 (1977)
29. Scheinberg, S.: Power series in one variable. J. Math. Anal. Appl. **31**, 321–333 (1970)
30. Sternberg, S.: Infinite Lie groups and formal aspects of dynamical systems. J. Math. Mech. **10**, 451–474 (1961)

Chapter 5
Fischer–Muszély Additivity: A Half Century Story

Roman Ger

Abstract This is an extended version of my talk presented at *the 30th International Summer Conference on Real Functions Theory* that was held in Stará Lesná (Slovakia) from September 4 to 9, 2016.

Keywords Fischer–Muszély equation (additivity) • Strictly convex spaces • General solution • The hierarchy of (non)commutativity • Pexiderization • Fischer–Muszély type inequalities • Stability

Mathematics Subject Classification (2010) Primary 39B52; Secondary 39B82, 49B99

5.1 Background

In the beginning was the word (of Fischer and Muszély in Hungarian and English: *A Cauchy-féle függvényegyenletek bizonyos típusú általánosí-tásai* (see [11]) and *On some new generalizations of the functional equation of Cauchy* (see [12]):

Examining certain problems in physics M. Hosszu (Észrevételek a relativitáselméleti időfogalom Reichenbach-féle értelmezéséhez, NME magyarnyelvű Közleményi Miskolc (1964), 223–233) *obtained the functional equation*

$$f(x+y)^2 = [f(x)+f(y)]^2, \tag{$*$}$$

where x, y, f are real.

R. Ger (✉)
Institute of Mathematics, University of Silesia, Bankowa 14, 40-007 Katowice, Poland
e-mail: roman.ger@us.edu.pl

J. Brzdęk et al. (eds.), *Developments in Functional Equations and Related Topics*, Springer Optimization and Its Applications 124,
DOI 10.1007/978-3-319-61732-9_5

In another paper of M. Hosszu (Egy alternativ fűggvényegyenletrő, Mat. Lapok 14 (1963), 98–102) *proved that Equation* $(*)$ *is equivalent to the functional equation of Cauchy, i.e. to the equation*

$$f(x+y) = f(x) + f(y) \tag{$**$}$$

H. Światak examined in (On the equation $\varphi(x+y)^2 = [\varphi(x)g(y) + \varphi(y)g(x)]^2$, Zeszyty Naukowe Uniwersytetu Jagiellońskiego, Nr II. Prace Matematyczne, Zeszyt 10 (1965), 97–104) *a generalization of Equation* $(*)$ *in the class of continuous functions.*

A similar alternative functional equation is considered in a paper of J. Aczél, K. Fladt and M. Hosszu (Lösungen einer mit dem Doppelverhältnis zusamenhängender Funktionalgleichung, MTA Mat. Kut. Int. Közl 7A (1962), 335–352).

At the end of his paper M. Hosszu puts the question: what is the general solution of Equation $(*)$*?*

E. Vincze was the first to give an answer to this question in his papers

- *Alternativ fűggvényegyenletek megoldásairól,* Mat Lapok 14 (1963), 179–195*;*
- *Beitrag zur Theorie der Cauchyschen Funktionalgleichungen,* Arch. Mat. 15 (1964), 132–135*;*
- *Über eine Verallgemeinerung der Cauchyschen Funktionalgleichung, Funkcialaj Ekvacioj 6 (1964), 55–62.*

He proved that the functional equation

$$f(x+y)^n = [f(x) + f(y)]^n$$

is equivalent to the functional equation of Cauchy, where x, y *are in an additive Abelian semigroup,* f *is an arbitrary complex-valued function and* n *is a natural number.*

5.2 Fischer–Muszély Equation

Plainly, Equation $(*)$ may equivalently be written in the form

$$|f(x+y)| = |f(x) + f(y)|$$

and, if so, why not to replace the absolute value sign by the norm?

Throughout the years the functional equation

$$\|f(x+y)\| = \|f(x) + f(y)\| \tag{FM}$$

has extensively been studied by many authors, see, e.g., Fischer and Muszély [11, 12], Dhombres [9], Aczél and Dhombres [1], Berruti and Skof [4], Skof [28], Ger [16–22], Schöpf [27], Ger and Koclęga [23], Száz [29]. The reason why this functional equation was attracting so much attention is, on the one side, the facts established in the papers spoken of in the Background and, on the other side, because of its links with the theory of isometries; moreover, it leads to some characterizations of strictly convex normed linear spaces as well as to some of their generalizations. The main result from [18] states that any map f from a (not necessarily commutative) group into a strictly convex space has to be additive, i.e. to satisfy the Cauchy equation

$$f(x+y) = f(x) + f(y). \tag{C}$$

On the other hand, already in 1979 Dhombres [9] exhibited an example of a continuous solution $f : \mathbb{R} \to X$ of Equation $(*)$ that fails to satisfy (C).

In the case where the domain $\mathbb{R}$ is replaced by the halfline $[0, \infty)$ one may "produce" a rich family of C^∞-nonadditive solutions of Equation (FM).

This inspired Schöpf [27] to look for a description of all continuous (resp. differentiable) solutions of (FM) mapping the real line $\mathbb{R}$ into a not necessarily strictly convex normed linear space $(X, \|\cdot\|)$. Looking for some alternative representations Ger and Koclęga [23] have shown that any function f of that kind fulfilling merely very mild regularity assumptions has to be proportional to an odd isometry mapping $\mathbb{R}$ into X.

Last but not least, in 2003, Tabor [31] has obtained the additivity of surjective solutions to (FM).

Theorem 5.1 (Fischer and Muszély [11]) *Let* $(X, +)$ *be a semigroup and* $(Y, (\cdot|\cdot))$ *be a unitary space. Let further* $f : X \to Y$ *be a solution to functional equation* (FM). *Then* f *is additive.*

Problem Is it possible to replace the unitary target space by a strictly convex one?

Numerous characterizations of strictly convex spaces are known (see, e.g., the monograph of Day [6]). Among them the following one was given by Dhombres in [9]

A normed space (real or complex) $(X, \|\cdot\|)$ *is strictly convex if and only if each function* $f : \mathbb{R} \to X$ *belonging to the class*

$$\mathscr{F} := \{g : \mathbb{R} \to X : \ g \text{ has a measurable majorant on a set of positive measure}\}$$

and satisfying the functional equation (FM) *has to be additive.*

Moreover, Dhombres writes (p. 2.28 in [9]): *The problem of determining those normed spaces characterized by the equivalence of Equation* (FM) *and the equation of additivity, even in the case of the domain being some group like the additive* $\mathbb{R}$, *remains open.*

Actually, to show that the space considered is strictly convex it suffices to consider only continuous solutions of Equation (FM) (see Aczél and Dhombres [1] and Theorem 5.4 below). But while studying logical connections between (FM) and additivity it seems desirable indeed to get rid of the class $\mathscr{F}$. This is actually possible; namely, we have the following:

Theorem 5.2 *Let* $(G,+)$ *be a group (not necessarily commutative) and let* $(X, \|\cdot\|)$ *be a strictly convex space. Then every function* $f : G \to X$ *satisfying Equation* (FM) *for all* $x, y \in G$ *is additive.*

Proof Without the use of strict convexity one may show [see Dhombres (p. 2.23 in [9])] that the equality

$$\|f(2x)+f(x)\| = \|f(2x)\| + \|f(x)\|$$

holds true for all $x \in G$. Then strict convexity implies that for every $x \in G$ such that $f(x) \neq 0 \neq f(2x)$ there exists a positive number $\lambda(x)$ such that $f(2x) = \lambda(x)f(x)$. Since we obviously have

$$\|f(2x)\| = 2\|f(x)\|, \qquad x \in G, \tag{5.1}$$

we infer that $\lambda(x) = 2$ whenever $f(x) \neq 0 \neq f(2x)$. However, in view of (5.1), if one of the values $f(x)$ or $f(2x)$ vanishes, then so does the other; consequently, the equality $f(2x) = 2f(x)$ is fulfilled for all elements x from G.

Putting $y = -x$ in (FM) and taking into account that (5.1) implies the equality $f(0) = 0$, we derive the oddness of f. Now, observe that for all $x, y \in G$ one has

$$\| f(x+y) - \frac{1}{2} f(x) \| = \| \frac{1}{2} f(x) + f(y) \| = \| \frac{f(x+y)+f(y)}{2} \| . \tag{5.2}$$

In fact,

$$\begin{aligned} 2 \| f(x+y) - \frac{1}{2} f(x) \| &= \| 2f(x+y) - f(x) \| = \| f(x+y+x+y) + f(-x) \| \\ &= \| f(y+x+y) \| = \| f(x+y) + f(y) \|, \end{aligned}$$

and, on the other hand,

$$\begin{aligned} 2 \| \frac{1}{2} f(x) + f(y) \| &= \| f(x) + 2f(y) \| = \| f(x) + f(2y) \| \\ &= \| f(x+2y) \| = \| f(x+y) + f(y) \|, \end{aligned}$$

which ends the proof of (5.2). Fix arbitrarily x and y from G and put $u := f(x+y) - \frac{1}{2} f(x)$ and $v := \frac{1}{2}f(x) + f(y)$; then (5.2) states that

$$\| u \| = \| v \| = \| \frac{u+v}{2} \|,$$

which, in view of the strict convexity of X, gives $u = v$. Thus

$$f(x+y) = f(x) + f(y),$$

which was to be proved. □

Remark 5.1 Under the assumption that the group considered is uniquely 2-divisible this result was presented by the author at the 26-th International Symposium on Functional Equations (Catalonia, 1988); see [16]. A year later, during the 27-th ISFE, the present version as well as its detailed proof was presented; see [17]. Assuming that the domain of the function in question yields a real linear space, in 1991 Berruti and Skof (Lemma fondamentale in [4]) proved the analogous assertion. Their proof relies essentially on Baker's lemma from [3].

Below we derive Baker's main result of [3] from ours.

Theorem 5.3 (Baker) *Let $(E, \|\cdot\|)$ and $(X, \|\cdot\|)$ be two real normed linear spaces and let $f : E \to X$ be an isometry. If the target space is strictly convex, then f has to be an affine function, i.e. there exists a constant $c \in X$ and a linear map $L : E \to X$ such that $f(x) = L(x) + c$ for all $x \in E$.*

Proof Put $c := f(0)$ and $L := f - c$. Then L is an isometry as well and $L(0) = 0$. Consequently,

$$\| L(x) - L(y) \| = \| x - y \| = \| L(x-y) \| \tag{5.3}$$

for all $x, y \in E$. Putting here $y = -x$, one gets

$$\| L(x) - L(-x) \| = 2 \| x \| = \| L(x) \| + \| -L(-x) \|$$

which, by means of the strict convexity of X, implies the oddness of L. This, jointly with (5.3), implies that the equality

$$\| L(x+y) \| = \| L(x) + L(y) \|$$

holds true for all $x, y \in E$. An appeal to Theorem 5.2 gives now the additivity of L which, being continuous, has to be linear. This ends the proof. □

The following characterization of strictly convex spaces in terms of the equivalence of Equation (FM) and the Cauchy functional equation yields a slight refinement of a result given by Aczél and Dhombres (p. 138 in [1]).

Theorem 5.4 *A normed linear space $(X, \|\cdot\|)$ is strictly convex if and only if for every its two-dimensional subspace $Y \subset X$ the functions*

$$f_c(x) = x \cdot c, \qquad x \in \mathbb{R},$$

where c stands for an arbitrarily fixed element of Y, are the only continuous solutions $f : \mathbb{R} \to Y$ of Equation (FM).

Proof Necessity. Fix a two-dimensional subspace Y of X and a continuous solution $f : \mathbb{R} \to Y$ of Equation (FM). Obviously, $(Y, \|\cdot\|)$ is strictly convex; therefore, by means of Theorem 5.2, f is additive and being continuous has to have the form f_c for some $c \in Y$.

Sufficiency. Assume, for the indirect proof, that $(X, \|\cdot\|)$ is not strictly convex. Then there exist elements $a, b \in X$, $a \neq b$ such that

$$\|a\| = \|b\| = \| \frac{a+b}{2} \| = 1.$$

Such vectors are linearly independent; in fact, if we had $b = \lambda a$ for some scalar λ (real or complex) we would get $|\lambda| = 1$ and $|1+\lambda| = 2$ implying the equality $\lambda = 1$, which is impossible. Consequently, the space $Y := \mathrm{Lin}\{a, b\}$ is two-dimensional. A continuous function

$$f(x) := \begin{cases} x \cdot a & \text{for } x \in [-1, 1] \\ a + (x-1) \cdot b & \text{for } x \in (1, \infty) \\ -a + (x+1) \cdot b & \text{for } x \in (-\infty, -1), \end{cases}$$

mapping $\mathbb{R}$ into Y yields a solution to (FM) (see Dhombres [9] or Aczél and Dhombres [1]) which obviously fails to be an f_c function. This contradiction completes the proof. □

Now, we are going to show that our Theorem 5.2 carries over to the case of linear topological spaces, topologized through families of suitable seminorms. To this aim, we shall first recall the definition introduced by Diminnie and White Jr. in [7]. Let X be a linear space and let $\mathscr{P}$ be a nonempty family of nonzero seminorms on X. For $p \in \mathscr{P}$ we put $N_p := \{x \in X : p(x) = 0\}$. The pair $(X, \mathscr{P})$ is said to be strictly convex if and only if for every $p \in \mathscr{P}$ and every $a, b \in X$ the conditions

$$p(a) = p(b) = p\left(\frac{a+b}{2}\right) = 1 \qquad \text{and} \qquad N_p \cap \mathrm{Lin}\{a, b\} = \{0\}$$

imply that $a = b$. Without loss of generality, in what follows, we shall be assuming that the family $\mathscr{P}$ consists of just a single seminorm: $\mathscr{P} = \{p\}$.

Theorem 5.5 *Let $(G, +)$ be a group (not necessarily commutative) and let X be a linear space endowed with a nonzero seminorm p such that the pair $(X, \{p\})$ is strictly convex. Suppose that $f : G \to X$ satisfies the functional equation*

$$p(f(x+y)) = p(f(x) + f(y)), \qquad x, y \in G. \tag{5.4}$$

Then there exists exactly one additive function $a : G \to X$ and exactly one function $n : G \to N_p$ such that

$$f(x) = a(x) + n(x), \qquad x \in G;$$

in particular,

$$p(f(x+y)-f(x)-f(y))=0, \qquad x,y\in G.$$

Proof One of the four equivalent conditions for a pair $(X,\{p\})$ to be strictly convex given by Diminnie and White Jr. in [8] states that there exists a strictly convex normed space $(Y,\|\cdot\|)$ and a linear mapping $F: X\to Y$ such that $p(x)=\|F(x)\|$ for all $x\in X$. Consequently, Equation (5.4) says that

$$\|F(f(x+y))\|=\|F(f(x)+f(y))\|=\|F(f(x))+F(f(y))\|, \quad x,y\in G.$$

Putting $g:=F\circ f$ we obtain

$$\|g(x+y)\|=\|g(x)+g(y)\|$$

for all $x,y\in G$ and, by the strict convexity of the space $(Y,\|\cdot\|)$, Theorem 5.2 implies the additivity of the map g; in other words

$$F(f(x+y))=F(f(x))+F(f(y)), \qquad x,y\in G.$$

Now, the additivity of F gives

$$C_f(x,y):=f(x+y)-f(x)-f(y)\in \ker F,$$

whence

$$p(C_f(x,y))=\| F(C_f(x,y)) \|=0,$$

i.e. $C_f(x,y)\in N_p$ for all $x,y\in G$.

Let N_p^c denote the complementary space to the linear subspace N_p of the space X. Then, for every $x\in G$, the value $f(x)$ can uniquely be factorized as $a(x)+n(x)$, where $a(x)\in N_p^c$ and $n(x)\in N_p$. Since, for any $x,y\in G$, one has

$$N_p^c \ni a(x+y)-a(x)-a(y)=C_f(x,y)-n(x+y)+n(x)+n(y)\in N_p,$$

the function a is additive, which finishes the proof. □

Now we are going to present an example illustrating the utility of Theorem 5.2 while solving some functional equations.

Assume that we are given a (not necessarily commutative) group $(G,+)$ and real numbers α,β,γ such that

$$\alpha>0 \qquad \text{and} \qquad \beta^2-4\alpha\gamma<0. \tag{5.5}$$

We will find the general solution of the functional equation

$$\alpha\left[\varphi(x+y)^2-(\varphi(x)+\varphi(y))^2\right]+\beta\left[\varphi(x+y)\psi(x+y)\right.$$
$$\left.-(\varphi(x)+\varphi(y))(\psi(x)+\psi(y))\right]+\gamma\left[\psi(x+y)^2-(\psi(x)+\psi(y))^2\right]=0 \qquad \text{(e)}$$

in the class of all functions $\varphi, \psi : G \to \mathbb{R}$. An easy calculation shows that Equation (e) may equivalently be written in the form

$$\left(\begin{bmatrix}\alpha, & \frac{1}{2}\beta\\ \frac{1}{2}\beta, & \gamma\end{bmatrix}\cdot\begin{bmatrix}\varphi(x+y)\\ \psi(x+y)\end{bmatrix}\,\middle|\,\begin{bmatrix}\varphi(x+y)\\ \psi(x+y)\end{bmatrix}\right)$$
$$=\left(\begin{bmatrix}\alpha, & \frac{1}{2}\beta\\ \frac{1}{2}\beta, & \gamma\end{bmatrix}\cdot\begin{bmatrix}\varphi(x)+\varphi(y)\\ \psi(x)+\psi(y)\end{bmatrix}\,\middle|\,\begin{bmatrix}\varphi(x)+\varphi(y)\\ \psi(x)+\psi(y)\end{bmatrix}\right)$$

for all $x, y \in G$; here $(\cdot|\cdot)$ stands for the usual inner product in $\mathbb{R}^2$. Let us put

$$A := \begin{bmatrix}\alpha, & \frac{1}{2}\beta\\ \frac{1}{2}\beta, & \gamma\end{bmatrix} \qquad \text{and} \qquad f(x) := (\varphi(x), \psi(x)), \qquad x \in G.$$

Then the latter equation states that

$$(A\cdot f(x+y)\mid f(x+y)) = (A\cdot(f(x)+f(y))\mid f(x)+f(y))$$

for all $x, y \in G$. Since conditions (5.5) guarantee that the matrix A is positive definite the formula

$$\langle u|v\rangle := (A\cdot u\,|v), \qquad u, v \in \mathbb{R}^2,$$

produces a new inner product in $\mathbb{R}^2$ and the equation considered assumes the form

$$\| f(x+y)\|^2 = \| f(x)+f(y)\|^2, \qquad x, y \in G,$$

where $\|u\|^2 = \langle u|u\rangle, u \in \mathbb{R}^2$. Since any inner product space is obviously strictly convex, Theorem 5.2 establishes the additivity of f and hence that of the component functions φ and ψ. Conversely, every pair of additive functions $\varphi, \psi : G \to \mathbb{R}$ yields a solution to Equation (e).

5.3 General Solution

In what follows, we are presenting a factorization of the general solution of Equation (FM) for functions mapping a commutative group into a real normed linear space (with no regularity assumptions whatsoever), into isometric and additive mappings. We believe that, in this way, we have finally achieved a clear explanation of seemingly divergent earlier approaches focused on different endeavours either to show that (FM) implies additivity or to express the solutions of (FM) in terms of isometries.

5.3.1 Preliminary Results

Given an Abelian group $(X, +)$ we call a function p mapping X into the set $\mathbb{R}$ of all real numbers *sublinear* provided that p is *subadditive*, i.e.

$$p(x + y) \leq p(x) + p(y), \qquad x, y \in X,$$

and satisfies a homogeneity condition

$$p(nx) = np(x),$$

for all $x \in X$ and all $n \in \mathbb{N}_0$ (nonnegative integers).

The following Hahn–Banach type theorem is a special case of Kranz's result (Theorem 2 in [24]).

Lemma 5.1 *Let $(X, +)$ be an Abelian group and let $(X_0, +)$ stand for a subgroup of $(X, +)$. Assume that we are given a sublinear functional $p : X \to \mathbb{R}$ and an additive functional $a_0 : X_0 \to \mathbb{R}$ such that*

$$a_0(x) \leq p(x), \qquad x \in X_0.$$

Then there exists an additive extension $a : X \to \mathbb{R}$ of a_0 such that

$$a(x) \leq p(x), \qquad x \in X.$$

As a matter of fact, the sublinearity assumption on the functional p above might simply be replaced by subadditivity alone but, in the sequel, we will need the following corollary in which sublinearity is actually essential.

Corollary 5.1 *Let $(X, +)$ be an Abelian group and let $x_0 \in X$. Given an even sublinear functional $p : X \to \mathbb{R}$ there exists an additive functional $a : X \to \mathbb{R}$ such that $a \leq p$ and $a(x_0) = p(x_0)$.*

Proof Denote by $\mathbb{Z}$ the set of all integers and put $X_0 := \{nx_0 : n \in \mathbb{Z}\}$. Obviously, a functional $a_0 : X_0 \to \mathbb{R}$ is unambiguously defined by the formula

$$a_0(nx_0) := np(x_0), \qquad n \in \mathbb{Z};$$

moreover, a_0 is additive and majorized by p on X_0 since p, being even, has to be nonnegative. Now, it suffices to apply Lemma 5.1 to complete the proof. □

In what follows, we are going to show that the validity of Fischer's conjecture (see [13] and Kuczma [25]) stating that an (even!) sublinear functional p admits a representation of the form $p = \|\cdot\| \circ A$ where $A : X \to Z$ stands for an additive map with values in a suitable real normed linear space $(Z, \|\cdot\|)$, carries over to groups. The idea of the proof is based on the paper of Berz [5]; we have only to ensure that the passage to commutative group domains is possible. To proceed we need yet another lemma.

Lemma 5.2 *Let $(X, +)$ be an Abelian group and let $p : X \to \mathbb{R}$ be an even sublinear functional. Then the equality*

$$p(x) = \sup\{a(x) : a : X \to \mathbb{R} \textit{ is additive and } a \le p\}$$

holds true for all $x \in X$.

Proof By virtue of Corollary 5.1, the family T of all additive real functionals a on X majorized by p is nonvoid.

Therefore, the formula

$$\tilde{p}(x) = \sup\{a(x) : a \in T\}, \quad x \in X,$$

correctly defines a functional $\tilde{p} : X \to \mathbb{R}$. Plainly, we have $\tilde{p} \le p$. On the other hand, by means of Corollary 5.1 again, for an arbitrarily fixed $x_0 \in X$ there exists an $a \in T$ such that $p(x_0) = a(x_0) \le \tilde{p}(x_0)$. Thus, $p \le \tilde{p}$, which finishes the proof. □

In the sequel, as usual, given a nonempty set T by $B(T, \mathbb{R})$ we denote a Banach space of all bounded real functions on T, equipped with the uniform convergence norm $\|\cdot\|_\infty$.

Theorem 5.6 *Let $(X, +)$ be an Abelian group and let $p : X \to \mathbb{R}$ be an even sublinear functional. Then there exists a nonempty set $T \subset \mathbb{R}^X$ and an additive operator $A : X \to B(T, \mathbb{R})$ such that*

$$p(x) = \|A(x)\|_\infty, \qquad x \in X.$$

Proof Let $T \subset \mathbb{R}^X$ stand for the family of all additive real functionals a on X majorized by p. According to Lemma 5.2, we have

$$p(x) = \sup\{a(x) : a \in T\}, \qquad x \in X.$$

In view of the evenness of p as well as the oddness of the members of T we obtain the estimation $|a(x)| \le p(x)$ valid for every $x \in X$ and every $a \in T$. Therefore, the formula

$$A(x)(a) := a(x), \qquad a \in T, x \in X,$$

correctly defines a map $A : X \to B(T, \mathbb{R})$. Clearly, A yields an additive operator and, moreover, the equality

$$p(x) = \sup\{|A(x)(a)| : a \in T\} = \|A(x)\|_\infty,$$

is satisfied for all $x \in X$. Thus the proof has been completed. □

5.3.2 Main Result

Now, we are in a position to prove a factorization theorem announced at the beginning of the present section.

Theorem 5.7 *Let* $(X, +)$ *be an Abelian group and let* $(Y, \|\cdot\|)$ *be a real normed linear space. Let further* $f : X \to Y$ *be a solution to functional equation* (FM). *Then there exist: a nonempty set* $T \subset \mathbb{R}^X$, *an additive operator* $A : X \to B(T, \mathbb{R})$ *and an odd isometry* $I : A(X) \to Y$ *such that*

$$f(x) = I(A(x)), \qquad x \in X.$$

Conversely, for an arbitrary real normed linear space $(Z, \|\cdot\|_Z)$, *any additive operator* $A : X \to Z$ *and any odd isometry* $I : A(X) \to Y$ *the superposition* $f := I \circ A$ *yields a solution of Equation* (FM).

Proof Let f be a solution of Equation (FM) and let a functional $p : X \to \mathbb{R}$ be given by the formula

$$p(x) := \|f(x)\|, \qquad x \in X.$$

Equation (FM) implies easily the subadditivity of p as well as the relationship

$$p(2x) = 2p(x), \qquad x \in X.$$

A simple induction shows that then $p(nx) = np(x)$ holds true for every $x \in X$ and every positive integer n. In other words, the functional p is sublinear. Observe that $f(0) = 0$ [by putting $x = y = 0$ in (FM)] whence the oddness of f results by setting $y = -x$ in (FM). Consequently the sublinear functional p is even. Therefore, by virtue of Theorem 5.6, there exist: a nonempty set $T \subset \mathbb{R}^X$ and an additive operator $A : X \to B(T, \mathbb{R})$ such that

$$p(x) = \|A(x)\|_\infty, \qquad x \in X.$$

Denote by $\hat{X}$ the quotient space $X/\ker A$ and define an operator $\hat{A} : \hat{X} \to B(T, \mathbb{R})$ by the formula

$$\hat{A}(x + \ker A) := A(x), \qquad x \in X.$$

Obviously, the operator $\hat{A}$ is both additive and injective. Now, observe that the formula

$$\hat{f}(x+\ker A) := f(x), \qquad x \in X,$$

correctly defines a map $\hat{f} : \hat{X} \to Y$. Indeed, once we have $x + \ker A = y + \ker A$ for some x, y from X, then $x - y \in \ker A$ whence by means of (FM) and the oddness of f we get

$$0 = \|A(x-y)\|_\infty = p(x-y) = \|f(x-y)\| = \|f(x) - f(y)\|$$

and, a fortiori, $f(x) = f(y)$.

Clearly, the image $G := \hat{A}(\hat{X}) = A(X)$ yields a subgroup of the additive group $(B(T,\mathbb{R}), +)$ and the formula

$$I(u) := \hat{f}\left(\hat{A}^{-1}(u)\right), \qquad u \in G,$$

establishes a map from the group $(G, +)$ into the normed space $(Y, \|\cdot\|)$. We are going to show that

(i) $\|I(u) + I(v)\| = \|I(u+v)\|, \qquad u, v \in G,$
(ii) $\|I(u)\| = \|u\|_\infty, \qquad u \in G.$

In fact, to see that (i) holds true, fix arbitrarily u, v from G. Then there exist x, y in X such that $u = \hat{A}(x + \ker A)$ and $v = \hat{A}(y + \ker A)$. Thus $u + v = \hat{A}(x + y + \ker A)$ whence

$$\begin{aligned}
\|I(u) + I(v)\| &= \|\hat{f}\left(\hat{A}^{-1}(u)\right) + \hat{f}\left(\hat{A}^{-1}(v)\right)\| \\
&= \|\hat{f}(x+\ker A) + \hat{f}(y+\ker A)\| = \|f(x) + f(y)\| = \|f(x+y)\| \\
&= \|\hat{f}(x+y+\ker A)\| = \|\hat{f}\left(\hat{A}^{-1}(u+v)\right)\| = \|I(u+v)\|.
\end{aligned}$$

To check (ii), observe that for every $u \in G$ one has

$$\begin{aligned}
\|I(u)\| &= \|\hat{f}\left(\hat{A}^{-1}(u)\right)\| = \|\hat{f}(x+\ker A)\| \\
&= \|f(x)\| = p(x) = \|A(x)\|_\infty = \|\hat{A}(x+\ker A)\|_\infty = \|u\|_\infty.
\end{aligned}$$

Since, as we have seen already, (i) implies the oddness of I, we infer that for every $u, v \in G$ one has

$$\|I(u) - I(v)\| = \|I(u) + I(-v)\| = \|I(u-v)\| = \|u - v\|$$

because of (i) and (ii). Thus the map I yields an odd isometry mapping G into Y.

Finally, for any $x \in X$ we have

$$I(A(x)) = I\left(\hat{A}(x + \ker A)\right) = (I \circ \hat{A})(x + \ker A) = \hat{f}(x + \ker A) = f(x),$$

which completes the necessity part of the proof.

Conversely, given a real normed linear space $(Z, \|\cdot\|_Z)$, an additive operator $A : X \to Z$ and an odd (hence also norm preserving) isometry $I : A(X) \to Y$, we see that the superposition $f := I \circ A$ satisfies Equation (FM) because for all $x, y \in X$ one gets

$$\|f(x) - f(y)\| = \|I(A(x)) - I(A(y))\|$$
$$= \|A(x) - A(y)\|_Z = \|A(x - y)\|_Z = \|I(A(x - y))\| = \|f(x - y)\|;$$

now, since f itself is odd as a superposition of an odd and additive mapping, it remains to replace here y by $-y$ to get (FM). This finishes the proof. □

In the case where the domain group $(X, +)$ is uniquely 2-divisible, it is worthwhile to note that actually the functional $p = \|\cdot\| \circ f$ discussed above is not merely sublinear but also *Jensen-convex*, i.e. it satisfies the functional inequality

$$p\left(\frac{x + y}{2}\right) \le \frac{p(x) + p(y)}{2}$$

for all points x, y from X. In particular, assuming that $(X, +)$ is simply the additive group of a normed real linear space $(X, \|\cdot\|_X)$ we see that very mild regularity assumption imposed upon p (for instance, continuity at a single point, Baire measurability, boundedness on a second category Baire subset of X, etc.; see Kuczma's monograph [25] for numerous further much more delicate instances) implies its continuity. Consequently, we get easily the following:

Theorem 5.8 *Let $(X, \|\cdot\|_X)$ and $(Y, \|\cdot\|_Y)$ be two real normed linear spaces. Let further $f : X \to Y$ be a solution to the functional equation* (FM) *such that the functional $p : X \to \mathbb{R}$ defined by the formula*

$$p(x) := \|f(x)\|_Y, \qquad x \in X,$$

satisfies any regularity condition that forces a Jensen-convex functional to be continuous. Then there exist: a nonempty set $T \subset \mathbb{R}^X$, a continuous linear operator $L : X \to B(T, \mathbb{R})$ and an odd isometry $I : L(X) \to Y$ such that

$$f(x) = I(L(x)), \qquad x \in X.$$

Conversely, for an arbitrary real normed linear space $(Z, \|\cdot\|_Z)$, any continuous linear operator $L : X \to Z$ and any odd isometry $I : L(X) \to Y$ the superposition $f := I \circ L$ yields a solution of Equation (FM) *and the corresponding functional p is continuous.*

Proof As we have already observed the functional p being Jensen-convex has to be continuous. Therefore the additive operator $A : X \to B(T, \mathbb{R})$ such that

$$p(x) = \|A(x)\|_{\infty}, \qquad x \in X,$$

is continuous as well. Therefore, since it is well known that additivity implies rational homogeneity, jointly with continuity it forces A to be linear (recall that we deal with real normed linear spaces).

Since the latter assertion is obvious, this ends the proof. □

5.3.3 Derivation of Earlier Results

We shall first derive the main result of [18] (cf. Theorem 5.2 above) from Theorem 5.9. To this end, we shall prove two propositions which, I believe, may present an interest of their own.

Proposition 5.1 (A Modified Version of Baker's Theorem; See [3]) *Let* $(Z, \|\cdot\|_Z)$ *and* $(Y, \|\cdot\|_Y)$ *be two real normed linear spaces and let* $(Y, \|\cdot\|_Y)$ *be strictly convex. Let further* $(G, +)$ *be a subgroup of the additive group* $(Z, +)$ *such that* $G = 2G$. *If* $I : G \to Y$ *is an isometry vanishing at zero, then* I *is additive.*

Proof Fix arbitrarily elements $u, v \in G$. Then

$$\left\|I\left(\frac{u+v}{2}\right) - I(u)\right\|_Y = \left\|\frac{u+v}{2} - u\right\|_Z = \frac{1}{2}\|u - v\|_Z = \frac{1}{2}\,\|I(u) - I(v)\|_Y$$

as well as

$$\left\|I\left(\frac{u+v}{2}\right) - I(v)\right\|_Y = \left\|\frac{u+v}{2} - v\right\|_Z = \frac{1}{2}\|u - v\|_Z = \frac{1}{2}\,\|I(u) - I(v)\|_Y,$$

whence, in view of the uniqueness of the midpoint of a metric segment in a strictly convex space, implies the equality

$$I\left(\frac{u+v}{2}\right) = \frac{I(u) + I(v)}{2}.$$

Hence, on account of the assumption that $I(0) = 0$, we obtain the additivity of I. This ends the proof. □

It turns out that the assumption $G = 2G$ is superfluous whenever the isometry in question is odd. Namely, we have the following:

Proposition 5.2 *Let* $(Z, \|\cdot\|_Z)$ *and* $(Y, \|\cdot\|_Y)$ *be two real normed linear spaces and let* $(Y, \|\cdot\|_Y)$ *be strictly convex. Let further* $(G, +)$ *be a subgroup of the additive group* $(Z, +)$. *If* $I : G \to Y$ *is an odd isometry, then* I *is additive.*

Proof Put

$$\tilde{G} := \bigcup\{2^{-n}G : n \in \mathbb{N}_0\}.$$

It is easily seen that the structure $(\tilde{G}, +)$ yields a subgroup of the group $(Z, +)$ and that $G \subset \tilde{G}$. Moreover, we have $\tilde{G} = 2\tilde{G}$. Therefore, by means of Proposition 5.1, to finish the proof, it suffices to show that I admits an isometric extension onto $\tilde{G}$. This is actually the case, because I being an odd isometry satisfies Equation (FM) whence, in particular, $I(2u) = 2I(u)$ (see, e.g., [1, p. 139]). Consequently, the formula

$$\tilde{I}\,(2^{-n}u) := 2^{-n}I(u), \qquad u \in G, n \in \mathbb{N}_0,$$

unambiguously defines a map $\tilde{I} : \tilde{G} \to Y$ which, obviously, yields an extension of I. To see that $\tilde{I}$ itself is an isometry, fix arbitrarily $u, v \in G$ and $n, m \in \mathbb{N}_0$. Then

$$\begin{aligned}
&\|\tilde{I}\,(2^{-n}u) - \tilde{I}\,(2^{-m}v)\,\|_Y = \|2^{-n}I(u) - 2^{-m}I(v)\|_Y \\
&= 2^{-n-m}\|2^m I(u) - 2^n I(v)\|_Y = 2^{-n-m}\|I(2^m u) - I(2^n v)\|_Y \\
&= 2^{-n-m}\|2^m u - 2^n v\|_Z = \|2^{-n}u - 2^{-m}v\|_Z,
\end{aligned}$$

which completes the proof. □

Corollary 5.2 *Any solution of Equation* (FM) *mapping an Abelian group into a strictly convex real normed linear space* $(Y, \|\cdot\|_Y)$ *satisfies the Cauchy functional equation* (C).

Proof An appeal to Theorem 5.7 shows that $f = I \circ A$ where $A : X \to B(T, \mathbb{R})$ is an additive operator and $I : A(X) \to Y$ is an odd isometry. Plainly, $A(X)$ is a subgroup of the additive structure $(B(T, \mathbb{R}), +)$ whence, on account of Proposition 5.2, I is additive; therefore so is also the composition $f = I \circ A$. □

Remark 5.2 The main result of [12] (i.e. Theorem of Fischer and Muszély here) cannot, however, be derived from Corollary 5.2 (even with semigroups replaced by groups) because the commutativity of the domain was not assumed there. On the other hand, the only place in the proof of our Theorem 5.7, requiring commutativity of the domain was an indirect appeal to Corollary 5.1 via Lemma 5.2 and Theorem 5.6. Therefore, the following question arises in a natural way.

Problem Does Lemma 5.1 carry over to non-Abelian groups? An essential step towards a positive answer to that question will be discussed in Section 5.4.

Corollary 5.3 *Any solution* $f : \mathbb{R} \to Y$ *of Equation* (FM), *where* $(Y, \|\cdot\|_Y)$ *stands for a real normed linear space, such that the function* $p : \mathbb{R} \to \mathbb{R}$ *defined by the formula*

$$p(x) := \|f(x)\|_Y, \qquad x \in \mathbb{R},$$

satisfies any regularity condition that forces a Jensen-convex function to be continuous, has to be proportional to an odd isometry mapping $\mathbb{R}$ into Y.

Proof An appeal to Theorem 5.8 shows that $f = I \circ L$ where $L : \mathbb{R} \to B(T, \mathbb{R})$ is a continuous linear operator and $I : L(\mathbb{R}) \to Y$ yields an odd isometry. Clearly, we simply have

$$L(x) = x \cdot c, \qquad x \in \mathbb{R},$$

where c is a fixed element of $B(T, \mathbb{R})$. Without loss of generality we may assume that $c \neq 0$. Setting

$$\tilde{I}(x) := \frac{1}{\|c\|_\infty} I(x \cdot c), \qquad x \in \mathbb{R},$$

we infer that

$$\|\tilde{I}(x) - \tilde{I}(y)\|_Y = \frac{1}{\|c\|_\infty} \|I(x \cdot c) - I(y \cdot c)\|_Y = \frac{1}{\|c\|_\infty} \|x \cdot c - y \cdot c\|_\infty = |x - y|$$

for all $x, y \in \mathbb{R}$ stating that $\tilde{I}$ yields an isometry. The oddness of $\tilde{I}$ results from that of I. Finally,

$$f(x) = I(x \cdot c) = \|c\|_\infty \tilde{I}(x), \qquad x \in \mathbb{R},$$

i.e. f is proportional to the odd isometry $\tilde{I}$, as claimed. □

The subsequent corollary (the main result in Schöpf's paper [27]) does not follow directly from our Theorem 5.9. The derivation of condition (iii) below is possible via a structural result of Jacek Tabor describing the form of odd isometries on the real line (see Ja. Tabor, Isometries from $\mathbb{R}$ to a Banach space, oral communication). We omit the details here.

Corollary 5.4 *Any continuous solution $f : \mathbb{R} \to Y$ of Equation* (FM), *where* $(Y, \|\cdot\|_Y)$ *stands for a real normed linear space, satisfies the following conditions:*

(i) *f is odd,*
(ii) $\|f(xy)\| = |x|\|f(y)\|$ *for all* $x, y \in \mathbb{R}$,
(iii) $\operatorname{conv}\left\{\frac{f(y)-f(x)}{\|f(y)-f(x)\|} : x, y \in \mathbb{R},\ x < y\right\}$ *is contained in the unit sphere $S \subset X$.*

Conversely, any function $f : \mathbb{R} \to Y$ that enjoys properties (i), (ii) *and*

(iii′) *for every quadruple x, y, u, v of real numbers such that $x < y$ and $u < v$ the segment joining the points $\frac{f(y)-f(x)}{\|f(y)-f(x)\|}$ and $\frac{f(v)-f(v)}{\|f(v)-f(v)\|}$ is contained in S,*

is necessarily continuous and satisfies Equation (FM).

Corollary 5.5 *Let $(X, +)$ be an Abelian group with uniquely performable division by 2 and 3 and let $(Y, \|\cdot\|_Y)$ be a real Banach space. Then any surjective solution $f : X \to Y$ of Equation* (FM) *is additive.*

Proof An appeal to Theorem 5.7 shows that $f = I \circ A$ where $A : X \to B(T, \mathbb{R})$ is an additive map and I stand for an odd isometry mapping the set $G := A(X)$ into Y. Clearly, the subgroup $(G, +)$ of the additive group $(B(T, \mathbb{R}), +)$ enjoys the following property:

$$G = \lambda G, \qquad \lambda \in D := \{\pm 2^n 3^m : n, m \in \mathbb{Z}\}.$$

On the other hand, the surjectivity of f implies that I yields a surjective isometry of G onto the Banach space Y. Therefore G is a closed subset of the space $B(T, \mathbb{R})$ and, a fortiori,

$$G = \lambda G, \qquad \lambda \in \mathbb{R},$$

because of the density of the set D in $\mathbb{R}$. Hence, the isometry I yields a surjection of the Banach space $(G, +)$ onto Y and being odd has to be linear by means of the well-known Mazur–Ulam theorem. Consequently, f is additive as a composition of two additive maps. □

Remark 5.3 Corollary 5.5 is, however, a considerably weaker version of Tabor's result from [27] where neither commutativity nor divisibility assumptions were imposed upon the domain group.

Two further questions might be asked:

- what about the uniqueness of the factorization spoken of in Theorem 5.7?
- does the result carry over to the case of Abelian semigroups?

The first question has a negative answer; actually we are pretty far from any kind of uniqueness. This is visible already from the last part of the statement of Theorem 5.7 the platform space $(Z, \|\cdot\|)$ occurring in the "only if" part, whichever it could be, may always be replaced by the space $B(T, \mathbb{R})$ considered in the "if" part of the theorem.

The other question remained open for many years and finally has been partially answered by Badora who has shown in [2] that commutativity may be replaced by the requirement that the group in question is a so-called $\mathscr{G}$-group. We shall discuss this problem in the next section.

5.4 The Hierarchy of (Non)Commutativity

Recall that the essential part of the proof of Lemma 5.2 was to show that

the family of all additive real functionals a on X majorized by p is nonvoid.

In that connection Badora [2] decided to introduce the notion of $\mathscr{G}$-groups, as those enjoying this property. More exactly:

Definition 5.1 We say that a group $(G, +)$ belongs to the class $\mathscr{G}$ if and only if for each subadditive functional $p : G \to \mathbb{R}$ there exists an additive functional $a : G \to \mathbb{R}$ such that $a \le p$.

It turns out that that notion is closely connected with the validity of Hahn–Banach extension theorem for groups. Namely, the following characterization of the class of $\mathscr{G}$-groups holds true.

Theorem 5.9 (Badora [28]) *Let $(G, +)$ be a group. Then $(G, +) \in \mathscr{G}$ if and only if for each subgroup $(G_0, +)$ of the group $(G, +)$ and for every subadditive functional $p : G \to \mathbb{R}$ such that*

$$M(x) := \sup\{p(-a + x + a) - p(x) : a \in G_0\} < \infty$$

and

$$\liminf_{n\to\infty} \frac{1}{n} M(nx) = 0,$$

for all $x \in G$, and for every additive functional $a_0 : G_0 \to \mathbb{R}$ with $a_0 \le p|_{G_0}$, there exists an additive functional $a : G \to \mathbb{R}$ such that $a|_{G_0} = a_0$ and $a \le p$.

Corollary 5.6 *Let $(G, +)$ be a group from the class $\mathscr{G}$ and let $p : G \to \mathbb{R}$ be a subadditive functional such that*

$$p(2x) = 2p(x), \qquad x \in G.$$

Then for every subgroup $(G_0, +)$ of the group $(G, +)$ and for every additive functional $a_0 : G_0 \to \mathbb{R}$ enjoying the property $a_0 \le p|_{G_0}$, there exists a functional $a : G \to \mathbb{R}$ such that $a|_{G_0} = a_0$ and $a \le p$.

Moreover, Badora has shown in [2] that the following classes of groups $(G, +)$ are contained in class $\mathscr{G}$:

- Abelian groups
- *amenable groups*, i.e. those admitting a positive, translation invariant linear functional $M : B(G, \mathbb{R}) \to \mathbb{R}$ with $M(1) = 1$;
- *weakly commutative groups*, i.e. those enjoying the following property: for each $x, y \in G$ there exists a positive integer n such that $2^n(x + y) = 2^n x + 2^n y$.

By *Hyers groups* we comprehend those enjoying the following property: for each functional $f : G \to \mathbb{R}$ with bounded Cauchy difference $G \times G \ni (x, y) \mapsto f(x + y) - f(x) - f(y) \in \mathbb{R}$, there exists a homomorphism $a : G \to \mathbb{R}$ such that $f - a$ is bounded.

The following chain of inclusions holds true:

$$\begin{array}{c} Abel \subset Amen \subset \mathscr{G} \subset Hyers \\ \qquad\qquad\quad \cup \\ \qquad\qquad\quad \textit{weak commutativity} \end{array}$$

It is known that free groups with two free generators fail to be Hyers ones (see Forti's remark [14]); consequently, such groups stay off the class $\mathscr{G}$. Till now it is not known whether anyone of the inclusions

$$Amen \subset \mathscr{G} \subset Hyers$$

is strict.

Undoubtedly, Badora's idea of introducing the class $\mathscr{G}$ proved to be extremely useful. In particular, in all corresponding results concerning Equation (FM), the commutativity assumption of the group considered may now be replaced by the requirement that this group belongs to the class $\mathscr{G}$. Above all, it holds true in the case of the factorization Theorem 5.7 which now, without any changes in the proof, may be improved as follows:

Theorem 5.10 *Let a group $(X, +)$ be a member of class $\mathscr{G}$ and let $(Y, \|\cdot\|)$ be a real normed linear space. Let further $f : X \to Y$ be a solution to functional equation* (FM). *Then there exist: a nonempty set $T \subset \mathbb{R}^X$, an additive operator $A : X \to B(T, \mathbb{R})$ and an odd isometry $I : A(X) \to Y$ such that*

$$f(x) = I(A(x)), \qquad x \in X.$$

Conversely, for an arbitrary real normed linear space $(Z, \|\cdot\|_Z)$, any additive operator $A : X \to Z$ and any odd isometry $I : A(X) \to Y$ the superposition $f := I \circ A$ yields a solution of Equation (FM).

5.5 Pexiderization

The results presented in the present section are published with detailed proofs in paper [19] of mine in which an answer to a question posed by Ludwig Reich during my stay at the Karl-Franzens Universität (Graz, Austria, Autumn 1995) gives a description of solutions to the functional equation

$$\|f(x + y)\| = \|g(x) + h(y)\|. \tag{PFM}$$

Surprisingly, in contrast to the preceding results, even in the case of strictly convex ranges, the pexiderized Equation (FM), i.e. Equation (PFM) fails to be equivalent to the Pexider functional equation

$$f(x + y) = g(x) + h(y). \tag{P}$$

Indeed, let $(X, +)$ be a groupoid and let $(Y, \|\cdot\|)$ be a normed linear space with $\dim Y \geq 2$. Fix arbitrarily a positive real number ϱ and a $d \in Y$. Denoting by $S(a, \varrho)$ the sphere $\{u \in Y : \|u - a\| = \varrho\}$, $a \in Y$, one can easily check that the triple

(f, g, d) yields a solution to (PFM) for quite arbitrary mappings $f : X \to S(0, \varrho)$ and $g : X \to S(-d, \varrho)$. Therefore, in general, Equation (PFM) enjoys an abundance of solutions being far away from translations of an additive map which are the only ones satisfying the Pexider equation (cf. Aczél and Dhombres [1] or Kuczma [25], for instance). As we shall see later on such a phenomenon is caused by the lack of zeros of the map f. If f vanishes at at least one point of its domain, then all the triples (f, g, h) fulfilling (PFM) may be expressed in terms of mappings G fulfilling the equation

$$\|G(x - y)\| = \|G(x) - G(y)\|. \tag{5.6}$$

5.5.1 Solutions Admitting Zeros

Assuming that either f or, equivalently, the two-place function $(x, y) \mapsto g(x) + h(y)$ vanishes at some point we shall reduce Equation (PFM) to (5.6). Namely we have the following:

Theorem 5.11 *Let $(X, +)$ be a group (not necessarily commutative) and let $(Y, \|\cdot\|)$ be a (real or complex) normed linear space. Assume that functions $f, g, h : X \to Y$ satisfy the functional equation* (PFM) *for all $x, y \in X$ and $f(x_0) = 0$ for some $x_0 \in X$. Then there exists a solution $G : X \to Y$ of Equation* (5.6) *and a vector $a \in Y$ such that*

$$g(x) = G(x) + a, \qquad x \in X, \tag{5.7}$$

$$h(x) = -G(x_0 - x) - a, \qquad x \in X, \tag{5.8}$$

and f is a selection of the multifunction

$$X \ni x \mapsto S(0, \|G(x) - G(x_0)\|) \subset Y. \tag{5.9}$$

Conversely, for every solution $G : X \to Y$ of Equation (5.6), *for every vector $a \in Y$, for every point $x_0 \in X$ and for every selection f of the multifunction* (5.9), *the triple (f, g, h) with g and h given by* (5.7) *and* (5.8), *respectively, yields a solution to* (PFM) *with $f(x_0) = 0$.*

Remark 5.4 The assumption on f to possess a zero in X may equivalently be replaced by the requirement

$$h^{-1}(-g(X)) \neq \emptyset \qquad \text{or} \qquad g^{-1}(-h(X)) \neq \emptyset.$$

In particular, this is the case provided that at least one of the maps g and h is surjective.

Theorem 5.12 *Let $(X, +)$ be a group (not necessarily commutative) and let $(Y, \|\cdot\|)$ be a (real or complex) strictly convex normed linear space. Assume that functions $f, g, h : X \to Y$ satisfy the functional equation* (PFM) *for all $x, y \in X$ and $f(x_0) = 0$ for some $x_0 \in X$. If either the even part of g is constant or the function $X \ni x \mapsto h(x + x_0) \in Y$ has constant even part, then there exists an additive map $G : X \to Y$ and constants $a, b \in Y$ such that*

$$g(x) = G(x) + a, \qquad x \in X,$$

$$h(x) = G(x) + b, \qquad x \in X,$$

and f is a selection of the multifunction

$$X \ni x \mapsto S(0, \|G(x) + a + b\|) \subset Y.$$

Conversely, for every additive function $G : X \to Y$, for every vectors $a, b \in Y$ and for every selection f of the above multifunction, the triple (f, g, h) with g and h given by the above formulae yields a solution to (PFM).

Remark 5.5 A particular selection

$$f(x) := G(x) + a + b, \qquad x \in X,$$

of the multifunction considered in Theorem 5.12 leads to a solution (f, g, h) of the Pexider equation (P). However, in general, Theorem 5.12 shows that even in the case of strictly convex ranges, a solution (f, g, h) of (PFM) may still be far from any triple solving (P) because of multitude of possible selections f. Nevertheless, remarkable is the fact that functions g and h in any such triple are exactly those occurring in solutions of the Pexider equation (translations of an additive function).

5.5.2 *Basic Equation and Additivity*

As we have seen, Equation (5.6) happened to be basic while studying (PFM). Obviously, each odd solution of (5.6) satisfies (FM) and every solution of (FM) is easily checked to be odd. Therefore

Remark 5.6 Equations (5.6) and (FM) are equivalent in the class of odd functions mapping a group into a normed linear space.

Replacing x by $x + y$ in (5.6) we arrive at

$$\|G(x)\| = \|G(x + y) - G(y)\|,$$

which, in case of Abelian domains, is equivalent to

$$\|G(x + y) - G(x)\| = \|G(y)\|. \tag{S}$$

Equally simple is the way back whence

Remark 5.7 Equations (5.6) and (S) are equivalent in the class of functions mapping a commutative group into a normed linear space.

Equation (S) was examined by Skof [28] in the case where the unknown function G is defined on a real linear space. Her principal goal was to give sufficient conditions for a solution of (S) to be additive. As we shall see later on, the main results (Theorems 1 and 2 in [28]) are special cases of our Theorem 5.13 (ii) and Corollary 5.8, respectively.

We proceed with the following:

Theorem 5.13 *Let $(X, +)$ be an Abelian group and let $(Y, \|\cdot\|)$ be a strictly convex normed linear space. If $G : X \to Y$ is a solution to the equation*

$$\|G(x - y)\| = \|G(x) - G(y)\|, \qquad x, y \in X,$$

then the following conditions are pairwise equivalent:

(i) *G is additive;*
(ii) *$G(X) = -G(X)$;*
(iii) *G is odd;*
(iv) *$\|G(2x)\| = 2\|G(x)\|$ for all $x \in X$.*

Remark 5.8 The commutativity of $(X, +)$ was used exclusively to show that (ii) $\Rightarrow$ (iii). Even in this case the relationship

$$\|G(x + y)\| = \|G(y + x)\|, \qquad x, y \in X, \tag{5.10}$$

is sufficient to conduct that part of the proof of Theorem 5.13. Indeed, having (5.10) we replace y by $y - x$ to get

$$\|G(y)\| = \|G(x + y - x)\| = \|G(x + y) - G(x)\|$$

and that is what was really needed. The question whether or not Equation (5.6) implies (5.10) in non-Abelian groups remains open.

Remark 5.9 Unlike (FM) Equation (5.6) always admits nonadditive solutions (no matter whether or not the target space is strictly convex) provided that the domain constitutes a group possessing subgroups of index 2. If that is the case, $(K, +)$ is a subgroup of index 2 of the group $(X, +)$ and $c \neq 0$ is an arbitrarily fixed vector of the normed linear space $(Y, \|\cdot\|)$, then any function $G : X \to Y$ given by the formula

$$G(x) = \begin{cases} 0 \text{ if } x \in K \\ c \text{ if } x \in X \setminus K \end{cases} \tag{5.11}$$

yields a nonadditive solution of Equation (5.6). Indeed, G being even and nonzero cannot be additive since, otherwise, it would be odd. To check that it satisfies

Equation (5.2) fix arbitrarily a pair $(x, y) \in X^2$. The following three possibilities have to be distinguished:

(a) $x, y \in K$: then so does $x - y$ and both sides of (5.6) are equal to 0;
(b) $x, y \in X \setminus K$: then $x - y$ is in K and we have the equalities

$$G(x - y) = 0 = c - c = G(x) - G(y);$$

(c) exactly one of the arguments x, y is in K : then $x-y \in X\setminus K$ whence $G(x-y) = c$ and $G(x) - G(y) \in \{-c, c\}$; thus (5.6) is satisfied as well.

Remark 5.10 Functions of the form (5.11) are, jointly with the additive solutions, the only ones that satisfy Mikusiński's functional equation

$$G(x + y) \neq 0 \quad \text{implies} \quad G(x + y) = G(x) + G(y) \tag{M}$$

(cf. Dubikajtis et al. [10] or Kuczma [25]). Therefore, in the light of Remark 5.9, each solution of Equation (M) satisfies the basic equation (5.6). In the sequel we shall show, among others, that the converse is true in the case of real functionals on groups.

5.5.3 Solutions with Values in Inner Product Spaces

Except for Theorem 5.14 below, in the present section we deal with solutions to the basic equation (5.6) which map a given group into an inner product space. So, we replace the assumption of strict convexity upon the target space by a stronger requirement: the norm comes from an inner product structure.

Theorem 5.14 *Let $(X, +)$ be a group (not necessarily commutative) such that $X = 2X$ and let $(Y, (\| \cdot \|)$ be a normed linear space (real or complex). Then any even solution of Equation* (5.6)*, mapping X into Y vanishes identically on X.*

Proof Let $G : X \to Y$ be an even solution of (5.6). Replacing y by $-y$ in (5.6) leads to

$$\|G(x + y)\| = \|G(x) - G(y)\|, \qquad x, y \in X,$$

whence, by putting here $y = x$ we obtain the equality $G(2x) = 0$ valid for all $x \in X$. Since, by assumption, $2X = X$ this completes the proof. □

Remark 5.11 In view of Remark 5.10 the 2-divisibility assumption is essential because each function of the form (5.11) is even.

In what follows we wish to realize how far are the solutions of (5.6) from those of (FM). The following two results jointly with Corollary 5.7 provide some information in that direction.

Theorem 5.15 *Let* $(X,+)$ *be a group (not necessarily commutative) and let* $(Y,(\cdot|\cdot))$ *be an inner product space (real or complex). Then* $G : X \to Y$ *is a solution of Equation* (5.6) *if and only if*

$$\|G(x)+G(y)\|^2 = \|G(x+y)\|^2 + 4\Re(G(x)|G_e(y))$$

for all $x, y \in X$, *where* G_e *stands for the even part of* G.

Theorem 5.16 *Let* $(X,+)$ *be a commutative group and let* $(Y,(\cdot|\cdot))$ *be a real inner product space. Then Equation* (5.6) *is equivalent to the system*

$$\|G(x)+G(y)\|^2 = \|G(x+y)\|^2 + \|G(x)+G(y)-G(x+y)\|^2$$
$$\|G(x)+G(y)-G(x+y)\|^2 = 4\,(G(x)|G_e(y))$$

assumed for all $x, y \in X$. *In particular, any solution* $G : X \to Y$ *of* (5.6) *enjoys the property*

$$G(x+y) \perp G(x)+G(y)-G(x+y).$$

Observe that due to the commutativity of the group $(X,+)$ the assertion of Theorem 5.15 implies the equality

$$(G(x)|G_e(y)) = (G(y)|G_e(x))$$

valid for all $x, y \in X$. Plainly, we have also

$$(G(-x)|G_e(y)) = (G(y)|G_e(x)), \qquad x, y \in X,$$

which, by subtracting these two equalities side by side, we deduce the following:

Corollary 5.7 *Under the assumptions of Theorem 5.16 every solution* $G : X \to Y$ *of Equation* (5.6) *has the following property:*

$$G_o(x) \perp G_e(y), \qquad x, y \in X,$$

where G_o *and* G_e *stand for the odd and even part of* G, *respectively. In particular, if the set* $\{G_o(x) : x \in X\}$ *is total, then* G *is additive.*

Finally, we shall show that in the case of real functionals the basic equation (5.6) and Mikusiński's equation (M) are equivalent.

Theorem 5.17 *Let* $(X,+)$ *be a commutative group. Then a function* $G : X \to \mathbb{R}$ *satisfies the equation*

$$|G(x-y)| = |G(x)-G(y)|, \qquad x, y \in X, \tag{5.12}$$

if and only if G is a solution to Mikusiński's equation

$$G(x+y)\;[G(x+y)-G(x)-G(y)]=0, \qquad x,y\in X. \tag{5.13}$$

Proof Let $G : X \to \mathbb{R}$ be a solution of (5.12). An appeal to Theorem 5.16 shows that

$$G(x+y) \perp G(x+y)-G(x)-G(y)$$

for all $x, y \in X$ which, in the real case, states nothing else but (5.13).

As to the converse Remark 5.10 may directly be applied. This ends the proof. □

Remark 5.10 jointly with Theorem 5.17 immediately implies the following:

Corollary 5.8 *If* $(X, +)$ *is a commutative group with no subgroups of index 2, then a function* $G : X \to \mathbb{R}$ *satisfies Equation* (5.12) *if and only if G is additive.*

5.6 Inequality Case

Is there any chance to obtain nontrivial results for the case where the equality sign in Equation (FM) would be replaced by that of inequality? More precisely, there are two possibilities:

- to assume that for every x, y from the domain (semigroup, at least, written additively) of a function f whose codomain is a normed linear space, one has

$$\|f(x+y)\| \leq \|f(x)+f(y)\|;$$

- to assume that for every x, y from the domain (semigroup, at least, written additively) of a function f whose codomain is a normed linear space, one has

$$\|f(x+y)\| \geq \|f(x)+f(y)\|.$$

The first possibility seems to be pointless because of the abundance of solutions that might be expected. For instance, given any normed linear space $(E, \|\cdot\|)$ the function $f : E \to \mathbb{R}$ defined by $f(x) = \|x\|,\ x \in E$, is a solution. For any nonnegative increasing subadditive function $\varphi : [0,\infty) \to \mathbb{R}$ the function $f : \mathbb{R} \to \mathbb{R}$ defined by the formula $f(x) = \varphi(|x|),\ x \in [0,\infty)$, is a solution as well.

What concerns the other possibility, the following very interesting result of Gyula Maksa and Peter Volkmann has been obtained in their paper [26]. In what follows, the details will be reported on.

Theorem 5.18 (Maksa and Volkmann [26]) *Let* $(X, +)$ *be a group and* $(Y, (\cdot|\cdot))$ *be a real or complex inner product space. Let further* $f : X \to Y$ *be a solution to the functional inequality*

$$\|f(x+y)\| \geq \|f(x)+f(y)\|, \qquad x,y\in X. \tag{MV}$$

Then f is additive.

Proof Putting $x = y = 0$ in (MV) we infer that $f(0) = 0$. Consequently, on setting $y = -x$ in (MV) we get $f(-x) = -f(x),\ x \in X$. Squaring both sides of (MV) we arrive at

$$\|f(x+y)\|^2 \geq \|f(x)\|^2 + 2\Re(f(x)|f(y)) + \|f(y)\|^2. \tag{5.14}$$

Replacing here x and y by $x + y$ and $-y$, respectively, and taking into account the oddness of f, we obtain the inequality

$$\|f(x)\|^2 \geq \|f(x+y)\|^2 - 2\Re(f(x+y)|f(y)) + \|f(y)\|^2$$

whence

$$-\|f(x+y)\|^2 \geq -\|f(x)\|^2 - 2\Re(f(x+y)|f(y)) + \|f(y)\|^2.$$

Now, adding the latter inequality to (5.14) side by side we infer that

$$2\Re(f(x)|f(y)) - 2\Re(f(x+y)|f(y)) + 2\|f(y)\|^2 \leq 0,$$

or, equivalently,

$$\Re(f(x) + f(y) - f(x+y)|f(y)) \leq 0. \tag{5.15}$$

Replacing in (5.14) x and y by $-x$ and $x + y$, respectively, and taking into account the oddness of f, we obtain the inequality

$$\|f(y)\|^2 \geq \|f(x)\|^2 - 2\Re(f(x)|f(x+y)) - \|f(x+y)\|^2$$

whence

$$-\|f(x+y)\|^2 \geq \|f(x)\|^2 - 2\Re(f(x)|f(x+y)) - \|f(y)\|^2.$$

Now, adding the latter inequality to (5.14) side by side we infer that

$$2\|f(x)\|^2 + 2\Re(f(x)|f(y) - f(x+y)) \leq 0,$$

or, equivalently,

$$\Re(f(x) + f(y) - f(x+y)|f(x)) \leq 0. \tag{5.16}$$

Replacing here x and y by $x + y$ and $-y$, respectively, and taking into account the oddness of f, we get

$$\Re(f(x+y) - f(y) - f(x)|f(x+y)) \leq 0,$$

or, equivalently,

$$\Re(f(x) + f(y) - f(x+y)| - f(x+y)) \leq 0. \tag{5.17}$$

Now, adding (5.15)–(5.17), side by side, we deduce finally that the inequality

$$\|f(x) + f(y) - f(x+y)\|^2 \leq 0,$$

holds true for all elements x, y from X. This implies the additivity of f and finishes the proof. □

That kind result fails to hold in the case where the group domain is replaced by a semigroup one. In fact, take $(X, +) = ([0, \infty), +)$, $(Y, (\cdot|\cdot)) = (\mathbb{R}, \cdot)$, and $f : [0, \infty) \to \mathbb{R}$ given by the formula $f(x) = x^2$, $x \in [0, \infty)$. Then

$$|f(x+y)| = (x+y)^2 = x^2 + y^2 + 2xy \geq x^2 + y^2 = |f(x) + f(y)|.$$

The authors of [26] have posed also the following:

Problem Is it possible to replace the unitary target space by a strictly convex one?

The aforesaid result of Maksa and Volkmann has recently been generalized by Száz in [29]. The generalization consists in replacing the target inner product space by a group $(Y, +)$ endowed with an inner product $Q : Y \times Y \to \mathbb{C}$ subjected to satisfy the following conditions:

(a) $Q(x, x) \geq 0$ and $Q(x, x) = 0$ forces x to be 0;
(b) $Q(y, x) = \overline{Q(x, y)}$;
(c) $Q(x + y, z) = Q(x, z) + Q(y, z)$,

for all x, y, z from Y.

Theorem 5.19 (Száz [29, 30]) *Let $(X, +)$ be a group and $(Y+)$ be a group endowed with an inner product Q. Put*

$$q(u) := \sqrt{Q(u, u)}, \qquad u \in Y.$$

Then for every map $f : X \to Y$ the following conditions are pairwise equivalent:

- *f is additive;*
- *$q(f(x+y)) \geq q(f(x) + f(y))$ for all $x, y \in X$;*
- *f is odd and*

$$\Re Q(f(x), f(y)) \leq \frac{1}{2} \left(q(f(x+y))^2 - q(f(x))^2 - q(f(y))^2\right)$$

for all $x, y \in X$.

In a final Remark 3.4 of his paper spoken of, Száz emphasizes that his proof of the above theorem "does not requires particular tricks" (author's spelling) and therefore it is “more simple” than that presented by Maksa and Volkmann (see the proof of Theorem 5.18 above).

In a feature article of Száz *Remarks and Problems at the Conference on Inequalities and Application* [30], containing 228 references, item nr [207] is a self-citation and reads as follows:

[207] Á. Száz, *A generalization of a theorem of Maksa and Volkmann on additive functions*, Tech. Rep., Inst. Math., Univ. Debrecen 2016/5, 6 pp. (The publication of an improved and enlarged version of this work in the Anal. Math. was probably prevented by a close colleague of Ger.)

No comments.

5.7 Stability

We shall present two single results in two categories:

- Hyers–Ulam stability of the Fischer–Muszély equation;
- Fischer–Muszély equation postulated almost everywhere.

It turns out that Fischer–Muszély equation is stable in the sense of Hyers and Ulam. More precisely we have the following result established by Tabor in his paper [31] for the class of surjective mappings.

Theorem 5.20 (Tabor [31]) *Let* $(G, +)$ *be a group and let* $(X, \|\cdot\|)$ *be a Banach space. If a surjective map* $f : G \to X$ *satisfies the inequality*

$$\big|\|f(x+y)\| - \|f(x)+f(y)\|\big| \leq \varepsilon, \qquad x, y \in G,$$

with a given $\varepsilon \geq 0$*, then*

$$\|f(x+y) - f(x) - f(y)\| \leq 13\varepsilon, \qquad x, y \in G.$$

In particular ($\varepsilon = 0$)*, any surjective solution of Equation* (FM) *is additive.*

Corollary 5.9 *If* $(G, +)$ *is amenable, or more generally, if* $(G, +)$ *happens to be a* $\mathcal{G}$*-group, then there exists exactly one additive map* $a : G \to X$ *such that* $\|f(x) - a(x)\| \leq 13\varepsilon$ *for all* $x \in G$*. Consequently, in that case, the Fischer–Muszély functional equation is stable in the class of surjective mappings.*

Now we want to exhibit another stability property: we shall show that under suitable assumptions a function satisfying the Fischer–Muszély functional equation postulated almost everywhere has to coincide with an additive map almost everywhere.

In what follows the symbol $(G, +)$ will stand for an additively written group. Recall that a nonempty family $\mathcal{J} \subset 2^G \setminus \{G\}$ is called a *proper linearly invariant ideal* (briefly: p.l.i. ideal) in G provided that it satisfies the following conditions:

(i) if $A, B \in \mathscr{J}$, then $A \cup B \in \mathscr{J}$;
(ii) if $A \in \mathscr{J}$ and $B \subset A$, then $B \in \mathscr{J}$;
(iii) if $A \in \mathscr{J}$ and $x \in G$, then $x - G \in \mathscr{J}$.

We say that a property $\mathscr{P}(x)$ holds *$\mathscr{J}$-almost everywhere in G* whenever $\mathscr{P}(x)$ is valid for all $x \in G \setminus U$ for some set $U \in \mathscr{J}$.

For a subset $M \subset G^2$ and $x \in G$ we define a *section*

$$M[x] := \{y \in G : (x, y) \in M\}.$$

An ideal $\widehat{\mathscr{J}}$ in G^2 is said to be *conjugate* with an ideal $\mathscr{J}$ in G if and only if for every set $M \in \widehat{\mathscr{J}}$ the appurtenance $M[x] \in \mathscr{J}$ takes place $\mathscr{J}$-almost everywhere in G.

The family

$$\Omega(\mathscr{J}) := \{M \subset G^2 : M[x] \in \mathscr{J} \text{ for } \mathscr{J}\text{-almost all } x \in G\}$$

yields the largest (in sense of the set inclusion) p.l.i. ideal in G^2 being conjugate to $\mathscr{J}$ [see, e.g., Kuczma [25, Ch. XVII, §5]].

Our main result reads as follows.

Theorem 5.21 *Given a p.l.i. ideal $\mathscr{J}$ in a group $(G, +)$ and a real or complex inner product space $(H, (\cdot|\cdot))$, assume that a map $f : G \to H$ satisfies Equation* (FM) *for all pairs $(x, y) \in G^2$ off a set $M \in \Omega(\mathscr{J})$ such that $T_1(M)$ and $T_2(M)$ stay in $\Omega(\mathscr{J})$ for $T_1(x, y) := (y, x)$ and $T_2(x, y) := (y, x - y)$, $(x, y) \in G^2$.*

If, moreover, for any set U from $\mathscr{J}$ the set $\frac{1}{2}\,U := \{x \in G : 2x \in U\}$ belongs to $\mathscr{J}$ and there exists a member E of $\mathscr{J}$ such that

$$M \cap \bigcup_{k=1}^{3} \{(x, kx) \in G^2 : x \notin E\} = \emptyset,$$

then there exists a unique additive map $a : G \to H$ such that

$$\{x \in G : f(x) \neq a(x)\} \in \mathscr{J}.$$

Proof To apply the technique used by Fischer and Muszély in [12] (see also p. 139 in Aczél and Dhombres [1], fix arbitrarily an $x \in G \setminus (E \cup \frac{1}{2}\,E)$; then all the pairs (x, x), $(x, 2x)$ and $(x, 3x)$ as well as $(2x, 2x)$ are off M and we have

$$\|f(2x)\| = 2\|f(x)\|, \ \|f(3x)\| = \|f(x) + f(2x)\|,$$
$$4\|f(x)\| = \|f(4x)\| = \|f(x) + f(3x)\|,$$

which like in [12], forces the equality

$$f(2x) = 2f(x), \qquad x \in G \setminus (E \cup \frac{1}{2}\, E). \tag{5.18}$$

Since M is supposed to be a member of $\Omega(\mathscr{J})$, there exists a set $U \in \mathscr{J}$ such that for every $x \in G \setminus U$ the section $M[x]$ falls into $\mathscr{J}$.

Let N stand for the set-theoretical union of the following seven sets: $M, (E \cup (\frac{1}{2}\, E)) \times G, G \times (E \cup (\frac{1}{2} E))$ and

$$M_1 := \{(x,y) \in G^2 : x \in \frac{1}{2}\, U \text{ or } y \in M[2x]\}, \;\; M_2 := \{(x,y) \in G^2 : x \in U \text{ or } y \in \frac{1}{2} M[x]\},$$
$$M_3 := \{(x,y) \in G^2 : x \in U \text{ or } y \in -x + M[x]\}, \;\; M_4 := (T_1 \circ T_2)(M).$$

Each one of these seven sets yields a member of the ideal $\Omega(\mathscr{J})$. Indeed, this is obvious for the first three sets as well as, by the invariance assumptions, for the set M_4. To check that $M_1 \in \Omega(\mathscr{J})$ note that for every $x \notin \frac{1}{2}\, U \in \mathscr{J}$ the section

$$M_1[x] = \{y \in G : (x,y) \in M_1\} = \{y \in G : y \in M[2x]\} = M[2x] \text{ belongs to } \mathscr{J}.$$

Similarly, since for every $x \notin U \in \mathscr{J}$ the section

$$M_2[x] = \{y \in G : (x,y) \in M_2\} = \{y \in G : y \in \frac{1}{2}\, M[2x]\} = \frac{1}{2} M[2x] \text{ belongs to } \mathscr{J},$$

we infer that $M_2 \in \Omega(\mathscr{J})$. Finally, for every $x \notin U \in \mathscr{J}$ the section

$$M_3[x] = \{y \in G : (x,y) \in M_3\} = \{y \in G : y \in -x + M[x]\} = -x + M[x] \text{ belongs to } \mathscr{J},$$

which shows that $M_3 \in \Omega(\mathscr{J})$.

Consequently, the union N of all the sets spoken of yields a member of the ideal $\Omega(\mathscr{J})$ as well. Now, fix arbitrarily a pair $(x,y) \in G^2 \setminus N$. Then:

1. $\|f(x+y)\| = \|f(x) + f(y)\|$ because $(x,y) \notin M$;
2. $f(2x) = 2f(x)$ and $f(2y) = 2f(y)$ because of (5.18) and the fact that $x, y \notin E \cup \frac{1}{2}\, E$;
3. $\|f(2x+y)\| = \|f(2x)+f(y)\|$ because $(x,y) \notin M_1$ which forces the pair $(2x, y)$ to stay off the set M;
4. $\|f(2x+y)\| = \|f(x) + f(x+y)\|$ because $(x,y) \notin M_3$ which forces the pair $(x, x+y)$ to stay off the set M;
5. $\|f(x+2y)\| = \|f(x)+f(2y)\|$ because $(x,y) \notin M_2$ which forces the pair $(x, 2y)$ to stay off the set M;
6. $\|f(x+2y)\| = \|f(x+y) + f(y)\|$ because $(x,y) \notin M_4$ which forces the pair $(x+y, y)$ to stay off the set M.

Relations 3. and 4. jointly with 2. imply that

$$\|f(x) + (f(x) + f(y))\| = \|f(x) + f(x + y)\|, \tag{5.19}$$

whereas a similar conclusion

$$\|(f(x) + f(y)) + f(y)\| = \|f(x + y) + f(y)\|, \tag{5.20}$$

can be drawn from relations 5. and 6. jointly with 2. By means of 1., after squaring both sides of (5.19) and (5.20), by a simple calculation, we derive the equalities

$$\Re((f(x)|f(x + y) - f(x) - f(y))) = 0 = \Re((f(y)|f(x + y) - f(x) - f(y))),$$

respectively, which immediately imply that

$$\Re((f(x) + f(y)|f(x + y) - f(x) - f(y))) = 0. \tag{5.21}$$

Along the same lines as in the paper [12] of Fischer and Muszély, from the trivial equality

$$\|f(x + y)\|^2 = \|(f(x) + f(y)) + (f(x + y) - f(x) - f(y))\|^2$$

with the aid of 1. and (5.21) we derive the relationship

$$\|f(x + y) - f(x) - f(y)\|^2 = 0.$$

This clearly forces the additivity relation

$$f(x + y) = f(x) + f(y)$$

that remains valid for all pairs $(x, y) \in G^2 \setminus N$, i.e. $\Omega(\mathscr{J})$-almost everywhere in G^2. Now, it remains to apply a de Bruijn's type result from [15]: there exists a unique additive function $a : G \to H$ such that the equality $f(x) = a(x)$ holds for $\mathscr{J}$-almost all $x \in G$, i.e.

$$\{x \in G : f(x) \neq a(x)\} \in \mathscr{J}.$$

Thus the proof has been completed. □

Remark 5.12 The leading idea of the proof above was to run along the lines of the proof presented in [12] treating it as the obstacle race. However, the set of obstacles, although basically caused by the fact that the validity of the (FM) equation is postulated merely almost everywhere, was enlarged by another one; namely, close to the bottom of page 199 in [12] the authors write:

If we interchange the variables x and y in Equation (16) *we get*

$$[\Re(f(y), f(x+y) - (f(x) + f(y))] = 0], \tag{17}$$

which is wrong; actually, we get then

$$[\Re(f(y), f(y+x) - (f(x) + f(y))] = 0],$$

and not (17) because of the lack of the commutativity of the domain semigroup.

In what follows we shall present a few corollaries illustrating some consequences of the theorem just proved.

Corollary 5.10 *Let $(X, \|\cdot\|)$ stand for a normed linear space and let $(H, (\cdot|\cdot))$ be an inner product space. If a map $f : X \to H$ satisfies the Fischer–Muszély functional equation* (FM) *in a vicinity of infinity (outside an arbitrarily given ball centred at the origin), then there exist a unique additive map $a : X \to H$ and a bounded set $B \subset X$ such that $f(x) = a(x)$ for all $x \in X \setminus B$.*

Proof Let $\mathscr{J}$ stand for the p.l.i. ideal of all bounded subsets of the space X. Clearly, any bounded set and, in particular, any ball $M := B((0,0), r)$ in the product space X^2 yields a member of $\Omega(\mathscr{J})$. Assume that

$$\|f(x+y)\| = \|f(x) + f(y)\|, \qquad (x, y) \in X^2 \setminus M.$$

Put $T_1(x, y) := (y, x)$ and $T_2(x, y) := (y, x - y)$, $(x, y) \in X^2$. The images $T_1(M)$ and $T_2(M)$ are contained in M and $\sqrt{5}M$, respectively, so that they stay in $\Omega(\mathscr{J})$. Moreover, $\frac{1}{2}\,U$ is bounded for any bounded set U. Finally, since the set $E := \{x \in X : \|x\| \le r\}$ belongs to $\mathscr{J}$ and for every $x \in X \setminus E$ one has

$$\|(x, kx)\| = \sqrt{1 + k^2}\|x\| \ge \sqrt{2}r > r, \qquad k \in \{1, 2, 3\},$$

the condition

$$M \cap \bigcup_{k=1}^{3} \{(x, kx) \in G^2 : x \notin E\} = \emptyset$$

is satisfied. Thus all the assumptions of Theorem 5.21 are fulfilled which ends the proof. □

Corollary 5.11 *Let $(G, +)$ stand for a uniquely 2-divisible locally compact group and let $(H, (\cdot|\cdot))$ be an inner product space. Denote by h_1 and h_2 the left Haar measures in G and G^2, respectively, with $h_1(G) = \infty$; moreover, let h_1^* be the outer Haar measure associated with h_1. Assume that for every set $U \subset G$ one has $h^*(\{x \in G : 2x \in U\}) < \infty$ provided that $h^*(U) < \infty$. If a map $f : G \to H$ satisfies the Fischer–Muszély functional equation* (FM) *for all $(x, y) \in G^2 \setminus M$ where $M \subset G^2$ is a set of finite measure h_2 and such that*

$$M \cap \bigcup_{k=1}^{3} \{(x, kx) \in G^2 : x \in G\} = \emptyset,$$

then there exist a unique additive map $a : G \to H$ *and a set* $B \subset G$ *such that* $h_1^*(B) < \infty$ *and* $f(x) = a(x)$ *for all* $x \in G \setminus B$.

Proof Let $\mathscr{J}$ stand for the p.l.i. ideal of all subsets of G having finite outer measure h_1^*. Since, by Fubini's theorem, one has

$$\infty > h_2(M) = \int_G h_1(M[x])dh_1(x),$$

we infer that h_1-almost all sections $M[x]$ are of finite h_1 measure. This proves that M falls into the ideal $\Omega(\mathscr{J})$. Let T_1 and T_2 be defined as in the statement of Theorem 5.18. Directly from the definition of the product measure it follows that $h_2(T_1(M)) = h_2(M) < \infty$ and

$$h_2(T_2(M)) = \int_G h_1(T_2(M)[x])dh_1(x) = \int_G h_1(-x + T_1(M)[x])dh_1(x)$$

$$= \int_G h_1(T_1(M)[x])dh_1(x) = h_2(T_1(M)) = h_2(M) < \infty.$$

Therefore, h_1−almost all sections $T_2(M)[x]$ are of finite h_1 measure which forces the image $T_2(M)$ to fall into the ideal $\Omega(\mathscr{J})$. To finish the proof it suffices to apply Theorem 5.18. □

Corollary 5.12 *Let* $(G, +)$ *stand for a uniquely 2-divisible Polish topological group and let* $(H, (\cdot|\cdot))$ *be an inner product space. Assume that the map* $G \ni x \mapsto \frac{1}{2}x \in G$ *is a homeomorphism of G onto itself. If a map* $f : G \to H$ *satisfies the Fischer–Muszély functional equation* (FM) *for all* $(x, y) \in G^2 \setminus M$ *where* $M \subset G^2$ *is a first category (in the sense of Baire) subset of the group* G^2 *and such that*

$$M \cap \bigcup_{k=1}^{3} \{(x, kx) \in G^2 : x \in G\} = \emptyset,$$

then there exist a unique additive map $a : G \to H$ *and a first category set* $B \subset G$ *such and* $f(x) = a(x)$ *for all* $x \in G \setminus B$.

Proof Let $\mathscr{J}$ stand for the p.l.i. ideal of all first category sets in G. Then with the aid of the celebrated Kuratowski–Ulam theorem we establish the fact that M belongs to the ideal $\Omega(\mathscr{J})$. Since the maps $T_1(x, y) := (y, x)$ and $T_2(x, y) := (y, x-y)$, $(x, y) \in G^2$ yield homeomorphic self-mappings of G^2 we infer that both the images $T_1(M)$ and $T_2(M)$ stay in $\Omega(\mathscr{J})$. Moreover since, by assumption, the map $G \ni x \mapsto \frac{1}{2}x \in G$ is a homeomorphism of G onto itself, the set $\frac{1}{2}U$ is of the first Baire category provided that so is U. To finish the proof it remains to apply Theorem 5.18. □

Corollary 5.13 *Let $(\mathbb{Z}, +)$ be the additive group of all integers and let $(H, (\cdot|\cdot))$ be an inner product space. If a sequence $(a_n)_{n\in\mathbb{Z}}$ of elements of the space H satisfies the Fischer–Muszély equation*

$$\|a_{n+m}\| = \|a_n + a_m\| \tag{5.22}$$

for all but finite set of pairs $(n, m) \in \mathbb{Z}^2$, then there exists a unique vector $c \in H$ such that $a_n = nc$ for all but finite number of integers n.

Proof Let $\mathscr{J}$ stand for the p.l.i ideal of all finite subsets of $\mathbb{Z}$. Assuming that relation (5.22) holds for all $n, m \in \mathbb{Z}$ off a set $M := \{(n, m) \in \mathbb{Z}^2 : |n|, |m| \leq n_0\}$ where n_0 is a positive integer, we see that M belongs to the ideal $\Omega(\mathscr{J})$. Plainly the maps $T_1(n, m) := (m, n)$ and $T_2(n, m) := (m, n - m)$, $(n, m) \in \mathbb{Z}^2$ transform finite sets into finite sets, which forces the images $T_1(M)$ and $T_2(M)$ to stay in $\Omega(\mathscr{J})$. Moreover, for every finite set $U \subset \mathbb{Z}$ the set $\{n \in \mathbb{Z} : 2n \in U\}$ is finite as well. Finally, on setting $E := \{-n_0, \ldots, -1, 0, 1, \ldots, n_0\}$ we have $E \in \mathscr{J}$ and M is disjoint with the union

$$\bigcup_{k=1}^{3}\{(n, kn) \in \mathbb{Z}^2 : n \notin E\}$$

that is contained in $\mathbb{Z}^2 \setminus M$. Thus all the assumptions of Theorem 5.21 are fulfilled which implies the existence of a unique additive map $a : \mathbb{Z} \to H$ such that the set $\{n \in \mathbb{Z} : a(n) \neq a_n\}$ is finite. Since, obviously, $a(n) = na(1)$, $n \in \mathbb{Z}$, we get the equality $a_n = nc$ for all but finite number of integers n, with a unique $c := a(1) \in H$, as claimed. □

Remark 5.13 As it states, the formulation of Theorem 5.21 leaves room for improvements. For instance, it would be desirable to have

- the group considered replaced by a semigroup;
- the inner product space replaced by a strictly convex one;
- the assumption

$$M \cap \bigcup_{k=1}^{3}\{(x, kx) \in G^2 : x \notin E\} = \emptyset,$$

  removed.

Unfortunately, at present none of these three wishes can be accomplished because of the proof technique applied.

References

1. Aczél, J., Dhombres, J.: Functional Equations in Several Variables. Cambridge University Press, Cambridge (1989)
2. Badora, R.: On the Hahn-Banach theorem for groups. Arch. Math. (Basel) **86**, 517–528 (2006)
3. Baker, J.A.: Isometries in normed spaces. Am. Math. Mon. **78**, 655–658 (1971)
4. Berruti, G., Skof, F.: Risultati di equivalenza per un'equazione di Cauchy alternativa negli spazi normati. Atti Acad. Sci. Torino Cl. Sci. Fis. Mat. Natur. **125**, 154–167 (1991)
5. Berz, E.: Sublinear functions on $\mathbb{R}$. Aequationes Math. **12**, 200–206 (1975)
6. Day, M.M.: Normed Linear Spaces. Springer, New York (1973)
7. Diminnie, C., White, A. Jr.: Strict convexity in topological vector spaces. Math. Japon. **22**, 49–56 (1977)
8. Diminnie, C., White, A. Jr.: Strict convexity conditions for seminorms. Math. Japon. **24**, 489–493 (1980)
9. Dhombres, J.: Some Aspects of Functional Equations. Chulalongkorn University, Bangkok (1979)
10. Dubikajtis, L., Ferens, C., Ger, R., Kuczma, M.: On Mikusiński's functional equation. Ann. Polon. Math. **28**, 39–47 (1973)
11. Fischer, P., Muszély, Gy.: A Cauchy-féle függvényegyenletek bizonyos típusú általánosí-tásai. Mat. Lapok **16**, 67–75 (1965)
12. Fischer, P., Muszély, Gy.: On some new generalizations of the functional equation of Cauchy. Can. Math. Bull. **10**, 197–205 (1967)
13. Fischer, P.: Remarque 5 – probleme 23. Aequationes Math. **1**, 300 (1968)
14. Forti, G.L.: Remark 11, report of meeting. Aequationes Math. **29**, 90–91 (1985)
15. Ger, R.: Note on almost additive functions. Aequationes Math. **17**, 73–76 (1978)
16. Ger, R.: Remark presented at the 26-th International Symposium on Functional Equations, April 24–May 3, 1988, Sant Feliu de Guixols, Catalonia (Spain). In: Report of Meeting. Aequationes Mathematicae, vol. 37, p. 101 (1989)
17. Ger, R.: Talk presented at the 27-th International Symposium on Functional Equations, August 14–24, 1989, Bielsko-Biała - Katowice - Kraków (Poland). In: Report of Meeting. Aequationes Mathematicae, vol. 39, p. 325 (1990)
18. Ger, R.: On a characterization of strictly convex spaces. Atti Acad. Sci. Torino Cl. Sci. Fis. Mat. Natur. **127**, 131–138 (1993)
19. Ger, R.: A Pexider-type equation in normed linear spaces. Österreich. Akad. Wiss. Math.-Natur. Kl. Sitzungsber. **206**, 291–303 (1998)
20. Ger, R.: Fischer-Muszély additivity of mappings between normed spaces. In: The First Katowice-Debrecen Winter Seminar on Functional Equations, Report of Meeting. Annales Mathematicae Silesianae, vol. 15, pp. 89–90 (2001)
21. Ger, R.: Fischer-Muszély additivity on Abelian groups. Comment. Math. Tomus Specialis in Honorem Juliani Musielak, 83–96 (2004)
22. Ger, R.: Fischer-Muszély functional equation almost everywhere. Aequationes Math. **89**, 207–214 (2015)
23. Ger, R., Koclęga, B.: Isometries and a generalized Cauchy equation. Aequationes Math. **60**, 72–79 (2000)
24. Kranz, P.: Additive functionals on abelian semigroups. Comment. Math. Prace Mat. **16**, 239–246 (1972)
25. Kuczma, M. An Introduction to the Theory of Functional Equations and Inequalities. Polish Scientific Publishers & Silesian University, Warszawa/Kraków/Katowice (1985)
26. Maksa, Gy., Volkmann, P.: Characterization of group homomorphisms having values in an inner product space. Publ. Math. Debrecen **56**, 197–200 (2000)
27. Schöpf, P.: Solutions of $\|f(\xi+\eta)\|=\|f(\xi)+f(\eta)\|$. Math. Pannon. **8**, 117–127 (1997)
28. Skof, F.: On the functional equation $\|f(x+y)-f(x)\| = \|f(y)\|$. Atti Acad. Sci. Torino Cl. Sci. Fis. Mat. Natur. **127**, 229–237 (1993)

29. Száz, A.: A generalization of a Theorem Maksa and Volkmann on additive functions. Tech. Rep., Inst. Math., Univ. Debrecen 2016/5
30. Száz, A.: Remarks and problems at the Conference on Inequalities and Applications, Hajdúszoboszló, Hungary, 2016. Tech. Rep., Inst. Math., Univ. Debrecen 2016/9
31. Tabor, Ja.: Stability of the Fischer-Muszély functional equation. Publ. Math. Debrecen **62**, 205–211 (2003)

Chapter 6
Alien Functional Equations: A Selective Survey of Results

Roman Ger and Maciej Sablik

Abstract We present the (relatively short) history of the "alienation" in the theory of functional equations. The notion originally has been proposed by Dhombres (Aequationes Math 35:186–212,1988). Later the topic has been developed and generalized by many authors. We summarize the present state of the research in this area.

Keywords Alienation • Functional equations of Cauchy • Hosszú • d'Alembert • Jensen • Derivations • Inequalities

Mathematics Subject Classification (2010) Primary 39B22; Secondary 39B72

6.1 Introduction

Dhombres in his paper [9] considers the following four Cauchy equations:

$$C_1(f)(x,y) := f(x+y) - (f(x) + f(y)) = 0; \tag{6.1}$$

$$C_2(f)(x,y) := f(xy) - f(x)f(y) = 0; \tag{6.2}$$

$$C_3(f)(x,y) := f(x+y) - f(x)f(y) = 0; \tag{6.3}$$

$$C_4(f)(x,y) := f(xy) - (f(x) + f(y)) = 0. \tag{6.4}$$

Dhombres introduced the following definitions.

Definition 6.1 Equations (i) and (j), $i, j \in \{1, \dots, 4\}$, $i \neq j$, are *s-independent* on X, Y if the only common solutions $f : X \to Y$ of (i), (j) are $f = 0$ or $f(x) \equiv x$.

R. Ger (✉) • M. Sablik
Institute of Mathematics, University of Silesia, Bankowa 14, 40-007 Katowice, Poland
e-mail: roman.ger@us.edu.pl; maciej.sablik@us.edu.pl

J. Brzdęk et al. (eds.), *Developments in Functional Equations and Related Topics*, Springer Optimization and Its Applications 124,
DOI 10.1007/978-3-319-61732-9_6

Remark 6.1 The notion of s-independence may depend on the sets X and Y. Indeed, (6.1) and (6.2) are s-independent as long as we are concerned with $X = Y = \mathbb{R}$ but there exists a nontrivial solution $f : \mathbb{C} \to \mathbb{C}$ of the system (6.1) and (6.2). Hence (6.1) and (6.2) are not s-independent on $\mathbb{C}$.

Remark 6.2 One can easily check that

- (6.1) and (6.3) are s-independent on a ring
- (6.1) and (6.4) are s-independent on $\mathbb{R}_+$, $\mathbb{R}$, i.e. for $f : \mathbb{R}_+ \to \mathbb{R}$.

Remark 6.3 Let us observe the following.

- (6.1) and (6.4) are a fortiori s-independent on $\mathbb{R}$ or $\mathbb{C}$. However, (6.4) has only $f = 0$ as solution, when considered for functions $f : \mathbb{R} \to \mathbb{R}$. This is why we restrict ourselves to $f : \mathbb{R}_+ \to \mathbb{R}$.
- (6.2) and (6.4) are s-independent on $\mathbb{R}_+$, $\mathbb{R}$, hence they are a fortiori s-independent on $\mathbb{R}$ or $\mathbb{C}$.

Since the notion of s-independence seemed somehow restrictive, Dhombres continued with the following definition.

Definition 6.2 Let $E_1(f) = 0$ and $E_2(f) = 0$ be two functional equations for a function $f : X \to Y$, where X and Y are non-empty sets. The equations E_1 and E_2 are *alien* with respect to X and Y, if any solution $f : X \to Y$ of

$$E_1(f) + E_2(f) = 0, \tag{6.5}$$

is a solution of the system

$$\begin{cases} E_1(f) = 0 \\ E_2(f) = 0. \end{cases} \tag{6.6}$$

If there is no risk of confusion we note $E_1 \perp E_2$ if the two equations are alien. In [9] we also find the following result:

Proposition 6.1 *Let X be a unitary ring divisible by* 2 *and let Y be a unitary ring with the following two properties:*

(i) $y^3 = y \in Y$ *implies* $y \in \{1, -1, 0\}$,
(ii) $y^2 = 0$ *and* $y \in Y$ *imply* $y = 0$.

Then Equations (6.1) *and* (6.2) *are alien with respect to X and Y.*

It may happen that $E_1 \perp E_2$ but $\sim (-E_1 \perp E_2)$. Indeed, the equation

$$f(x) + f(y) + f(xy) = f(x + y) + f(x)f(y) \tag{6.7}$$

$f : K \to K$, where K is a field of characteristic different from 2 has $f = 2$ as a solution, which is not a field homomorphism. Thus $C_1 \perp C_2$ but not $\sim (-C_1 \perp C_2)$.

This observation leads to the following new definition.

Definition 6.3 Two equations $E_1(f) = 0$ and $E_2(f) = 0$ are *weakly alien* with respect to X and Y if any non-constant solution $f : X \to Y$ of (6.5) solves the system (6.6).

Following Dhombres we admit yet another definition.

Definition 6.4 Two equations $E_1(f) = 0$ and $E_2(f) = 0$ are *strongly alien* with respect to X and Y if any couple (f, g) of functions mapping X into Y and solving

$$E_1(f) + E_2(g) = 0,$$

solves also the system

$$\begin{cases} E_1(f) = 0 \\ E_2(g) = 0. \end{cases}$$

Dhombres has also noted the following.

One can imagine many more kinds of dependence between functional equations C_i, $i \in \{1, 2, 3, 4\}$ other than s-independence or being weakly or strongly alien. In connection with conditional equations we shall study elsewhere the m-independence, i.e. the case where

$$E_1(f)E_2(f) = 0$$

implies $E_1(f) = 0$ or $E_2(f) = 0$. *The Pexider analogue is the functional equation*

$$E_1(f)E_2(g) = 0.$$

Actually, Dhombres in [9] dealt with the following equation:

$$af(xy) + bf(x)f(y) + cf(x + y) + d(f(x) + f(y)) = 0, \tag{6.8}$$

for mapping f defined on a unitary ring with the uniquely performed division by 2 and with values in a unitary ring and a field, respectively. Here a, b, c and d are some constants from the range of f. Actually, the main result concerning Equation (6.8) concerns the situation where X is a unitary ring divisible by 2 and Y is a field. In [9] (Theorem 11) one can see the table of all solutions of (6.8) depending on the behaviour of constants a, b, c and d which are assumed to belong to the centre of Y. The solution is expressed in terms of solutions to (6.1), (6.2) or (6.3) or is arbitrary constant, or a specific constant, or vanishes everywhere.

A particular case of (6.8) is ($a = c = 1$, $b = d = -1$)

$$f(x + y) + f(xy) = f(x) + f(y) + f(x)f(y). \tag{6.9}$$

In [9] we find the following result.

Theorem 6.1 ([9], Théorème 5) *Let X be unitary ring divisible by 2 and let Y be a unitary ring. Let $f : X \to Y$ be any function with $f(0) = 0$. Then f solves (6.9) for all $x, y \in X$ if, and only if, f solves the system*

$$\begin{cases} f(x+y) = f(x) + f(y), \\ f(xy) = f(x)f(y), \end{cases} \tag{6.10}$$

for all $x, y \in X$ (in other words, f is a ring homomorphism).

The crucial part of Dhombres's proof was to get the oddness of a solution f of (6.9). However, even in the very simple case of unitary rings $X = \mathbb{Z}$ (the integers) and $Y = \mathbb{R}$ (the reals) Equation (6.9) admits non-odd (actually even) and hence non-homomorphic solutions of the form

$$f(x) = \begin{cases} 0 & \text{for } x \in 2\mathbb{Z} \\ -1 & \text{for } x \in 2\mathbb{Z} + 1 \, . \end{cases}$$

More generally, it is not hard to check that for any two elements c, d from the target ring Y such that $c = c^3$ and $cd = dc = d^2 = 0$ a map $f : \mathbb{Z} \to Y$ given by the formula

$$f(x) = \begin{cases} \frac{1}{2}x(c + c^2) + d & \text{for } x \in 2\mathbb{Z} \\ \\ \frac{1}{2}(x-1)c^2 + \frac{1}{2}(x+1)c + d & \text{for } x \in 2\mathbb{Z} + 1 \, . \end{cases}$$

yields a non-homomorphic solution of (6.9) unless $c = c^2$ and $d = 0$.

Therefore it is most desirable to relax the assumptions upon the rings considered.

6.2 Extension of Dhombres's Results

The first author of the present survey was dealing with the question in papers [13] and [14]. However, any attempt to do that presented in both papers shows that omitting the divisibility hypothesis and/or the existence of unit elements causes essential difficulties and requires some developed techniques. To give you a flavour, let us present the following:

Theorem 6.2 ([14], Theorem 3) *Let X be a unitary ring with e as a unit and let Y be an arbitrary ring. If $f : X \to Y$ is a solution of Equation (6.9) such that $f(0) = 0$, then the ring Y_0 generated by $f(X)$ in Y is unitary with c^2 as a unit, where $c := f(e)$. Moreover, $c^3 = c$ and f satisfies the following system of functional equations:*

$$\begin{cases} f(2x+y) = f(2x) + f(y) \\ f(2xy) = f(2x)f(y) \\ f(2z)\,[f(x+y) - f(x) - f(y)] = 0 \\ f(2z)\,[f(xy) - f(x)f(y)] = 0 \end{cases} \tag{6.11}$$

for all $x, y, z \in X$.

In particular, if the ring X *is either 2-divisible or* $f(2a) \in \{c, c^2\}$ *for some* $a \in X$ *or* $f(2a) \neq 0$ *is not a zero divisor for some* $a \in X$*, then* f *yields a ring homomorphism between* X *and* Y.

Another result in that spirit:

Theorem 6.3 ([14], Theorem 4) *Under the assumptions and denotations of the previous theorem the sets*

$$I := \{x \in X : f(2x) = 0\} \quad \text{and} \quad J := \{u \in Y_0 : uc = cu = -u\}$$

form two-sided *ideals in the rings* X *and* Y_0*, respectively. The quotient ring* Y_0/J *is unitary with the unit element* $e_J := c + J$*. Moreover,*

$$\begin{cases} f(x+y) - f(x) - f(y) \in J \\ f(xy) - f(x)f(y) \in J \end{cases} \tag{6.12}$$

for all $x, y \in X$*. In other words, the map*

$$X \ni x \longmapsto F(x) := f(x) + J \in Y_0/J$$

establishes a homomorphism between the rings X *and* Y_0/J *fulfilling the condition* $F(e) = e_J$.

6.2.1 Even Solutions

Plainly, whatever has been told about solutions of Equation (6.9) till now applies, in particular, for even solutions. However, in this case, we are able to explain the occurrence of $\{0, -1\}$-solutions in the case of the ring $\mathbb{Z}$ of all integers. Namely, we have the following:

Theorem 6.4 ([14], Theorem 6) *Let* X *be a unitary ring with* e *as a unit and let* Y *be an arbitrary ring. If* $j : X \to Y$ *is an even solution of Equation* (6.7) *such that* $j(0) = 0$*, then the ring* Y_0 *generated by* $j(X)$ *in* Y *is unitary with* c^2 *as a unit, where* $c := j(e)$*. Moreover,* $c^2 = -c$ *and* $j\,|_{2X} = 0$*,* j *satisfies the functional equation of Hosszú:*

$$j(x+y-xy)+j(xy)=j(x)+j(y), \qquad x,y \in X, \tag{6.13}$$

and

$$2j(x)\,(j(x)-c)=0, \qquad x \in X\,. \tag{6.14}$$

If, in addition, the cardinality of the quotient ring $X/2X$ *does not exceed* 2*, then the set* $Z := \{x \in X : j(x) = 0\}$ *yields a two-sided ideal of the ring* X*,* $2X \subset Z$ *and*

$$j(x) = \begin{cases} 0 \; for\; x \in Z \\ c \; for\; x \in Z+e \end{cases} \tag{6.15}$$

Conversely, in that case, each function $j : X \to Y$ *of that form with* $-c^2 = c = j(e)$ *yields an even solution of Equation* (6.9)*, vanishing at* 0*.*

6.2.2 A Generalized Ring Homomorphisms Equation

In 2010 the first author, jointly with an Austrian mathematician Ludwig Reich, has established in [19] the general solution of the functional equation

$$af(xy) + bf(x)f(y) + cf(x+y) + df(x) + kf(y) = 0 \tag{6.16}$$

yielding a joint generalization of equations that has been studied by Dhombres, Alzer, Hammer, Benz, Halter-Koch and Ger. Around 2004, Alzer (private communication), motivated by an entirely different type of problems was asking about solutions of the equation

$$f(x+y)-f(xy)=f(x)+f(y)-f(x)f(y),$$

which, however, may simply be viewed as the result of *subtraction* (instead of *summation*) of the additivity and multiplicativity equations side by side. It turned out that actually Alzer was interested in the inequality

$$f(x)f(y)-f(xy) \le f(x)+f(y)-f(x+y).$$

A similar inequality has earlier been studied by Hammer in [23].

We also mention two different approaches to the problem of characterizing field homomorphisms by means of functional equations. The first one uses, e.g., the functional equation

$$f\left(x(x+y)^{-1}\right)(f(x)+f(y)) = f(x)$$

to characterize homomorphisms of skew fields (see Benz [3]). The second approach characterizes field homomorphisms in the class of additive functions by a functional equation in a single variable (cf. Halter-Koch [22]).

The emphasis is given upon the dropping of the 2-divisibility assumption in X and replacing the range Y by an integral domain; this, however, by definition, requires the commutativity of the multiplication. Ger and Reich believed that, in this way, they had also achieved a greater uniformity of the presentation as well as that their approach allows one to have a better insight into the reasons of the occurrence of non-homomorphic solutions.

The crucial result reads as follows:

Theorem 6.5 ([19], Theorem) *Let X and Y be two unitary rings and let Y be commutative with no zero divisors. Given five elements a, b, c, d and $k \in Y$, denote by $\mathscr{S}$ the family of all functions $f : X \to Y$ such that $f \neq 0, f(0) = 0$, and satisfying Equation* (6.16) *for all $x, y \in X$. If $\mathscr{S} \neq \emptyset$, then $k = d = -c$. If $c = 0$ in Equation* (6.16) *reduced to*

$$af(xy) + bf(x)f(y) = c\,(f(x) + f(y) - f(x + y))\,, \tag{6.17}$$

then

- *$b = 0$ implies that either $\mathscr{S} = \emptyset$ provided that $a \neq 0$ or, otherwise, $\mathscr{S}$ coincides with the family of all nonzero functions mapping X into Y vanishing at* 0*;*
- *$b \neq 0$ and $f \in \mathscr{S}$ imply that $f(e) \neq 0$ and $g := f(e)^{-1}f$ yields a multiplicative mapping from X into $\tilde{Y}$—the field of fractions of the ring Y.*

If $c \neq 0$ in Equation (6.16) *reduced to* (6.17) *and $f \in \mathscr{S}$, then the following four cases are the only possible ones:*

(i) *$a = b = 0$ and cf is additive;*
(ii) *$a = 0 \neq b$ and there exists an exponential map $g : X \to \tilde{Y}$ such that*

$$bf(x) = c(1 - g(x)), \qquad x \in X;$$

(iii) *$a \neq 0 = b$ and either f is an arbitrary nonzero constant function provided that $a = c$, or f is even, $f(2x) \equiv 0$ on X and f is constant on the cosets forming the elements of the quotient ring $X/2X$, provided that $a = 2c$. In the latter case the formula*

$$F(x + 2X) := f(x), \qquad x \in X.$$

correctly defines an even map $F : X/2X \to Y$ which solves the corresponding equation on $X/2X$ and vanishes at zero ;
(iv) *$a \neq 0 \neq b$ and either*

(j) *f is additive and $af(xy) = -bf(x)f(y)$ for all $x, y \in X$ or*
(jj) *$bf(x) \equiv c - a$ on X or*

(jjj) *f is even, $f(2x) \equiv 0$ on X and f is constant on the cosets forming the elements of the quotient ring $X/2X$ with F defined and behaving like in* (iii).

The result is completed with following remarks.

Remark 6.4 The assumption that $f(0) = 0$ is by no means restrictive. Indeed, while dealing with Equation (6.16), with the aid of the substitutions $g(x) := f(x) - f(0), x \in X$, $d' := bf(0) + d$ and $k' := bf(0) + k$ we get

$$\begin{aligned} &ag(xy) + bg(x)g(y) + cg(x+y) + d'g(x) + k'g(y) \\ &+[af(0) + bf(0)^2 + cf(0) + df(0) + kf(0)] = 0 \end{aligned}$$

and the constant term in square brackets vanishes [just put $x = y = 0$ in (6.16)]. Therefore

$$ag(xy) + bg(x)g(y) + cg(x+y) + d'g(x) + k'g(y) = 0$$

and obviously $g(0) = 0$.

Remark 6.5 ([19], Remark 3) It follows from this theorem that except for the trivial case where all five coefficients in (10) are vanishing (then, plainly, (6.16) is satisfied for all functions mapping X into Y), a map $f \in \mathscr{S}$ yields a nonzero ring homomorphism if and only if $k = d = -c \neq 0$ and $b = -a \neq 0$ (see the case (iv)(j)).

Remark 6.6 ([19], Remark 4) It is to be observed that in the family $\mathscr{S}$ the solutions of three Cauchy equations: additivity, exponentiality and multiplicativity occur in some cases, but not logarithmic functions. The latter effect is caused by the fact that to have nontrivial logarithmic functions we have to remove zero from the domain.

6.2.3 The Alienation of Additivity and Exponentiality

In contrast to Equation (6.9) or

$$f(x+y) + f(xy) = f(x) + f(y) + f(x)f(y),$$

where two Cauchy functional equations (additivity and multiplicativity of the same function f) have been summed up side by side, trying to examine possible alienation of additivity and exponentiality, Ger (cf. [16]) decided to discuss a Pexider version of the problem:

$$f(x+y) + g(x+y) = f(x) + f(y) + g(x)g(y), \qquad x, y \in S,$$

from the very beginning. This could not be avoided because, otherwise, for $f = g$, i.e. in the case of equation

$$2f(x+y) = f(x) + f(y) + f(x)f(y), \qquad x, y \in S,$$

the problem becomes trivial. Indeed, one may easily check that constant solutions f are the only possible ones (at least under the assumption that the binary law "+" in S admits a neutral element). On the other hand, it seems hardly likely that given two maps f, g the validity of the equation

$$f(x+y) + g(x+y) = f(x) + f(y) + g(x)g(y), \qquad x, y \in S,$$

brings us back to the additivity of f and hence the exponentiality of g (or, in the language of Dhombres, Equations (6.1) and (6.3) are strongly alien, cf. Definition 6.4).

Note also that in the case where the target ring $(R, +, \cdot)$ yields an integral domain (a commutative unitary ring with no zero divisors), no nontrivial linear combination of an additive map a from a groupoid $(S, +)$ to R and an exponential map $e : S \to R$ is quadratic unless $e(x) \equiv 0$ or $e(x) \equiv 1$. Here and in the sequel a map $f : S \to R$ is termed *quadratic* whenever

$$\Delta_y^3 f(x) := f(x+3y) - 3f(x+2y) + 3f(x+y) - f(x) = 0, \qquad x, y \in S.$$

Let us introduce a new definition.

Definition 6.5 We say that mappings $a : S \to \mathbb{R}$ and $e : S \to \mathbb{R}$ are *quadratically equivalent* if for some non-vanishing constants $\alpha, \beta \in \mathbb{R}$ we have

$$\Delta_y^3(\alpha\, a + \beta\, e)(x) \equiv 0 \quad \text{on} \quad S \times S,$$

Now, if our additive a and exponential e were quadratically equivalent, i.e. if for some non-vanishing constants $\alpha, \beta \in R$ we had

$$\Delta_y^3(\alpha\, a + \beta\, e)(x) \equiv 0 \quad \text{on} \quad S \times S,$$

then in view of the linearity of the operator Δ_y^3 we would get

$$\beta\, [e(x+3y) - 3e(x+2y) + 3e(x+y) - e(x)] = 0, \qquad x, y \in S,$$

which due to the exponentiality of e states that

$$e(x)\,(e(y) - 1)^3 \equiv 0 \quad \text{on} \quad S \times S.$$

Thus $e(x) \equiv 0$ or $e(y) \equiv 1$, as claimed.

Our main result establishes the alienation of additivity and exponentiality up to the quadratical equivalence.

6.2.3.1 Some Lemmas

In what follows the minimal requirement upon the domain groupoid is that the binary law in question is associative.

Lemma 6.1 ([16], Lemma 1) *Let* $(S,+)$ *be a semigroup and let* $(R,+,\cdot)$ *be a ring. If functions* $f,g:S\to R$ *satisfy functional equation*

$$f(x+y)+g(x+y)=f(x)+f(y)+g(x)g(y), \qquad x,y\in S, \tag{6.18}$$

then for all $x,y,z\in S$ *one has*

$$g(x+y)-g(x)g(y)=g(x+y)g(z)-g(x)g(y+z)+g(y+z)-g(y)g(z). \tag{6.19}$$

Proof It is well known that the Cauchy difference

$$F(x,y):=f(x+y)-f(x)-f(y), \qquad x,y\in S,$$

satisfies the cocycle equation

$$F(x+y,z)+F(x,y)=F(x,y+z)+F(y,z), \qquad x,y,z\in S. \tag{6.20}$$

Since (6.18) states that $F(x,y)=g(x)g(y)-g(x+y)$, $x,y\in S$, the equality (6.19) follows as a result of a simple calculation. □

The following lemma is also almost evident.

Lemma 6.2 ([16], Lemma 2) *Let* $(S,+)$ *be a semigroup and let* $(R,+,\cdot,1)$ *be a unitary ring. If a function* $g:S\to R$ *satisfies Equation* (6.19) *for all* $x,y,z\in S$, *then the function* $h:=1-g$ *yields a solution to the equation*

$$[h(x)-h(y+x)]h(z)=h(y)[h(x)-h(x+z)], \qquad x,y,z\in S. \tag{6.21}$$

Proof After inserting $g=1-h$ into (6.19) it suffices to interchange the roles of the variables x and y. □

The last of the lemmas reads as follows.

Lemma 6.3 ([16], Lemma 3) *Let* $(S,+)$ *be an Abelian semigroup and let* $(R,+,\cdot,1)$ *be an integral domain. Denote by* $\widetilde{R}$ *the field of quotients generated by* R. *If a function* $h:S\to R$ *satisfies Equation* (6.21) *for all* $x,y,z\in S$, *then there exists a function* $\varphi:S\to\widetilde{R}$ *such that*

$$h(x+y)=\varphi(x)h(y)+h(x), \qquad x,y\in S. \tag{6.22}$$

Proof In the trivial case where $h(x)\equiv 0$ any function φ satisfies (6.22). Therefore, in what follows, we may assume that

$$Z:=\{x\in S:\ h(x)=0\}\neq S.$$

Then, Equation (6.21) implies that for every $x \in S$ and all $y, z \in S \setminus Z$ one has

$$\frac{h(x+y)-h(x)}{h(y)} = \frac{h(x+z)-h(x)}{h(z)} =: \varphi(x).$$

Consequently, for all pairs $(x, y) \in S \times (S \setminus Z)$, we obtain the relationship

$$h(x+y) = \varphi(x)h(y) + h(x). \tag{6.23}$$

Note that having z in Z it follows from (6.21) that z yields a period of h. Hence, Equation (6.23) is satisfied unconditionally (i.e. the assertion (6.22) holds true), which completes the proof. □

6.2.3.2 Main Results

Now, we are in a position to derive the general description of solutions to the functional equation in question in various domain and ranges.

Theorem 6.6 ([16], Theorem 1) *Let* $(S, +, 0)$ *be an Abelian monoid and let* $(R, +, \cdot, 1)$ *stand for an integral domain. If functions* $f, g : S \to R$ *satisfy equation*

$$f(x+y) + g(x+y) = f(x) + f(y) + g(x)g(y) \tag{6.24}$$

for all $x, y \in S$, *then there exist constants* $p, q \in R$, $q \neq 0$, *additive maps* $a, A : S \to R$ *and a function* $r : S \to R$ *such that*

$$p\,r(x+y) = r(x)r(y), \qquad x, y \in S,$$

and either

$$\begin{cases} q^2 f(x) = a(x) + (p-q)r(x) + p(q-p) & \text{for } x \in S, \\ q\,g(x) = r(x) + q - p & \text{for } x \in S, \end{cases} \tag{6.25}$$

or

$$\begin{cases} 2f(x) = a(x)^2 + A(x) & \text{for } x \in S, \\ g(x) = 1 - a(x) & \text{for } x \in S. \end{cases} \tag{6.26}$$

Conversely, each pair of functions $f, g : S \to R$ *satisfying either of the systems* (6.25), (6.26) *yields a solution to Equation* (6.24).

Proof Let, as previously, $\tilde{R}$ stand for the field of quotients generated by R and let functions $f, g : S \to R$ satisfy Equation (6.24). On account of Lemmas 6.1–6.3, the function $h := 1 - g$ generates then a map $\varphi : S \to \tilde{R}$ satisfying (6.22). Due to the commutativity of the addition in S and the multiplication in $\tilde{R}$ we have also

$$h(x+y) = h(x)\varphi(y) + h(y), \qquad x, y \in S. \tag{6.27}$$

Now, an appeal to Theorem 1 from Aczél's and Dhombres's monograph [1, p. 242] leads to the following three possibilities: either h is additive, or $h = b(1 - e)$ with an exponential function $e : S \to \tilde{R}$, or $h(x) \equiv b \in \tilde{R}$.

The first possibility gives $g = 1 - a$ with an additive map $a : S \to R$, which inserted into (6.24) implies that

$$f(x+y) - f(x) - f(y) = a(x)a(y), \qquad x, y \in S.$$

Put $A := 2f - a^2$ to get (by means of the additivity of a),

$$A(x+y) - A(x) - A(y) = 2\,[f(x+y) - f(x) - f(y) - a(x)a(y)] = 0,$$

for all $x \in S$, which states that the map $A : S \to R$ is additive and we arrive at formulas (6.26).

In the case where $h = b(1 - e)$ there exist constants $p, q \in R$, $q \neq 0$, such that

$$h = \frac{p}{q}\,(1 - e) \quad \text{i.e.} \quad q\,(1 - g) = q\,h = p\,(1 - e)$$

whence

$$r := p\,e = q\,g + p - q.$$

Clearly, r maps S into R and due to the exponentiality of e, for all $x, y \in S$, one has

$$r(x+y) = pe(x+y) = pe(x)e(y) = r(x)e(y)$$

whence

$$p\,r(x+y) = r(x)r(y),$$

as claimed. To prove the first of formulas (6.25) in the case discussed, note that we have $g = b\,e + 1 - b$ whence by (6.24) and the exponentiality of e, for all $x, y \in S$, one obtains

$$\begin{aligned} f(x+y) - f(x) - f(y) &= b(b-1)[e(x+y) - e(x) - e(y) + 1] \\ &= b(b-1)\,[(e(x+y) - 1) - (e(x) - 1) - (e(y) - 1)] \end{aligned}$$

or, equivalently,

$$\begin{aligned} &f(x+y) - b(b-1)\,[(e(x+y) - 1)] \\ &\quad = f(x) - b(b-1)\,[(e(x) - 1)] + f(y) - b(b-1)\,[(e(y) - 1)]\,, \end{aligned}$$

which proves that the map $a_0 := f - b(b-1)(e-1)$ is additive. With $b = p/q$, $p, q \in R$, $q \neq 0$, we deduce now that

$$a(x) := q^2 a_0(x) = q^2 f(x) - p(p-q)(e(x)-1)$$
$$= q^2 f(x) - (p-q)(r(x)-p) \in R, \qquad x \in S.$$

Thus, $q^2 f = a + (p-q)r + p(q-p)$ with an additive map $a : S \to R$, which was to be shown.

Finally, the last possibility $h(x) \equiv b \in \tilde{R}$, forces b to belong to the ring R and $g(x) \equiv c := 1 - b \in R$. Then Equation (6.24) implies that the map $a := f + c^2 - c : S \to R$ has to be additive and we arrive at $f = a + c(1-c)$, $g = c$ which is the special case of (6.25) with $r = 0$ and $q = 1$, $p = 1 - c$.

Thus the proof has been completed. □

We have the following corollary.

Corollary 6.1 ([16], Corollary) *Let $(S, +)$ be a commutative group and let $(\mathbb{Z}, +, \cdot)$ stand for the ring of all integers. If functions $f, g : S \to \mathbb{Z}$ satisfy Equation* (6.24) *for all $x, y \in S$, then either*

$$\begin{cases} f(x) = A(x) + c(1-c) \text{ for } x \in S, \\ g(x) = c \qquad\qquad\qquad \text{ for } x \in S, \end{cases} \tag{6.28}$$

with an additive map $A : S \to \mathbb{Z}$ and some constant $c \in \mathbb{Z}$, or the pair (f, g) is given by formulas (6.26) *with additive maps $a, A : S \to \mathbb{Z}$.*

Proof We apply Theorem 6.6 assuming that $R = \mathbb{Z}$. Having (6.25) with some constants $p, q \in \mathbb{Z}$, $q \neq 0$, additive maps $a, A : S \to \mathbb{Z}$ and a function $r : S \to \mathbb{Z}$ such that

$$p\, r(x+y) = r(x) r(y), \qquad x, y \in S,$$

since the domain S is endowed with a group structure, we infer that $p\, r(0) = r(0)^2$ and $p\, r(0) = r(x) r(-x)$ for all $x \in S$. In particular, we have either $r(0) = 0$ or $r(0) = p \neq 0$. In the first case we have $r(x) \equiv 0$ on S whence for all $x \in S$ we get

$$q^2 f(x) = a(x) + p(q-p) \qquad \text{and} \qquad q\, g(x) = q - p.$$

On setting $c_0 := 1 - g(0)$ the latter equality implies that $p = c_0 q$ and, a fortiori,

$$q^2 f(x) = a(x) + c_0(1-c_0) q^2, \qquad x \in S.$$

This forces the function $A := f - c_0(1-c_0)$ to be additive, giving (6.28) with $c := 1 - c_0$.

In the case where $r(0) = p \neq 0$ we see that $0 \neq r(x)$ divides p^2 for every $x \in S$; in particular, $Z := r(S)$ forms a finite subset of $\mathbb{Z} \setminus \{0\}$ and the map $e : S \to \frac{1}{p} Z$

given by the formula $e(x) := \frac{1}{p}r(x)$, $x \in S$, is a nonzero exponential function with a finite number of values. Since, obviously, $e(nx) = e(x)^n$ for all $x \in S$ and all positive integers n we have to have $e(x) \equiv 1$ which implies that $r(x) \equiv p$ on S. Thus $q^2 f(x) = a(x)$ as well as $qg(x) = q$ for all $x \in S$, stating that f itself has to be additive whereas $g = 1$, i.e. we have (6.28) with $c = 1$. This completes the proof. □

In the case where we assume additionally that the range is field we obtain

Theorem 6.7 ([16], Theorem 2) *Let $(S, +, 0)$ be an Abelian monoid and let $(F, +, \cdot)$ stand for a field. If functions $f, g : S \to F$ satisfy Equation* (6.24) *i.e.*

$$f(x+y) + g(x+y) = f(x) + f(y) + g(x)g(y)$$

for all $x, y \in S$, then there exist a constant $\lambda \in F$, additive maps $a, A : S \to F$ and an exponential function $e : S \to F$ such that either

$$\begin{cases} f(x) = a(x) + \lambda(1-\lambda)[1-e(x)] & \text{for } x \in S, \\ g(x) = (1-\lambda)e(x) + \lambda & \text{for } x \in S, \end{cases} \tag{6.29}$$

or

$$\begin{cases} f(x) = \frac{1}{2}a(x)^2 + A(x) & \text{for } x \in S, \\ g(x) = 1 - a(x) & \text{for } x \in S. \end{cases} \tag{6.30}$$

Conversely, each pair of functions $f, g : S \to F$ satisfying either of the systems (6.29), (6.30) *yields a solution to Equation* (6.24).

Proof From Theorem 6.6 we infer that either (6.25)

$$\begin{cases} f(x) = \frac{1}{q^2}a(x) + \left(\frac{p}{q} - 1\right)\frac{1}{q}r(x) + \frac{p}{q}\left(1 - \frac{p}{q}\right) & \text{for } x \in S, \\ g(x) = \frac{1}{q}r(x) + 1 - \frac{p}{q} & \text{for } x \in S, \end{cases}$$

holds or (6.26)

$$\begin{cases} f(x) = \frac{1}{2}a(x)^2 + \frac{1}{2}A(x) & \text{for } x \in S, \\ g(x) = 1 - a(x) & \text{for } x \in S. \end{cases}$$

is valid with

$$p\,r(x+y) = r(x)r(y), \qquad x, y \in S.$$

Now, if we had $p = 0$, then r must be the zero function and we get (6.29) with a standing for the additive function $\frac{1}{q^2}a$ and with $\lambda := 1$. For $p \neq 0$ the map $e := \frac{1}{p}r$ becomes exponential and, again, we arrive at (6.29) with $\lambda := 1 - \frac{p}{q}$ and with a standing for the additive function $\frac{1}{q^2}a$. To finish the proof it remains to observe that (6.30) results from (6.26) on replacing $\frac{1}{2}A$ by A. □

6.2.3.3 Quadratic Equivalence and the Crucial Result

Looking at the assertion of Theorem 6.6 we see that for some constants $p, q \in R$, $q \neq 0$, and additive maps $a, A : S \to R$, one has either

$$q^2 f(x) + q(q-p)g(x) = a(x) + q(q-p), \qquad x \in S, \tag{6.31}$$

or

$$2f(x) + g(x) = a(x)^2 + A(x) - a(x) + 1, \qquad x \in S, \tag{6.32}$$

depending on whether formulas (6.25) or (6.26) are valid, respectively. Therefore, in the sense of the Definition 6.5, f and g are not quadratically equivalent if and only if $q = p$. In fact, the right-hand sides of both (6.31) and (6.32) are special quadratic maps from S into R. Consequently, if the pair (f, g) of functions $f, g : S \to R$ that are not quadratically equivalent yields a solution to Equation (6.24) then $q^2 f$, and hence f itself, is additive. Then (6.24) forces g to be exponential.

Thus we have proved the following:

Theorem 6.8 ([16] Theorem 3) *Let $(S, +, 0)$ be an Abelian monoid and let $(R, +, \cdot, 1)$ stand for an integral domain. If functions $f, g : S \to R$ are not quadratically equivalent, then they satisfy equation*

$$f(x+y) + g(x+y) = f(x) + f(y) + g(x)g(y)$$

for all $x, y \in S$, if and only if f is additive and g is exponential.

A straightforward verification shows that quadratical equivalence yields an equivalence relation in the space of all mappings from S into R. Therefore, Theorem 6.8 states nothing else than that *additivity and exponentiality are alien to each other modulo quadratical equivalence.*

6.2.4 Alienation of Additive and Logarithmic Equations

The first author published in 2013 the paper [18] in which he dealt with (6.1) and (6.4) in context of their strong alienation. More exactly, if $(C, +, \cdot)$ is the cone of all positive elements in an Archimedean totally ordered unitary ring $(R, +, \cdot)$ and $(H, +)$ is an Abelian group then Ger was dealing with the question whether or not the equations

$$a(x+y) = a(x) + a(y)$$

and

$$\ell(xy) = \ell(x) + \ell(y),$$

are strongly alien in the sense of Dhombres. Although, at first glance, it seems hardly likely, bearing in mind the results obtained by Dhombres [9] and by Ger in [13] and [14] (see also Ger and Reich [19]), such a conjecture becomes more reasonable. In contrast to the papers just quoted, following the case of additivity and exponentiality dealt with in Ger's paper [16], it was decided to discuss strong alienation rather than alienation. Indeed, the case where $a = \ell$, i.e. in the case of equation

$$a(x+y) + a(xy) = 2a(x) + 2a(y),$$

we are faced to a very special form of the general functional equation studied in [19]; on the other hand, there are no nontrivial mappings that would be both additive and logarithmic. The result from [18] reads as follows.

Theorem 6.9 ([18] Theorem) *Given an Archimedean totally ordered unitary ring $(R, +, \cdot)$ and an Abelian group $(H, +)$ denote by C the positive cone in R. Then functions $f, g : C \to H$ satisfy equation*

$$f(x+y) + g(xy) = f(x) + f(y) + g(x) + g(y) \tag{6.33}$$

for all $x, y \in C$, if and only if there exist: an additive map $a : R \to H$, a logarithmic map $\ell : S \to H$ and a constant $c \in H$ such that

$$f(x) = a(x) + c \quad \textit{and} \quad g(x) = \ell(x) - c, \qquad x \in C.$$

Proof Assume that functions $f, g : C \to H$ satisfy Equation (6.33) for all $x, y \in C$ and put $h := f + g$. Let $F : C \times C \to H$ stand for the Cauchy kernel of h, i.e.

$$F(x, y) = h(x+y) - h(x) - h(y), \qquad x, y \in C.$$

Then F satisfies the cocycle equation (6.20) for all $x, y, z \in C$. On the other hand, by means of (6.33) and the definition of h, one has

$$F(x, y) = h(x+y) - f(x+y) - g(xy) = g(x+y) - g(xy)$$

provided that x, y are in C. Inserting that form of F into (6.20) we get the equality

$$g((x+y)z) + g(xy) - g(x+y) = g(x(y+z)) + g(yz) - g(y+z) \tag{6.34}$$

valid for every triple $(x, y, z) \in C^3$. Now, setting here $y = x$ gives

$$g(2xz) + g\left(x^2\right) - g(2x) = g(x(x+z)) + g(xz) - g(x+z), \qquad x, z \in C, \tag{6.35}$$

and putting $z = e$, the identity element of R, into (6.34) leads to

$$g(xy) = g(xy+x) + g(y) - g(y+e), \qquad x, y \in C.$$

With $y = e$ this implies the equality

$$g(x) = g(2x) + g(e) - g(2e),$$

and on setting $\alpha := g(2e) - g(e)$ we arrive at

$$g(2x) = g(x) + \alpha, \qquad x \in C. \tag{6.36}$$

Applying (6.36) in (6.35) we infer that

$$u(x) := g\left(x^2\right) - g(x) = g(x(x+z)) - g(x+z)$$

or, equivalently,

$$g(x(x+z)) = u(x) + g(x+z), \qquad x, z \in C. \tag{6.37}$$

With $y = x + z$ Equation (6.37) may equivalently be rewritten in the form

$$x \preceq y \rightarrow g(xy) = u(x) + g(y), \qquad x, y \in C.$$

Now, going back to (6.33), we deduce that

$$x \preceq y \rightarrow f(x+y) + u(x) = f(x) + f(y) + g(x).$$

In other words,

$$x \preceq y \rightarrow f(x+y) = A(x) + f(y), \qquad x, y \in C, \tag{6.38}$$

where we have put $A := f - u + g$. Now, we are going to show that map A is additive. To this end, observe that due to the inequality $x \preceq x + y$ valid for all $x, y \in C$, relation (6.38) implies that

$$f(2x+y) = A(x) + f(x+y), \qquad x, y \in C. \tag{6.39}$$

Replacing here y by $y + z$ we get

$$f(2x+y+z) = A(x) + f(x+y+z), \qquad x, y, z \in C,$$

whence, by setting here $2y$ in place of y one obtains

$$f(2(x+y)+z) = A(x) + f(2y+x+z), \qquad x, y, z \in C,$$

which, with the aid of a double use of (6.39), gives

$$A(x+y) + f(x+y+z) = f(2(x+y)+z) =$$

$$A(x) + f(2y+x+z) = A(x) + A(y) + f(x+y+z)$$

for all $x, y, z \in C$, proving the additivity of A, as claimed. It is well known and easily verifiable that the formula

$$a(x) := \begin{cases} A(x) & \text{whenever } x \in C \\ 0 & \text{for } x = 0 \\ -A(-x) & \text{whenever } x \in -C \end{cases}$$

uniquely extends A to an additive map $a : R \to H$.

Observe that, on account of (6.39) and the additivity of A, for arbitrary x, y from C one has

$$f(2x + y) - A(2x + y) = A(x) + f(x + y) - 2A(x) - A(y) = f(x + y) - A(x + y),$$

which on setting $c := f - A$ states that

$$c(2x + y) = c(x + y), \qquad x, y \in C. \tag{6.40}$$

Fix arbitrarily an $s \in C$ and a t such that $s \prec t \prec 2s$. With $x := t - s$ and $y := 2s - t$ we have then $x, y \in C$ as well as

$$s = x + y \quad \text{and} \quad t = 2x + y,$$

which jointly with (6.40) implies that $c(t) = c(s)$ for all t from the order segment $[s, 2s)$. As a matter of fact, we have also $c(2s) = c(s)$; indeed, fix a t_0 from the segment $(s, 2s)$ to get $s \prec t_0 \prec 2s \prec 2t_0$ (without loss of generality we may assume that $(s, 2s) \neq \emptyset$) whence $c(2s) = c(t_0) = c(s)$. Clearly, now we can conclude that for every $s \in C$ and each positive integer n the restriction $c|_{[s,2^n s]}$ is constant. Consequently, c is globally constant on C. In fact, fix arbitrarily two elements $a, b, a \prec b$, (recall that the ordering is total) from C. Since the ring R is supposed to be Archimedean there exists a positive integer n such that $a \prec b \prec 2^n a$ whence $c(b) = c(a)$.

To finish the proof, it remains to observe that the equality $f = A + c$ forces the map $g + c$ to be logarithmic, by means of (6.33). Since the reverse implication is fairly straightforward, the proof has been completed. □

6.3 Functional Equations Stemming from Actuarial Mathematics

In mathematical risk theory the notion of utility function plays a crucial role. The utility functions are used, e.g., to determine insurance premiums (cf. Bowers et al. [5], Gerber [20] or Tversky and Kahneman [42]). Roughly speaking, the notions of utility function and mathematical expectation or rather a special tool called *Choquet integral* make the insurance business go round. In particular, one determines the premium $H(X)$, X denotes a random variable associated with risk from the equation

$$u(w) = E_{gh}u\left(w + H(X) - X\right), \tag{6.41}$$

(cf. Kałuszka and Krzeszowiec [25]). Here u denotes a utility function, $g, h : [0,1] \to [0,1]$ are *probability distortion functions*, i.e. non-decreasing functions mapping $[0,1]$ into itself and keeping both ends fixed. E_{gh} stands for the expression

$$E_{gh}X = E_gX_+ - E_h(-X)_+$$

where E_g (and also E_h) is defined by

$$E_gX := \int_{-\infty}^{0} [g(P(X > t)) - 1]dt + \int_0^{\infty} g(P(X > t))dt,$$

provided both (Riemann) integrals on the left-hand side are finite. In the case where $h(p) = \overline{g}(p) = 1 - g(1-p)$ then $E_{g\overline{g}}X = E_gX$ and (6.41) reduces to

$$u(w) = E_g\left[u\left(w + H(X) - X\right)\right]. \tag{6.42}$$

Equation (6.42) has been considered by Heilpern in [24]. It is called sometimes the model of *rank-dependent utility*. Admitting some special forms of u one can determine H from (6.42). An interesting problem is when H is additive? More exactly, what are the conditions guaranteeing the following:

$$X, Y - \text{ independent} \to H(X+Y) = H(X) + H(Y). \tag{6.43}$$

In the paper of Heilpern we find the following result:

Theorem 6.10 ([24], Theorem 1(v)) *Let X and Y be independent risks.*

a) Let $g = id$. Then (6.43) *holds if, and only if, $u = id$ or $u(x) = \frac{1}{r}\left(1 - \exp(-rx)\right)$.*
b) If u is either identity or $u(x) = \frac{1}{r}\left(1 - \exp(-rx)\right)$, then (6.43) *holds if, and only if, $g = id$.*

In the proof Heilpern gets taking $u = id$

$$H(X) = E_{\overline{g}}(X),$$

where $\overline{g}$ is defined by

$$\overline{g}(x) = 1 - g(1-x), \qquad x \in [0,1].$$

In the case where u is exponential, i.e. $u(x) = \frac{1}{r}\left(1 - \exp(-rx)\right)$, where $r > 0$, he obtains

$$H(X) = \frac{1}{r}\ln E_{\overline{g}}\left(\exp(rX)\right).$$

It turns out that in the former case (6.43) is equivalent to

$$\overline{g}(p+q-pq)+\overline{g}(pq)=\overline{g}(p)+\overline{g}(q), \tag{6.44}$$

or the Hosszú's equation in [0, 1] (cf. also (6.13)). In the latter, (6.43) is equivalent to

$$\overline{g}(p)+\overline{g}(q)-\overline{g}(p+q-pq)=e^r\overline{g}(pq)-(e^r-1)\overline{g}(p)\overline{g}(q),$$

which can be rewritten as

$$\frac{\overline{g}(p+q-pq)-\overline{g}(p)-\overline{g}(q)+\overline{g}(pq)}{e^r-1}=\overline{g}(p)\overline{g}(q)-\overline{g}(pq),$$

or, still more generally,

$$h(p+q-pq)-h(p)-h(q)+h(pq)=\overline{g}(p)\overline{g}(q)-\overline{g}(pq), \qquad p,q\in[0,1]. \tag{6.45}$$

Heilpern solves (6.44) and (6.45) differentiating $\overline{g}$ twice and using other regularity properties. As we have seen, at least (6.44) can be solved using techniques of the theory of functional equations. Indeed, from Lajkó's result from [29] or [30] we get that $\overline{g}$ has to be of the form $A+c$, where A is an additive function, and c is a constant. Applying the definition of $\overline{g}$ and the properties of g, we arrive at $g=\mathrm{id}$, as claimed. As to Equation (6.45) we conjectured that it is actually an example of (strongly) alien functional equation, i.e. (6.45) can be split into a system

$$\begin{cases} g(xy)=g(x)g(y), \\ h(x+y-xy)+h(xy)=h(x)+h(y), \end{cases} \tag{6.46}$$

for $x,y\in[0,1]$. Unfortunately (or fortunately, as it turned out) Maksa (cf. [32]) disproved this conjecture giving an example of a nontrivial solution of (6.45) which does not satisfy (6.46). Here is his example. Let $M:[0,1]\to\mathbb{R}$ be a multiplicative function, i.e.

$$M(xy)=M(x)M(y).$$

Put

$$h(x)=M(1-x), \qquad x\in[0,1],$$

and

$$g(x)=1-M(1-x), \qquad x\in[0,1].$$

The pair (g,h) solves (6.45) but usually neither h nor g are solutions of (6.46).

6.3.1 *Positive Results on Alienation of Hosszú and Other Cauchy Equations*

Here is what we obtained jointly (cf. Maksa [32]).

Theorem 6.11 ([32] and [36]) *Hosszú equation* (6.44) *and the logarithmic Cauchy equation* (6.4) *are strongly alien for functions* $g, h : (0, 1) \to \mathbb{R}$.

Proof As Lajkó observed (oral communication) if a pair (f, g) of mappings defined on $(0, 1)$ satisfies

$$g(x) + g(y) - g(xy) = h(x + y - xy) - h(x) - h(y) + h(xy) \tag{6.47}$$

for $x, y \in (0, 1)$, then $(f = g + h, h)$ satisfies

$$h(x + y - xy) = f(x) + f(y) - f(xy), \qquad x, y \in (0, 1). \tag{6.48}$$

Using a theorem from [30], we get

$$h(x) = A(x) + c, \tag{6.49}$$

where $A : \mathbb{R} \to \mathbb{R}$ is an additive function. Now, from (6.48) and (6.49) it turns out that m given by

$$m(x) = f(x) - A(x) - c, \qquad x \in (0, 1),$$

satisfies

$$m(xy) = m(x) + m(y), \qquad x, y \in (0, 1).$$

But obviously $m = g$, and the proof is completed. □

We also obtained the following, using a method of Ger presented earlier and based on the use of cocycle equation.

Theorem 6.12 ([32] and [36]) *Hosszú equation* (6.44) *and the additive Cauchy equation* (6.1) *are strongly alien for functions* $g, h : \mathbb{R} \to \mathbb{R}$.

In the case of (6.45) we proved some partial results. Let $g : [0, 1] \to \mathbb{R}$ be any function and $\Psi = g(x)g(y) - g(xy),\ x, y \in [0, 1]$. Then

$$\Psi(xy, z) + g(z)\Psi(x, y) = \Psi(x, yz) + g(x)\Psi(y, z) \tag{6.50}$$

holds for all $x, y, z \in [0, 1]$. This shows that the values of g can be expressed by the values of Ψ provided that Ψ is not identically zero. Assuming that $\Psi \neq 0$ means that g is not multiplicative, or equivalently, h is not a solution of the Hosszú equation, in case of Equation (6.45). The following two results hold.

Theorem 6.13 ([32]) *If (g, h) is a solution of (6.45), $\Psi \neq 0$, and h is differentiable (on $[0, 1]$), then g is differentiable (on $[0, 1]$), too, and, if in addition, $g'(1) = 0$, then (and only then) there are real numbers $a, b, 1 < \alpha$ such that*

$$g(x) = 1 - (1-x)^\alpha \quad \textit{and} \quad h(x) = ax + b + (1-x)^\alpha, \qquad x \in [0, 1]. \tag{6.51}$$

Theorem 6.14 ([32]) *If (g, h) is a solution of (6.45), $\Psi \neq 0$, and h is Lebesgue integrable on $[0, 1]$ (or locally Lebesgue integrable on $(0, 1)$), then g and h are infinitely many times differentiable on $(0, 1)$.*

We also stated the following open problems.

Problem 6.1 Find the general solution of

$$g(x)g(y) - g(xy) = h(x + y - xy) - h(x) - h(y) + h(xy)$$

where $g, h : [0, 1] \to \mathbb{R}$ and the equation holds for all $x, y \in [0, 1]$.

Conjecture The pair (g, h) is a solution if and only if there exist an additive function $A : \mathbb{R} \to \mathbb{R}$, a multiplicative function $M : [0, 1] \to \mathbb{R}$, and $b \in \mathbb{R}$ such that either

$$g(x) = M(x), \ h(x) = A(x) + b, \qquad x \in [0, 1]$$

or

$$g(x) = 1 - M(1-x), \ h(x) = A(x) + b + M(1-x), \qquad x \in [0, 1].$$

Problem 6.2 Find all the solutions (g, h) of

$$g(x)g(y) - g(xy) = h(x + y - xy) - h(x) - h(y) + h(xy)$$

where $g, h : [0, 1] \to \mathbb{R}$, h is continuous (differentiable) on $[0, 1]$ and the equation holds for all $x, y \in [0, 1]$.

Problem 6.3 Find all the solutions (g, h) of

$$g(x)g(y) - g(xy) = h(x + y - xy) - h(x) - h(y) + h(xy)$$

where $g, h : [0, 1] \to \mathbb{R}$, g is continuous (differentiable) on $[0, 1]$ and the equation holds for all $x, y \in [0, 1]$.

6.3.2 Alienation of Hosszú and Exponential Equations

During Maksa's talk [32] the following question was also asked.

Problem 6.4 Find all the solutions (g, h) of

$$g(x)g(y) - g(x + y) = h(x + y - xy) - h(x) - h(y) + h(xy) \tag{6.52}$$

where $g, h : \mathbb{R} \to \mathbb{R}$ and the equation holds for all $x, y \in \mathbb{R}$.

The question has been answered under some additional assumptions by Maksa and Sablik in [34]. In that paper, we investigated the functional equation (6.52) supposing that the function h is continuous. First, define the function Γ on $\mathbb{R}^2$ by

$$\Gamma(x, y) = g(x)g(y) - g(x + y), \qquad x, y \in \mathbb{R}. \tag{6.53}$$

If Γ is identically zero, then we have that g is a solution of the exponential Cauchy equation (6.3) and h is a solution of the Hosszú equation (6.44). These equations are well-discussed and their general (and also their continuous) solutions are well known, e.g., from [28] and [7], respectively. Thus the interesting case now for us is the case $\Gamma \neq 0$. In this case, we can prove the following regularity improvement:

Theorem 6.15 ([34], Theorem 1) *Let $g, h : \mathbb{R} \to \mathbb{R}$ and define the function Γ by* (6.53). *Suppose that the pair (g, h) is a solution of* (6.52)*, Γ is not identically zero and h is continuous. Then g and h are differentiable on $\mathbb{R}$.*

The main result was the following.

Theorem 6.16 ([34], Theorem 2) *Suppose that the functions $g, h : \mathbb{R} \to \mathbb{R}$ satisfy functional equation* (6.52) *and h is continuous. Then g is a solution of the exponential Cauchy equation* (6.3) *and there exist $a, b \in \mathbb{R}$ such that*

$$h(x) = ax + b \tag{6.54}$$

holds for all $x \in \mathbb{R}$.

The following corollary of Theorem 6.16 is obvious.

Corollary 6.2 ([34], Corollary) *The exponential Cauchy equation* (6.3) *and the Hosszú equation* (6.44) *are strongly alien in the class of couples (g, h) such that h, solution of* (6.44)*, is continuous.*

6.4 Further Developments

6.4.1 Alienation of Exponential and Logarithmic Cauchy Equations

In their paper [27] Kominek and Sikorska looked at the equation

$$f(xy) - f(x) - f(y) = g(x + y) - g(x)g(y), \qquad x, y \in \mathbb{R}, \tag{6.55}$$

and looked for solutions $f, g : \mathbb{R} \to \mathbb{R}$ of the equation. They obtained the following:

Theorem 6.17 ([27], Theorem 1) *Let $f, g : \mathbb{R} \to \mathbb{R}$ be functions satisfying* (6.55). *Then*

$$f(x) \equiv 0 \text{ and } g \text{ is an arbitrary exponential function,}$$

or there exists a nonzero real constant α such that

$$f(x) \equiv \alpha(\alpha+1),\ g(x) \equiv \alpha+1,$$

or

$$f(x) = -\alpha x^2 + \alpha(\alpha+1),\ g(x) = -\alpha x + \alpha + 1, \qquad x \in \mathbb{R}.$$

Conversely, each of the above pairs of functions is a solution of (6.55) *with any $\alpha \in \mathbb{R}$.*

Investigating much more interesting case, $f : \mathbb{R} \setminus \{0\} \to \mathbb{R}$ and $g : \mathbb{R} \to \mathbb{R}$ and they satisfy

$$f(xy) - f(x) - f(y) = g(x+y) - g(x)g(y), \qquad x, y \in \mathbb{R} \setminus \{0\}, \tag{6.56}$$

they obtained the following result.

Theorem 6.18 ([27], Theorem 2) *Assume that $f : \mathbb{R} \setminus \{0\} \to \mathbb{R}$ and $g : \mathbb{R} \to \mathbb{R}$ satisfy* (6.56).
If $g(1) \neq 1$ or $f(1) \neq 0$, then

$$g(x+y) = g(x)g(y)\, x, y \in \mathbb{R} \text{ and } f(xy) = f(x) + f(y), \qquad x, y \in \mathbb{R} \setminus \{0\},$$

or there exist $\alpha \in \mathbb{R} \setminus \{0\}$ and a function $F : \mathbb{R} \setminus \{0\} \to \mathbb{R}$ satisfying $F(xy) = F(x) + F(y)$ for all $x, y \in \mathbb{R} \setminus \{0\}$ such that

$$g(x) = \alpha x + \alpha,\ x \in \mathbb{R}, \text{ and} f(x) = F(x) - \alpha x^2 + \alpha(\alpha+1), \qquad x \in \mathbb{R} \setminus \{0\},$$

or there exist $\beta \in \mathbb{R} \setminus \{1\}$ and a function $F : \mathbb{R} \setminus \{0\} \to \mathbb{R}$ satisfying $F(xy) = F(x) + F(y)$ for all $x, y \in \mathbb{R} \setminus \{0\}$ such that

$$g(x) = \beta,\ x \in \mathbb{R}, \text{ and} f(x) = F(x) + \beta^2 - \beta, \qquad x \in \mathbb{R} \setminus \{0\}.$$

If $g(1) = 1, f(1) = 0$ and g is continuous at the origin, then

$$g(x) \equiv 1,\ x \in \mathbb{R} \text{ and} f(xy) = f(x) + f(y), \qquad x, y \in \mathbb{R} \setminus \{0\}.$$

Conversely, each pair of functions described by the above formulae is a solution of (6.56).

Obviously, Kominek and Sikorska realized that the assumption about continuity of g should be relaxed and they formulated in [27] the following question:

Problem 6.5 Find all functions $f : \mathbb{R} \setminus \{0\} \to \mathbb{R}$, $g : \mathbb{R} \to \mathbb{R}$ satisfying the conditions

$$f(1) = 0, \qquad g(1) = 1 \tag{6.57}$$

and (6.56) holds for all $x, y \in \mathbb{R} \setminus \{0\}$.

Recently, during the 17th Katowice-Debrecen Winter Seminar in Zakopane, Maksa [33] presented the solution of this problem by showing that these functions are

$$f(x) = a(\ln|x|),\ x \in \mathbb{R} \setminus \{0\}, \qquad g(x) = \exp A(x), \qquad x \in \mathbb{R}$$

where $a, A : \mathbb{R} \to \mathbb{R}$ are additive functions with $A(1) = 0$. This result can be interpreted also in the way that, under the additional supposition (6.57), the logarithmic and the exponential Cauchy equations are alien.

6.4.2 *The Alienation Phenomenon and Associative Rational Operations*

The title of the present subsection is exactly the title of Ger's paper [17]. The first author of the present survey observed that (6.9) can be written as

$$f(x + y) + f(xy) = Q(f(x), f(y)), \tag{6.58}$$

with $Q(u, v) := u + v + uv$ being a rational associative operation. This observation motivated Ger to ask the following question: given an abstract rational associative operation Q does Equation (6.58) force f to be a ring homomorphism (with the target ring being a field)? The answer is negative in general, but under some additional assumptions, like 2-homogeneity of f and provided the range of f is large enough, Ger was able to get some sufficient and necessary conditions for the positive answer to his original question. In what follows X will stand for a unitary ring with unity e and F will denote a real closed field. In particular, F is formally real, i.e. a sum of squares of elements of F vanishes if and only if each of these elements is equal to zero. Moreover, for each element a of F either a or $-a$ is a square and $\mathrm{char}F = 0$. Chéritat [6] has shown that any nontrivial associative rational operation Q from a suitable subdomain of $F \times F$ into F admits a representation of the form

$$Q(u, v) = \varphi^{-1}\left(\frac{\varphi(u) + \varphi(v)}{1 + \omega\varphi(u)\varphi(v)}\right)$$

with some constant $\omega \in F$ and with a homography

$$\varphi(u) = \frac{au + b}{cu + d} \qquad \text{such that} \qquad ad - bc \neq 0\,.$$

It is not hard to check that the following forms of an associative rational operation Q spoken of are the only possible ones:

$$Q(u,v) = \varphi^{-1}(\varphi(u) + \varphi(v)) \tag{A}$$

or

$$Q(u,v) = \varphi^{-1}(\varphi(u)\varphi(v)) \tag{M}$$

or

$$Q(u,v) = \varphi^{-1}\left(\frac{\varphi(u) + \varphi(v)}{1 - \varphi(u)\varphi(v)}\right), \tag{T}$$

with a homography

$$\varphi(u) = \frac{au+b}{cu+d} \qquad \text{such that} \qquad ad - bc \neq 0\,. \tag{G}$$

The homography φ is then called a generator of the operation Q which, a fortiori, is termed additively, multiplicatively or tangentially generated provided that case (A), (M) or (T) does occur, respectively. Here are the results from [17].

Theorem 6.19 ([17], Theorem 1) *Given a rational associative operation Q assume that a map $f : X \to F$ satisfies Equation* (6.58) *for all $x, y \in X$ such that the pair $(f(x), f(y))$ falls into the domain of Q and* card $f(X) > 4$. *Then there exist constants λ, μ, ν and σ in F such that*

$$\sigma\,(f(x+y) + f(xy))f(x)f(y) = \lambda\,(f(x+y) - f(x) - f(y)) + \mu f(xy) + \nu f(x)f(y).$$

More precisely, if φ given by (G) *stands for the generator of Q, then*

(i) $a \neq 0 = b \neq d$, $\lambda = \mu = 1$, $\nu = -\frac{2c}{d}$ *and* $\sigma = \left(\frac{c}{d}\right)^2$ *provided that Q is additively generated;*
(ii) $b = d \neq 0$, $\lambda = \mu = 1$, $\nu = -\frac{a+c}{b}$ *and* $\sigma = \frac{ac}{b^2}$ *provided that Q is multiplicatively generated;*
(iii) $a \neq 0 = b \neq d$, $\lambda = \mu = d^2$, $\nu = -2cd$ *and* $\sigma = a^2 + c^2$ *provided that Q is tangentially generated.*

Theorem 6.20 ([17], Theorem 2) *Given a rational associative operation Q assume that a map $f : X \to F$ satisfies Equation* (6.58) *for all $x, y \in X$ such that the pair $(f(x), f(y))$ falls into the domain of Q and* card $f(X) > 4$. *Then f is 2-homogeneous, i.e.*

$$f(2x) = 2f(x), \qquad x \in X,$$

if and only if

$$Q(u,v) = \kappa uv + u + v, \qquad u, v \in X,\ \kappa \neq 0.$$

If that is the case, then f is additive and κf is multiplicative; moreover, Q is multiplicatively generated by a generator φ given by the formula

$$\varphi(u) = \kappa u + 1, \qquad u \in F.$$

6.4.3 Alienation of Cauchy and Leibniz Equations

Gselmann in [21] studied the question of (6.1) and

$$g(xy) = xg(y) + yg(x), \tag{6.59}$$

defining derivations, and sometimes called Leibniz equation. She has noticed that although the characterization of ring derivation has an extensive literature, most of the results are of the form: additivity along with an other algebraic property implies that the function in question is a derivation. The main purpose of the paper was to show that derivations can be characterized via a single equation. In the paper the author examines whether the equations occurring in the definition of derivations are independent. As a corollary of the main result, that concerns functional equation

$$f(x+y) - f(x) - f(y) = g(xy) - xg(y) - yg(x),$$

the following result is proved.

Theorem 6.21 ([21], Corollary 2.3) *Let $\mathbb{F}$ be a field and X be a linear space over $\mathbb{F}$, $\lambda, \mu \in \mathbb{F} \setminus \{0\}$. Then the function $f: \mathbb{F} \to X$ is a derivation if and only if*

$$\lambda\,[f(x+y) - f(x) - f(y)] = \mu\,[f(xy) - xf(y) - yf(x)]$$

is fulfilled for any $x, y \in \mathbb{F}$.

6.4.4 Exponential, Jensen and d'Alembert Equations

In 2016 Sobek published the paper [40] in which she presented results concerning mutual alienation of classical exponential Cauchy equation (6.3), Jensen equation and d'Alembert's. More precisely, assuming that F is a field of characteristic different from 2, $(S, +)$ is a commutative semigroup and σ is an endomorphism of S with $\sigma(\sigma(x)) = x$ for $x \in S$, Sobek has studied the equations

$$g(x+y) = g(x)g(y), \qquad x, y \in S,$$

$$f(x+y) + f(x+\sigma(y)) = 2f(x), \qquad x, y \in S, \tag{6.60}$$

and the following generalized version of the classical d'Alembert equation (cf. [39])

$$h(x+y)+h(x+\sigma(y))=2h(x)h(y), \qquad x,y\in S. \tag{6.61}$$

The solutions f, g and h are supposed to map S into F.

6.4.4.1 Exponential and Jensen Equations

The first equation considered by Sobek is the following.

$$f(x+y)+f(x+\sigma(y))+g(x+y)=2f(x)+g(x)g(y), \qquad x,y\in S. \tag{6.62}$$

In [40] it was shown that (6.62) forces f and g to solve the system (6.60)–(6.3), which means that Equations (6.60) and (6.3) are strongly alien.

Theorem 6.22 ([40], Theorem 2.1) *Assume that a pair of functions* (f,g), *where* $f,g:S\to F$, *satisfies Equation* (6.62). *Then f solves* (6.60) *and g satisfies* (6.3).
In a skillful proof the author applies some ideas from [38]. To give you a flavour we reproduce here the proof.

Proof Making use of (6.62), for every $x,y,z\in S$, we get

$$(f+g)(x+y+z)+f(x+y+\sigma z)=2f(x+y)+g(x+y)g(z), \tag{6.63}$$

$$(f+g)(x+\sigma y+z)+f(x+\sigma y+\sigma z)=2f(x+\sigma y)+g(x+\sigma y)g(z), \tag{6.64}$$

$$(f+g)(x+y+z)+f(x+\sigma y+\sigma z)=2f(x)+g(x)g(y+z) \tag{6.65}$$

and

$$(f+g)(x+\sigma y+z)+f(x+y+\sigma z)=2f(x)+g(x)g(\sigma y+z). \tag{6.66}$$

Summing up equalities (6.63) and (6.64) side by side, and subtracting from the equality thus obtained the sum of equalities (6.65) and (6.66), we infer that

$$\begin{aligned}2[f(x+y)+f(x+\sigma y)]&+[g(x+y)+g(x+\sigma y)]g(z)\\ &=4f(x)+g(x)[g(z+y)+g(z+\sigma y)], \qquad x,y,z\in S.\end{aligned}$$

Thus, applying (6.62) again, we obtain

$$\begin{aligned}&[g(x+y)+g(x+\sigma y)]g(z)-2g(x+y)\\ &\qquad=g(x)[g(z+y)+g(z+\sigma y)-2g(y)], \qquad x,y,z\in S.\end{aligned} \tag{6.67}$$

Replacing in (6.67) y by σy, we get

$$\begin{aligned}[g(x+\sigma y)+g(x+y)]g(z)-2g(x+\sigma y)\\ = g(x)[g(z+\sigma y)+g(z+y)-2g(\sigma y)], \qquad x,y,z\in S.\end{aligned} \tag{6.68}$$

Hence, subtracting (6.67) from (6.68) side by side, we arrive at

$$g(x+y)-g(x+\sigma y)=g(x)[g(y)-g(\sigma y)], \qquad x,y\in S, \tag{6.69}$$

whereas by summing up (6.67) and (6.68) side by side, we get

$$\begin{aligned}[g(x+y)+g(x+\sigma y)]\cdot[g(z)-1]\\ = g(x)[g(z+y)+g(z+\sigma y)-g(y)-g(\sigma y)], \qquad x,y,z\in S.\end{aligned} \tag{6.70}$$

Fix $x_0, z_0 \in S$ with $g(x_0)\neq 0$ and $g(z_0)\neq 1$ and define a function $U: S\to F$ in the following way:

$$U(y)=\frac{g(x_0+y)+g(x_0+\sigma y)}{g(x_0)}, \qquad y\in S.$$

Then

$$U(\sigma y)=U(y), \qquad y\in S \tag{6.71}$$

and, by (6.70),

$$g(x+y)+g(x+\sigma y)-g(y)-g(\sigma y)=U(y)[g(x)-1], \qquad x,y\in S. \tag{6.72}$$

Furthermore, in view of (6.70), we have

$$U(y)=\frac{g(z_0+y)+g(z_0+\sigma y)-g(y)-g(\sigma y)}{g(z_0)-1}, \qquad y\in S$$

and

$$g(x+y)+g(x+\sigma y)=g(x)U(y), \qquad x,y\in S. \tag{6.73}$$

So, from (6.69) and (6.73) it follows that

$$2g(x+y)=g(x)V(y), \qquad x,y\in S, \tag{6.74}$$

where the function $V: S\to F$ is defined by

$$V(y)=U(y)+g(y)-g(\sigma y), \qquad y\in S. \tag{6.75}$$

Since, in view of (6.71), $V(y) + V(\sigma y) = 2U(y)$ for $y \in S$, making use of (6.72) and (6.74), we obtain

$$\begin{aligned} 2g(x)U(y) &= g(x)V(y) + g(x)V(\sigma y) = 2g(x+y) + 2g(x+\sigma y) \\ &= 2U(y)[g(x) - 1] + 2g(y) + 2g(\sigma y), \qquad x, y \in S. \end{aligned}$$

Hence $U(y) = g(y) + g(\sigma y)$ for $y \in S$, which together with (6.75) gives $V = 2g$. Thus, taking into account (6.74), we conclude that g satisfies (6.3) and so, in view of (6.62), f solves (6.60). □

From Theorem 6.22 and [39] we derive the following result.

Corollary 6.3 ([40], Corollary 2.2) *Assume that a pair (f, g) of functions mapping S into F satisfies Equation* (6.62). *Then g satisfies Equation* (6.3) *and there exist a constant $c \in F$ and an additive function $a : S \to F$ such that $a(\sigma x) = -a(x)$ for $x \in S$ and $f(x) = a(x) + c$ for $x \in S$.*

6.4.4.2 Jensen and d'Alembert's Equations

The following result shows that the phenomenon of strong alienation takes place also in the case of the Jensen and the d'Alembert equations.

Theorem 6.23 ([40], Theorem 3.1) *Assume that a pair of functions (f, h), where $f, h : S \to F$, satisfies equation*

$$(f + h)(x + y) + (f + h)(x + \sigma y) = 2f(x) + 2h(x)h(y), \qquad x, y \in S. \tag{6.76}$$

Then f satisfies (6.60) *and h solves* (6.61).

Applying [39, Theorems 1–2], from Theorem 6.23 Sobek deduced the following result.

Corollary 6.4 ([40], Corollary 3.2) *Let F be a quadratically closed field of characteristic different from 2. Suppose that a pair of functions (f, h), where $f, h : S \to F$, satisfies Equation* (6.76). *Then there exist a function $g : S \to F$ satisfying Equation* (6.3), *an additive function $a : S \to F$ and a constant $c \in F$ such that $a(\sigma x) = -a(x)$ for $x \in S$, $f(x) = a(x) + c$ for $x \in S$ and*

$$h(x) = \frac{g(x) + g(\sigma x)}{2}, \qquad x \in S.$$

6.4.4.3 Exponential and d'Alembert's Equations

The alienation problem for the pair of Equations (6.3) and (6.61) is different. The following example shows that, in general, these equations are not strongly alien to each other.

Example 6.1 ([40], Example) Let $g, h : \mathbb{R} \to \mathbb{C}$ be the constant functions, say $g = c$ and $h = d$, where $c, d \in \mathbb{C} \setminus \{0, 1\}$ are such that $c(1-c) = 2d(d-1)$. Then, as one can easily check, the pair (f, g) satisfies Equation (6.77), but neither g fulfils (6.3), nor h satisfies (6.61).

However, under some additional assumptions, Equations (6.3) and (6.60) are strongly alien to each other. To this end, we will need the following simple result.

Lemma 6.4 ([40], Lemma 4.1) *Let $(S, +, 0)$ be a commutative monoid. Assume that a pair of functions (g, h), where $g, h : S \to F$, satisfies equation*

$$g(x+y) + h(x+y) + h(x+\sigma y) = g(x)g(y) + 2h(x)h(y), \qquad x, y \in S. \tag{6.77}$$

Then g satisfies (6.69) *and h is even with respect to σ, i.e.*

$$h(x) = h(\sigma x), \qquad x \in S. \tag{6.78}$$

Applying the above Lemma 6.4 Sobek proves

Theorem 6.24 ([40], Theorem 4.2) *Let $(S, +, 0)$ be a commutative monoid. Assume that a pair of functions (g, h), where $g, h : S \to F$, satisfies Equation* (6.77) *and $g(s_0) \neq g(\sigma s_0)$ for some $s_0 \in S$. Then g satisfies* (6.3) *and h satisfies* (6.61).
The paper [40] is concluded with a result which states that in the class of non-constant functions mapping a 2-divisible Abelian group into a field of characteristic different from 2, the exponential Cauchy equation and the d'Alembert equation are strongly alien to each other.

Corollary 6.5 ([40], Corollary 4.3) *Let $(S, +)$ be an Abelian group with $S = 2S$. Assume that a pair of functions (g, h), where $g, h : S \to F$, satisfies equation*

$$g(x+y) + h(x+y) + h(x-y) = g(x)g(y) + 2h(x)h(y), \qquad x, y \in S. \tag{6.79}$$

Then one of the following holds:

(i) either there exist $c, d \in F$ with $c(1-c) = 2d(d-1)$ such that $g = c$ and $h = d$;
(ii) or g satisfies (6.3) *and h satisfies equation*

$$h(x+y) + h(x-y) = 2h(x)h(y), \qquad x, y \in S. \tag{6.80}$$

6.4.5 Trigonometric Equations

Tyrala published in 2011 (cf. [44]) results concerning the alienation of Wilson's (sine, [1, 47]) and d'Alembert's (cosine, [1, 26, 43]) functional equations:

$$\left[f\left(\frac{x+y}{2}\right)\right]^2 - \left[f\left(\frac{x-y}{2}\right)\right]^2 = f(x)f(y), \qquad x, y \in G, \tag{6.81}$$

and

$$f(x+y)+f(x-y)=2f(x)g(y), \qquad x,y\in G. \tag{6.82}$$

We replace x by $2x$ and y by $2y$ in (6.81) and (6.82). Summing up these functional equations side by side, for all $x,y\in G$, we get

$$[f(x+y)]^2-[f(x-y)]^2+f(2x+2y)+f(2x-2y)=f(2x)[f(2y)+2g(2y)]. \tag{6.83}$$

Tyrala proved the following theorem (f_o and f_e stand for the odd and the even part of a function f).

Theorem 6.25 ([44], Theorem 1) *Let $(G,+)$ be a uniquely 2-divisible Abelian group. Then functions $f,g:G\to\mathbb{C}$ satisfy Equation (6.83) if and only if*

(i) $f=0$ *and g is arbitrary; or*
(ii) $f(x)=\alpha\neq 0,\ g(x)=1-\frac{1}{2}\alpha,\ x\in G$*; or*
(iii) there exists an additive function $A:G\to\mathbb{C}$ such that $f=A, g=1$; or
(iv) there exists an exponential function $m:G\to\mathbb{C}$ and some constant $\beta\in\mathbb{C}$ such that $f=\beta m_o,\ g=m_e$; or
(v) there exists an exponential function $m:G\to\mathbb{C}$ such that $f(x)=f(0)m_o(x)+f(0)m_e(x), g(x)=\frac{f(0)}{2}m_o(x)+\left(1-\frac{f(0)}{2}\right)m_e(x),\ x\in G$*; or*
(vi) there exists an exponential function $m:G\to\mathbb{C}$ such that $f(x)=-f(0)m_o(x)+f(0)m_e(x),\ g(x)=-\frac{f(0)}{2}m_o(x)+\left(1-\frac{f(0)}{2}\right)m_e(x),\ x\in G.$

6.4.6 Cauchy, Jensen and Lagrange Equations

Tyrala in [45] studied the dependence between Equation (6.1) and the so-called Lagrange equation

$$g(x)-g(y)=(x-y)f\left(\frac{x+y}{2}\right)$$

for all $x,y\in\mathbb{R}$. The latter was considered and solved, e.g., by Aczél (cf. Sahoo and Riedel book [37]). The main result of the article [45] reads as follows.

Theorem 6.26 ([45], Theorem 5) *Let $(R,+,\cdot)$ be a uniquely 2-divisible ring. Then functions $f,g:R\to R$ satisfy the generalization of the Lagrange functional equation*

$$f(x+y)+g(x)-g(y)=f(x)+f(y)+(x-y)f\left(\frac{x+y}{2}\right) \tag{6.84}$$

for all $x, y \in R$ *if and only if*

$$\begin{cases} f(x+y) = f(x) + f(y) \\ g(x) - g(y) = (x-y)f\left(\frac{x+y}{2}\right) \end{cases} \tag{6.85}$$

for each $x, y \in R$.

In 2016, Troczka-Pawelec and Tyrala went back to the problem of alienation of Cauchy and Lagrange equations. They published their results in [41]. Actually, they studied a generalization of the system (6.85), namely they replaced $g(y)$ on the left-hand side of (6.84) by $h(y)$. First, they proved the following:

Theorem 6.27 ([41], Theorem 4) *Let* $(R, +, \cdot)$ *be a uniquely 2-divisible ring. If functions* $f, g, h : R \to R$ *satisfy the functional equation*

$$f(x+y) + g(x) - h(y) = f(x) + f(y) + (x-y)f\left(\frac{x+y}{2}\right), \tag{6.86}$$

for all $x, y \in R$, *then there exists an additive function* $a : R \to R$ *such that*

$$\begin{cases} f(x) = a(x) + f(0) \\ g(x) = g(0) + \frac{1}{2}xa(x) + xf(0) \\ h(x) = g(0) + \frac{1}{2}xa(x) + xf(0) - f(0) \end{cases}$$

for all $x \in R$.

Theorem 6.28 ([41], Theorem 5) *Let* $(R, +, \cdot)$ *be a uniquely 2-divisible ring. If functions* $f, g, h : R \to R$ *satisfy the functional equation*

$$f(x+y) + g(x) - h(y) = f(x) + f(y) + (x-y)\left(\frac{f(x) + f(y)}{2}\right) \tag{6.87}$$

for all $x, y \in R$, *then there exists an additive function* $a : R \to R$ *such that*

$$\begin{cases} f(x) = a(x) + f(0) \\ g(x) = g(0) + \frac{1}{2}xa(x) + xf(0), \\ h(x) = g(0) + \frac{1}{2}xa(x) + xf(0) - f(0) \end{cases}$$

where $x \in R$.

Corollary 6.6 ([41], Corollary 1) *Let* $(R, +, \cdot)$ *be a uniquely 2-divisible ring and* $f(0) = 0$. *Functions* $f, g, h : R \to R$ *satisfy the functional equation* (6.86) *if and only if*

$$\begin{cases} f(x+y) = f(x) + f(y) \\ g(x) - h(y) = (x-y)f\left(\frac{x+y}{2}\right) \end{cases} \tag{6.88}$$

for all $x, y \in R$.

Corollary 6.7 ([41], Corollary 2) *Let* $(R, +, \cdot)$ *be a uniquely* 2*-divisible ring and* $f(0) = 0$. *Functions* $f, g, h : R \to R$ *satisfy for all* $x, y \in R$ *the functional equation* (6.87) *if and only if*

$$\begin{cases} f(x+y) = f(x) + f(y) \\ g(x) - h(y) = (x-y)\left(\frac{f(x)+f(y)}{2}\right), \end{cases} \tag{6.89}$$

where $x, y \in R$.

The authors dealt also with Jensen equation

$$f\left(\frac{x+y}{2}\right) = \frac{f(x)+f(y)}{2}, \qquad x, y \in R, \tag{6.90}$$

The following theorems are the main results of that paper.

Theorem 6.29 ([41], Theorem 6) *Let* $(R, +, \cdot)$ *be a uniquely* 2*-divisible ring. Functions* $f, g, h : R \to R$ *satisfy for all* $x, y \in R$ *the functional equation*

$$f\left(\frac{x+y}{2}\right) + g(x) - h(y) = \frac{f(x)+f(y)}{2} + (x-y)f\left(\frac{x+y}{2}\right) \tag{6.91}$$

if and only if

$$\begin{cases} f\left(\frac{x+y}{2}\right) = \frac{f(x)+f(y)}{2} \\ g(x) - h(y) = (x-y)f\left(\frac{x+y}{2}\right) \end{cases} \tag{6.92}$$

for all $x, y \in R$.

Theorem 6.30 ([41], Theorem 7) *Let* $(R, +, \cdot)$ *be a uniquely* 2*-divisible ring. Functions* $f, g, h : R \to R$ *satisfy for all* $x, y \in R$ *the functional equation*

$$f\left(\frac{x+y}{2}\right) + g(x) - h(y) = \frac{f(x)+f(y)}{2} + (x-y)\left(\frac{f(x)+f(y)}{2}\right) \tag{6.93}$$

if and only if

$$\begin{cases} f\left(\frac{x+y}{2}\right) = \frac{f(x)+f(y)}{2} \\ g(x) - h(y) = (x-y)\left(\frac{f(x)+f(y)}{2}\right) \end{cases} \tag{6.94}$$

for all $x, y \in R$.

Let us note in connection of the presented results the following.

Remark 6.7 It is noteworthy that in fact $h = g$ in results of Troczka-Pawelec and Tyrala. Indeed, if in

$$g(x) - h(y) = (x-y)f\left(\frac{x+y}{2}\right), \qquad x, y \in R,$$

[second equation in the system (6.92)] we interchange x and y, we get

$$g(y) - h(x) = (y-x)f\left(\frac{x+y}{2}\right), \qquad x, y \in R.$$

Adding the above equations side by side, we obtain

$$(g(x) - h(x)) + (g(y) - h(y)) = 0, \qquad x, y \in R,$$

and obviously

$$g(x) - h(x) = h(y) - g(y) = const = 0.$$

Similarly one can prove that $g = h$ in the case of the second equation of the system (6.94), as well as (6.91) or (6.93).

6.5 Inequalities

Dhombres's original idea was to characterize the ring homomorphisms which is defined by a system of equations with one equation only. But it is possible also to show equivalence of the system to a system of inequalities. Such was the idea of Rădulescu who proved in 1980 the following result.

Theorem 6.31 ([35]) *Let X stand for a compact Hausdorff topological space and let $C_{\mathbb{R}}(X)$ denote the space of all continuous real valued functions on X. If an operator $T: C_{\mathbb{R}}(X) \to C_{\mathbb{R}}(X)$ satisfies the following system:*

$$\begin{cases} T(f+g) \geq T(f) + T(g), \\ T(f \cdot g) \geq T(f) \cdot T(g), \end{cases} \tag{6.95}$$

for each $f, g \in C_{\mathbb{R}}(X)$, then there exist a clopen subset $B \subseteq X$ and a continuous function $\varphi: X \to X$ such that

$$T(f) = \chi_B \cdot f \circ \varphi$$

for all $f \in C_{\mathbb{R}}(X)$. In particular, T is linear, multiplicative and continuous and the system (6.95) *assumes the form:*

$$\begin{cases} T(f+g) = T(f) + T(g), \\ T(f \cdot g) = T(f) \cdot T(g). \end{cases} \tag{6.96}$$

The Rădulescu's result was later generalized by several mathematicians. Let us mention here research of Volkmann from [46]: *if A is a ring, then each solution $T\colon A \to \mathbb{R}$ of system* (6.95) *is additive and multiplicative.* Then Dhombres in [8] showed that *if A is a ring and R is an ordered ring in which nonzero elements have positive squares then each solution $T\colon A \to R$ of system* (6.95) *is additive and multiplicative.* In 2007, Ercan proved that (cf. [10]) *the Rădulescu's assumption that X is a compact Hausdorff space may be dropped.*

However, there exist also some counterexamples. In particular it turns out that the above-mentioned results fail to hold if we reverse one or both of the inequalities in system (6.95). Indeed,

- The absolute value of a real or complex number is subadditive and multiplicative.
- The function $-\chi_{\mathbb{R}\setminus\mathbb{Q}}$ is both superadditive and submultiplicative.

The assumptions upon the domain cannot be relaxed too much, as well, even if the mappings in question are smooth: the function

$$[0, +\infty) \ni x \mapsto -1 - x \in \mathbb{R}$$

is both superadditive and supermultiplicative.

6.5.1 Stability

Bourgin has shown in [4] that given a surjective map f from a ring into a Banach algebra such that both additivity and multiplicativity of f are assumed merely with some (ε, δ)-exactness, i.e.

$$\|f(x+y) - f(x) - f(y)\| \leq \varepsilon$$

and

$$\|f(xy) - f(x)f(y)\| \leq \delta,$$

then f has to be a ring homomorphism, i.e. f has to satisfy the system of two Cauchy functional equations (6.10), or

$$\begin{cases} f(x+y) = f(x) + f(y) \\ f(xy) = f(x)f(y) \end{cases}$$

exactly. This stability result has been then generalized by Badora in [2] who was applying different methods to get rid of, among others, the surjectivity assumption upon the map in question.

The functional equation we have been dealing with, i.e.

$$f(x+y) + f(xy) = f(x) + f(y) + f(x)f(y)$$

(6.9) may obviously be viewed as that characterizing ring homomorphism. This gives rise to a natural problem whether the Bourgin-Badora hyperstability result [(ε, δ)-exactness and the exact validity of the system are equivalent!] carries over to the case of the latter equation. The question was approached by Ger in [15]. An affirmative answer can hardly be expected, because given a positive ε a straightforward verification proves that an arbitrary map f from a ring into a normed algebra, enjoying the property that

$$\|f(x)\| \leq \eta \qquad \text{where} \quad 4\eta + \eta^2 \leq \varepsilon\,,$$

satisfies Equation (6.9) with ε-exactness. Moreover, it is worthy to observe also that taking arbitrary elements a and r from the domain and the range of the solution f of Equation (6.9), respectively, one can easily check that the map

$$x \longmapsto af(rx)$$

yields a solution as well, provided that $a^2 = a$ and $r^2 = r$. Therefore, the maps for which such shifts are bounded are, in a sense, uninteresting in the context discussed. What about the others? The following result provides an answer to that question.

Theorem 6.32 ([15], Theorem) *Let X be a unitary ring with a unit 1 and let $(\mathscr{A}, \|\cdot\|)$ stand for a commutative Banach algebra with a unit e. Given an $\varepsilon \geq 0$ assume let that a map $f : X \to \mathscr{A}$ is such that $f(0) = 0, f(1) = e, f(2) = 2e$, and*

$$\|f(x+y) + f(xy) - f(x) - f(y) - f(x)f(y)\| \leq \varepsilon, \qquad x, y \in X. \tag{6.97}$$

Then either there exist an $a \in \mathscr{A} \setminus \{0\}$ and an $r \in X \setminus \{0\}$ such that the map

$$X \ni x \longmapsto af(rx) \in \mathscr{A} \quad \textit{is bounded}$$

or

$$f \textit{ establishes a ring homomorphism between } X \textit{ and } \mathscr{A}.$$

Remark 6.8 The assumptions $f(0) = 0$ and $f(1) = e$ seem to be natural while dealing with homomorphisms. Note that none of them results from inequality (6.97). The same applies to $f(2) = 2e$; inequality (6.97) forces only the distance $\|f(2) - 2e\|$ to be majorized by ε. The question whether the commutativity of the target algebra is essential remains open.

Remark 6.9 The assertion of the theorem would certainly be more readable if we had simply the alternative: either f is bounded or f is a homomorphism (classical *superstability* effect). Plainly, that is actually the case whenever both the domain ring X and the Banach algebra $\mathscr{A}$ in question are fields. If $\mathscr{A}$ is a field, then f yields a homomorphism provided that no function of the form $x \longmapsto f(rx)$, $r \in X \setminus \{0\}$, is bounded. If X is a field, then f yields a homomorphism provided that no function af, $a \in \mathscr{A} \setminus \{0\}$, is bounded.

Laohakosol et al. in [31] obtained an analogue of Dhombres' theorem for mappings defined on $\mathbb{R}^+$.

6.5.2 More Inequalities

Rădulescu's result from [35] mentioned above may also be viewed as follows:

An operator $T\colon C(X) \to C(X)$ satisfies the system:

$$\begin{cases} T(f+g) \geq T(f) + T(g), \\ T(f \cdot g) \geq T(f) \cdot T(g), \end{cases}$$

for every $f, g \in C(X)$ if and only if

$$T(f+g) + T(f \cdot g) = T(f) + T(g) + T(f) \cdot T(g)$$

for all $f, g \in C(X)$.

What about possible equivalence of the system in question with the inequality

$$T(f+g) + T(f \cdot g) \geq T(f) + T(g) + T(f) \cdot T(g) \ ?$$

No hope, because of the following results.

Theorem 6.33 (Hammer, [23]) *A continuous function $f : \mathbb{R} \to \mathbb{R}$ that is differentiable at zero satisfies the functional inequality*

$$f(x+y) + f(xy) \geq f(x) + f(y) + f(x)f(y), \qquad x, y \in \mathbb{R} \tag{6.98}$$

if and only if f is constant and equal to 0 or

$$f(x) = x + \frac{a-1}{a}(e^{ax} - 1), \quad x \in \mathbb{R},$$

with $a = f'(0) \geq 1$.

With the aid of this result we infer that (under Hammer's assumptions) the "alienation phenomenon" holds true for inequality (6.98) if and only if $f'(0) = 0$ (which leads to $f = 0$) or $f'(0) = 1$ (which gives $f =$ id), i.e. merely for boundary cases.

Quite recently Fechner [11] has generalized Hammer's result in a few directions. First, he has started with two unknown functions instead of a single one. Second, he has taken a linear combination of inequalities from the system in question instead of the sum. His objective was to check whether the "alienation phenomenon" holds true for the functional inequalities discussed. The answer reads as follows.

Theorem 6.34 ([11], Theorem 1) *Let $f : \mathbb{R} \to \mathbb{R}$ be a differentiable at zero and continuous function and let $b, c \in \mathbb{R}$ be arbitrary nonzero constants. If f satisfies*

$$f(x+y) + bf(xy) \geq f(x) + f(y) + cf(x)f(y), \qquad x, y \in \mathbb{R},$$

jointly with $f(0) = 0$, *then* $f = 0$ *or*

$$f(x) = \frac{ac - b}{ac^2}\,[e^{acx} - 1] + \frac{b}{c}x, \qquad x \in \mathbb{R},$$

where $a = f'(0)$ *and moreover* $ac > 0$ *and* $(ac - b)bc \geq 0$.

Fechner proved also the following.

Corollary 6.8 ([11], Corollary 1) *Let* $f : \mathbb{R} \to \mathbb{R}$ *be a differentiable at zero and continuous function and let* $b, c \in \mathbb{R}$ *be arbitrary nonzero constants. If* f *satisfies the inequality*

$$f(x + y) + bf(xy) \geq f(x) + f(y) + cf(x)f(y), \qquad x, y \in \mathbb{R},$$

jointly with $f(0) = 0$, *then* f *solves the system*

$$\begin{cases} f(x + y) \geq f(x) + f(y), & x, y \in \mathbb{R}, \\ bf(xy) \geq cf(x)f(y), & x, y \in \mathbb{R}, \end{cases}$$

if and only if $f = 0$ *or* $f'(0) = \frac{b}{c}$.

In connection with his results, Fechner asked the following two questions. Consider the inequality

$$\alpha \cdot C_1 f(x, y) + \beta \cdot C_2 g(x, y) \geq 0, \qquad x, y \in \mathbb{R}, \tag{6.99}$$

where α, β are real constants, and C_1, C_2 are defined by (6.1) and (6.2). Then the following two problems arise.

Problem 6.6 ([12], Problem 3.5) Solve (6.99) completely.

Problem 6.7 ([12], Problem 3.6) Solve inequality (6.98) under weaker regularity assumptions and/or in a more general setting.

References

1. Aczél, J., Dhombres, J.: Functional Equations in Several Variables. Cambridge University Press, Cambridge (1989)
2. Badora, R.: On approximate ring homomorphisms. J. Math. Anal. Appl. **276**, 589–597 (2002)
3. Benz, W.: Characterizations of homomorphisms of skew fields. Aequationes Math. **64**, 315–320 (2002)
4. Bourgin, D.G.: Approximately isometric and multiplicative transformations on continuous function rings. Duke Math. J. **16**, 385–397 (1949)
5. Bowers, N.L. Jr., Gerber, H.U., Hickman, J.C., Jones, D.A., Nesbitt, C.J.: Actuarial Mathematics. The Society of Actuaries, Schaumburg (1997)
6. Chéritat, A.: Fractions rationnelles associatives et corps quadratiques. Revue des Mathématiques de l'Enseignement Supérieur **109**, 1025–1040 (1998–1999)

7. Daróczy, Z.: On the general solution of the functional equation $f(x + y - xy) + f(xy) = f(x) + f(y)$. Aequationes Math. **6**, 130–132 (1971)
8. Dhombres, J.: Sur les fonctions simultanement suradditives et surmultiplicatives. C.R. Math. Rep. Acad. Sci. Canada **5**, 207–210 (1983)
9. Dhombres, J.: Relations de dépendance entre les équations fonctionnelles de Cauchy. Aequationes Math. **35**, 186–212 (1988)
10. Ercan, Z.: A remark on supra-additive and supra-multiplicative operators on $C(X)$. Math. Bohem. **132**, 55–58 (2007)
11. Fechner, W.: A note on alienation for functional inequalities. J. Math. Anal. Appl. **385**, 202–207 (2012)
12. Fechner, W.: Problems 3.5 and 3.6., Report of Meeting. Aequationes Math. **89**, 1509–1551 (2015)
13. Ger, R.: On an equation of ring homomorphisms. Publ. Math. Debr. **52**, 397–412 (1998)
14. Ger, R.: Ring homomorphisms equation revisited. Rocznik Nauk.-Dydakt. Prace Mat. **17**, 101–115 (2000)
15. Ger, R.: Stability of the equation of ring homomorphisms. Sci. Issues, Jan Długosz Univ. Czest. Math. **12**, 29–35 (2007)
16. Ger, R.: Additivity and exponentiality are alien to each other. Aequationes Math. **80**, 111–118 (2010)
17. Ger, R.: The alienation phenomenon and associative rational operations. Annales Math. Sil. **27**, 75–88 (2013)
18. Ger, R.: Alienation of additive and logarithmic equations. Ann. Univ. Sci. Budapest. Sect. Comput. **40**, 269–274 (2013)
19. Ger, R., Reich, L.: A generalized ring homomorphisms equation. Monatsh. Math. **159**, 225–233 (2010)
20. Gerber, H.U.: An Introduction to Mathematical Risk Theory. Homewood, Philadelphia (1979)
21. Gselmann, E.: Notes on the characterization of derivations. Acta Sci. Math. (Szeged) **78**, 137–145 (2012)
22. Halter-Koch, F.: Characterizations of field homomorphisms by functional equations. Aequationes Math. **59**, 298–305 (2000)
23. Hammer, C.: Über die Funktionalungleichung $f(x + y) + f(xy) \geq f(x) + f(y) + f(x)f(y)$. Aequationes Math. **45**, 297–299 (1993)
24. Heilpern, S.: A rank-dependent generalization of zero utility principle. Insur. Math. Econ. **33**, 67–73 (2003)
25. Kałuszka, M., Krzeszowiec, M.: Pricing insurance contracts under Cumulative Prospect Theory. Insur. Math. Econ. **50**, 159–166 (2012)
26. Kannappan, Pl.: The functional equations $f(xy) + f(xy)^{-1} = 2f(x)f(y)$ for groups. Proc. Am. Math. Soc. **19**, 69–74 (1968)
27. Kominek, Z., Sikorska, J.: Alienation of the logarithmic and exponential functional equations. Aequationes Math. **90**, 107–121 (2016)
28. Kuczma, M.: An Introduction to the Theory of Functional Equations and Inequalities. Birkhäuser, Basel (2009)
29. Lajkó, K.: Applications of extensions of additive functions. Aequationes Math. **11**, 68–76 (1974)
30. Lajkó, K.: Generalized Hosszú functional equations. Ann. Acad. Crac. Studia Math. **4**, 259–268 (2001)
31. Laohakosol, V., Pimsert, W., Udomkavanich, P.: Dependence relations among solutions of a universal Cauchy's functional equation. East-West J. Math. **2008**, 99–108 (2008)
32. Maksa, Gy.: Talk presented at the 14th Debrecen-Katowice Winter Seminar, Hajdúszoboszló, Hungary, January 29-February 1, 2014. In: Report of Meeting. Annales Math. Sil. **28**, 97–118 (2014)
33. Maksa, Gy.: Talk presented at the 17th Katowice-Debrecen winter seminar, Zakopane, Poland, 1–4 February, 2017

34. Maksa, Gy., Sablik, M.: On the alienation of the exponential Cauchy equation and the Hosszú equation. Aequationes Math. **90**, 57–66 (2016)
35. Rădulescu, M.: On a supra-additive and supra-multiplicative operator of $C(X)$. Bull. Math. Soc. Sci. Math. Repub. Soc. Roum. Nouv. Ser. **24**(72), 303–305 (1980)
36. Sablik, M.: Talk presented at the 51st ISFE, Rzeszów. In: Report of Meeting. Aequationes Math. **88**, 291–320 (2014), and at Positivity VII, Zaanen Centennial Conference, Leiden, 2013
37. Sahoo, P.K., Riedel, T.: Mean Value Theorems and Functional Equations. World Scientific, River Edge, NJ (1998)
38. Sinopoulos, P.: Generalized sine equations. I. Aequationes Math. **48**, 171–193 (1994)
39. Sinopoulos, P.: Functional equations on semigroups. Aequationes Math. **59**, 255–261 (2000)
40. Sobek, B.: Alienation of the Jensen, Cauchy and d'Alembert equations. Annales Math. Sil. **30**, 181–191 (2016)
41. Troczka-Pawelec, K., Tyrala, I.: On the alienation of the Cauchy equation and the Lagrange equation. Sci. Issues, Jan Długosz Univ. Czest. Math. **21**, 105–111 (2016)
42. Tversky, A., Kahneman, D.: Advances in prospect theory: cumulative representation of uncertainty. J. Risk Uncertain. **5**, 297–323 (1992)
43. Tyrala, I.: A characterization of the sine function by functional inequalities. Math. Inequal. Appl. **14**, 13–33 (2011)
44. Tyrala, I.: Solutions of the Dhombres-type trigonometric functional equation. Sci. Issues, Jan Długosz Univ. Czest. Math. **16**, 87–94 (2011)
45. Tyrala, I.: The generalization of the Lagrange functional equation. Sci. Issues Catholic University in Ruzomberok, Mathematica **IV**, 177–181 (2012)
46. Volkmann, P.: Sur un système d'inéquations fonctionnelles. C.R. Math. Rep. Acad. Sci. Canada **4**, 155–158 (1982)
47. Wilson, W.H.: On certain related functional equations. Bull. Am. Math. Soc. **26**, 300–312 (1919)

Chapter 7
Remarks on Analogies Between Haar Meager Sets and Haar Null Sets

Eliza Jabłońska

Abstract In the paper some analogies between Haar meager sets and Haar null sets in abelian Polish groups are presented.

Keywords Abelian Polish group • Haar meager set • Haar null set • Meager set • Set of Haar measure zero

Mathematics Subject Classification (2010) Primary 28C10, 28E05, 54B30, 54E52; Secondary 39B52, 39B62

7.1 Introduction

It is well known [3] that a subset A of an abelian Polish group X is called *Haar null* if there are a universally measurable set $B \subset X$ with $A \subset B$ and a Borel probability measure μ on X such that

$$\mu(x + B) = 0$$

for all $x \in X$. In [5] Darji introduced another family of "small" sets in an abelian Polish group X; he called a set $A \subset X$ *Haar meager* if there is a Borel set $B \subset X$ with $A \subset B$, a compact metric space K and a continuous function $f : K \to X$ such that

$$f^{-1}(B + x) \text{ is meager in } K \text{ for every } x \in X.$$

In a locally compact group these two definitions are equivalent to definitions of Haar measure zero sets and meager sets, respectively. That is why we can say that

---

E. Jabłońska (✉)
Department of Discrete Mathematics, Rzeszów University of Technology,
Powstańców Warszawy 12, 35-959 Rzeszów, Poland
e-mail: elizapie@prz.edu.pl

J. Brzdęk et al. (eds.), *Developments in Functional Equations and Related Topics*, Springer Optimization and Its Applications 124,
DOI 10.1007/978-3-319-61732-9_7

the notion of a Haar meager set is a topological analog to the notion of a Haar null set. Since lots of similarities between meager sets and sets of Haar measure zero are well known in locally compact abelian Polish groups (see, e.g., [24]), we would like to find as many analogies between Haar meager sets and Haar null sets as possible.

For each abelian Polish group Y we introduce the following notations:

$$\mathscr{HN}_Y := \{A \subset Y : A \text{ is Haar null}\},$$

$$\mathscr{HM}_Y := \{A \subset Y : A \text{ is Haar meager}\},$$

$$\mathscr{M}_Y := \{A \subset Y : A \text{ is meager}\};$$

and, if additionally Y is locally compact,

$$\mathscr{N}_Y := \{A \subset Y : A \text{ has Haar measure zero}\}.$$

Moreover, in the whole paper X is an abelian Polish group.

7.2 Basic Similarities

Let us start with the fact that both families, $\mathscr{HM}_X$ and $\mathscr{HN}_X$, are "small".

Theorem 7.1 ([3, Theorem 1]) *The family $\mathscr{HN}_X$ is a σ-ideal and, if X is locally compact,*

$$\mathscr{HN}_X = \mathscr{N}_X.$$

Theorem 7.2 ([5, Theorems 2.4, 2.9]) *The family $\mathscr{HM}_X$ is a σ-ideal and, if X is locally compact,*

$$\mathscr{HM}_X = \mathscr{M}_X.$$

Moreover, Darji (see [5, Theorem 2.2]) proved that in the case, where X is not locally compact,

$$\mathscr{HM}_X \subsetneq \mathscr{M}_X.$$

Clearly an analogous inclusion for Haar null sets is impossible.

An important result obtained by Christensen [3] is a theorem of Steinhaus' type.

Theorem 7.3 ([3, Theorem 2]) *For every universally measurable subset A of X, with $A \notin \mathscr{HN}_X$, the set*

$$\{x \in X : (A + x) \cap A \notin \mathscr{HN}_X\}$$

is a neighbourhood of 0 *in X; consequently* $0 \in$ *int* $(A - A)$.

A topological analogue of the above theorem also holds.

Theorem 7.4 ([16, Theorem 2]) *For every Borel subset A of X, $A \notin \mathscr{HM}_X$, the set*

$$\{x \in X : (A + x) \cap A \notin \mathscr{HM}_X\}$$

is a neighbourhood of 0 in X; i.e., $0 \in int\,(A - A)$.

The following generalization of Theorem 7.3 has been proved by Gajda [13].

Theorem 7.5 ([13, Theorem 1]) *For every $n \in \mathbb{N}$ and every universally measurable set $A \notin \mathscr{HN}_X$ the set*

$$\{x \in X : \bigcap_{k \in \{-n,,\ldots,n\}} (A + kx) \notin \mathscr{HN}_X\}$$

is a neighbourhood of 0 in X.

The above theorem is a very useful tool in functional equations. An analogous result has been proved in [17].

Theorem 7.6 ([17, Theorem 4]) *For every $n \in \mathbb{N}$ and Borel set $A \notin \mathscr{HM}_X$ the set*

$$\{x \in X : \bigcap_{k \in \{-n,,\ldots,n\}} (A + kx) \notin \mathscr{HM}_X\}$$

is a neighbourhood of 0 in X.

Christensen and Fischer [4] generalized Theorem 7.5 as follows.

Theorem 7.7 ([4, Theorem 2]) *For every $N \in \mathbb{N}$ and every universally measurable set $A \notin \mathscr{HN}_X$ the set*

$$\{(x_1, \ldots, x_N) \in X^N : A \cap \bigcap_{i=1}^{N} (A + x_i) \notin \mathscr{HN}_X\}$$

is a neighbourhood of 0 in X^N.

It turns out that an analogy to Theorem 7.7 also exists.

Theorem 7.8 ([18, Theorem 2.2]) *For every $N \in \mathbb{N}$ and Borel set $A \notin \mathscr{HM}_X$ the set*

$$\{(x_1, \ldots, x_N) \in X^N : A \cap \bigcap_{i=1}^{N} (A + x_i) \notin \mathscr{HM}_X\}$$

is a neighbourhood of 0 in X^N.

From Theorems 7.3 and 7.4 we obtain that σ-compact sets in non-locally compact groups are "small" in both senses. More precisely we have the following.

Corollary 7.1 ([3], [16, Corollary 1]) *If X is not locally compact, then each σ-compact set is Haar null as well as Haar meager.*

One of the well-known results is the decomposition theorem stating that the real line can be decomposed into two disjoint "small" sets: a meager one and a Lebesgue measure zero one. Doležal, Rmoutil, Vejnar and Vlasák proved that some special spaces also can be decomposed into two disjoint "small" sets.

Theorem 7.9 ([10, Theorems 22 and 25]) *Each Banach space, or $\mathbb{R}^\omega$, can be decomposed into two disjoint sets: a Haar meager one and a Haar null one.*

Let us pay attention yet that the Kuratowski–Ulam Theorem and the Fubini Theorem, which are analogues of each other in the locally compact groups, fail in non-locally compact groups.

Example 7.1 ([10, Example 20]) The set

$$C := \{(s,t) \in \mathbb{Z}^\omega \times \mathbb{Z}^\omega : t_n \leq s_n \leq 0 \text{ for } n \in \omega\}$$

is neither Haar null nor Haar meager. But the set

$$C[t] := \{s \in \mathbb{Z}^\omega : (s,t) \in C\}$$

is Haar meager as well as Haar null for each $t \in \mathbb{Z}^\omega$ (because it is compact). On the other hand, the set

$$A := \{s \in \mathbb{Z}^\omega : s_n \leq 0 \text{ for } n \in \omega\}$$

is non-Haar meager and non-Haar null and, for each $s \in A$, the set

$$C[s] := \{t \in \mathbb{Z}^\omega : (s,t) \in C\}$$

is neither Haar meager nor Haar null.

From this example we see that there exists a non-Haar meager and non-Haar null set in $\mathbb{Z}^\omega \times \mathbb{Z}^\omega$ such that in one direction all its section are Haar meager, and in the other direction there are non-Haar meager many sections which are non-Haar meager.

It is rather obvious that every set containing a translation of each compact set is "large" in both senses; i.e., the following proposition is valid.

Proposition 7.1 *Every set containing a translation of each compact set is neither Haar null nor Haar meager.*

This proposition is very useful, because allows to observe some further similarities between Haar meager sets and Haar null sets.

In the paper [22] Matoušková and Zelený constructed closed sets A, B in a non-locally compact abelian Polish group X such that A, as well as B, includes a translation of each compact set and the set $(A + x) \cap B$ is compact for each $x \in X$. Consequently we obtain two analogies characterizations of locally compact groups.

Proposition 7.2 *An abelian Polish group X is locally compact if and only if*

$$int\,(A+B)\neq\emptyset$$

for each universally measurable non-Haar null sets $A, B\subset X$

Proposition 7.3 *An abelian Polish group X is locally compact if and only if*

$$int\,(A+B)\neq\emptyset$$

for each Borel non-Haar meager sets $A, B\subset X$.

Dodos [6] has used Matoušková's and Zelený's result from [22] to show that the invariance under bigger subgroups is not sufficient to establish a dichotomy. More precisely, he proved the following fact.

Proposition 7.4 ([6, Proposition 12]) *If X is not locally compact and G is a σ-compact subgroup of X, then there exists a G-invariant* F_σ *subset F of X such that neither F nor* $X\setminus F$ *is Haar null.*

In view of Proposition 7.1, in the same way as Dodos, we can prove that an another type of dichotomy also does not hold.

Proposition 7.5 ([18, Proposition 3.2]) *If X is not locally compact and G is a σ-compact subgroup of X, then there exists a G-invariant* F_σ *subset F of X such that neither F nor* $X\setminus F$ *is Haar meager.*

Let us also recall that each meager set is contained in an F_σ meager set, as well as each set of Lebesgue measure zero is contained in a G_δ set of Lebesgue measure zero. It turns out that both theorems cannot be generalized on the case of Haar null sets and Haar meager sets. More precisely, Elekes and Vindyánszky [11] proved the following.

Theorem 7.10 ([11, Theorem 4.1]) *Let* $1\leq\xi<\omega_1$. *If X is non-locally compact, then there exists a Borel Haar null set that is not contained in any Haar null set from* $\Pi^0_\xi(X)$ *(i.e., the ξth multiplicative Borel class in X).*

The same type result for a Haar meager set has been proved by Doležal and Vlásak in [9].

Theorem 7.11 ([9, Theorem 10]) *Let* $1\leq\xi<\omega_1$. *If X is non-locally compact, then there exists a Borel Haar meager set that is not contained in any Haar meager set from* $\Sigma^0_\xi(X)$ *(i.e., the ξth additive Borel class in X).*

Clearly, for $\xi=2$, we obtain the existence of a Borel Haar null set without any G_δ Haar null hull, as well as the existence of a Borel Haar meager set without any F_σ Haar meager hull.

In the same papers we can also find the following theorems analogies each other.

Theorem 7.12 ([11, Theorem 4.1]) *If X is non-locally compact, then there exists a coanalytic Haar null set without any Borel Haar null hull.*

Theorem 7.13 ([9, Theorem 10]) *If X is non-locally compact, then there exists a coanalytic Haar meager set without any Borel Haar meager hull.*

Matoušková and Stegall [21] proved that a separable Banach space X is non-reflexive if and only if there exists a closed convex subset of X with empty interior, which contains a translation of any compact subset of X. Consequently, by Proposition 7.1, we obtain the following result.

Theorem 7.14 *Every separable nonreflexive Banach space contains a closed convex set with empty interior, which is neither Haar null nor Haar meager.*

Moreover, Matoušková [20, Theorem 4] has showed that this is unlike the situation in superreflexive spaces, where closed, convex, nowhere dense sets are Haar null. In turn Banakh [1, Proposition 5.7] has proved that each closed Haar null set in a Polish group is Haar meager. Hence we have the next theorem.

Theorem 7.15 *In separable superreflexive Banach spaces closed, convex, nowhere dense sets are Haar null as well as Haar meager.*

7.3 Generically Haar Meager Sets and Generically Haar Null Sets

Let us recall once again definitions of Haar meager sets and Haar null sets.

Definition 7.1 A set $A \subset X$ is *Haar null* if there is a universally measurable set $B \supset A$ and a Borel probability measure μ on X such that

$$\mu(x + B) = 0 \quad \text{for all} \quad x \in X.$$

Definition 7.2 A set $A \subset X$ is *Haar meager* if there is a Borel set $B \supset A$, a compact metric space K and a continuous function $f : K \to X$ such that

$$f^{-1}(B + x) \in \mathcal{M}_K \quad \text{for all} \quad x \in X.$$

It means that:

- each Haar null set has the only one witness parameter—*a test measure*;
- each Haar meager set has two witness parameters—*a witness metric space* and *a witness function*.

The following result has been proved in [2].

Proposition 7.6 *A Borel set $B \subset X$ is Haar meager if and only if there is a continuous function $f : 2^\omega \to X$ such that $f^{-1}(B + x)$ is meager in 2^ω for all $x \in X$.*

It means that a Haar meager set and a Haar null set have both the only one witness parameter—*a witness function* and *a test measure*, respectively.

Now, let $P(X)$ be the space of all Borel probability measures on X; this is a Polish space with Lévy metric.

Following Dodos [7, 8], given a universally measurable set $A \subset X$, by $T(A)$ we mean the set of all test measures for A, i.e.

$$T(A) := \{\mu \in P(X) : \mu(x + A) = 0 \text{ for every } x \in X\}.$$

Dodos [7] has proved the following.

Theorem 7.16 ([7, Proposition 5]) *If $A \subset X$ is a universally measurable Haar null set, then:*

- *$T(A)$ is dense in $P(X)$;*
- *if A is analytic, then either $T(A)$ is meager or $T(A)$ is comeager in $P(X)$;*
- *if A is σ-compact, then $T(A)$ is comeager in $P(X)$.*

Using Theorem 7.16, Dodos [8] has introduced the notion of a generically Haar null set and next he has proved a theorem of Steinhaus' type.

Definition 7.3 A set $A \subset X$ is *generically Haar null* if $T(A)$ is comeager in $P(X)$.

Theorem 7.17 ([8, Proposition 11]) *If $A \subset X$ is analytic, non-generically Haar null, then $A - A$ is non-meager.*

Now, let $C(2^\omega, X)$ be the space of all continuous functions $f : 2^\omega \to X$; this is a Polish space with the supremum metric (similarly as the space $P(X)$ with Lévy metric). For every Borel set $A \subset X$ we define

$$W(A) := \{f \in C(2^\omega, X) : f^{-1}(x + A) \in \mathcal{M}_{2^\omega} \text{ for every } x \in X\},$$

i.e., the set of all witness functions for A. Clearly, if $A \in \mathcal{HM}_X$, then $W(A) \neq \emptyset$, so this notation is analogous to $T(A)$.

In [1] and [2] an analogous result to Theorem 7.16 has been proved.

Theorem 7.18 ([2]) *Let $A \subset X$ be a Borel Haar meager set. Then:*

- *$W(A)$ is dense in $C(2^\omega, X)$;*
- *either $W(A)$ is meager, or $W(A)$ is comeager in $C(2^\omega, X)$;*
- *if A is σ-compact, then $W(A)$ is comeager in $C(2^\omega, X)$.*

Theorem 7.19 ([1], [2]) *If $A \subset X$ is analytic, non-generically Haar meager (i.e., $W(A)$ is not comeager in $C(2^\omega, X)$), then $A - A$ is non-meager.*

7.4 Analogies in Functional Equations

In this part (only) we assume that X is a Polish real linear space to present some further similarities between Haar meager sets and Haar null sets, which are very important in functional equations.

Lemma 7.1 ([23, Lemma 5]) *Let $A \notin \mathcal{HN}$ be a universally measurable set and $x \in X \setminus \{0\}$. Then there exists a Borel set $B \subset A$ such that the set $k_x^{-1}(B + z)$ has a positive Lebesgue measure in $\mathbb{R}$ for each $z \in X$, where $k_x : \mathbb{R} \to X$ is given by $k_x(\alpha) = \alpha x$.*

Lemma 7.2 ([19, Lemma 1]) *Let $A \notin \mathcal{HM}$ be a Borel set and $x \in X \setminus \{0\}$. Then there exists a Borel set $B \subset A$ such that the set $k_x^{-1}(B + z)$ is non-meager with the Baire property in $\mathbb{R}$ for each $z \in X$.*

Due to those two lemmas t-Wright convex functions, that are bounded on a "large" set, can be characterized.

Theorem 7.20 ([23, Theorem 8]) *Let $D \subset X$ be a nonempty convex open set and $t \in (0, 1)$. Each t-Wright convex function $f : D \to \mathbb{R}$ bounded on a non-Haar null universally measurable set $T \subset D$ is continuous.*

Theorem 7.21 ([19, Theorem 4]) *Let $D \subset X$ be a nonempty convex open set and $t \in (0, 1)$. Each t-Wright convex function $f : D \to \mathbb{R}$ bounded on a non-Haar meager Borel set $T \subset D$ is continuous.*

Now, using a weaker version of Lemma 7.1, the additive functions, that are bounded above on a "large" set, can be characterized. More precisely, the following theorem is true.

Theorem 7.22 ([14, Corollary 1]) *If $f : X \to \mathbb{R}$ is additive and bounded above on a universally measurable set $C \notin \mathcal{HN}$, then f is linear.*

Replacing [14, Lemma 1] by Lemma 7.2 in the proof of the above theorem, we obtain an analogous result.

Theorem 7.23 *If $f : X \to \mathbb{R}$ is additive and bounded above on a Borel set $C \notin \mathcal{HM}$, then f is linear.*

Moreover, using a weaker version of Lemma 7.1 and Theorem 7.22, solutions of a generalized Gołąb–Schinzel equation, that are bounded on a "large" set, can be characterized.

Theorem 7.24 ([15, Theorem 1]) *Let $f : X \to \mathbb{R}$, $M : \mathbb{R} \to \mathbb{R}$ and $|f(D)| \subset (0, a)$ for a positive number a and a universally measurable set $D \notin \mathcal{HN}$. Then functions f and M satisfy the equation*

$$f(x + M(f(x))y) = f(x)f(y) \tag{7.1}$$

if and only if one of the following three conditions holds:

(i) $f = 1$;

(ii) $M|_{(0,\infty)} = 1$ *and there exists a nontrivial linear functional* $h : X \to \mathbb{R}$ *such that*

$$f(x) = \exp h(x) \ \text{for} \ x \in X;$$

(iii) there exists a nontrivial linear functional $h : X \to \mathbb{R}$ *and* $c \in \mathbb{R} \setminus \{0\}$ *such that either*

$$M(y) = |y|^{1/c} \operatorname{sgn} y \ \text{for} \ y \in \mathbb{R},$$

$$f(x) = \begin{cases} |h(x)+1|^c \operatorname{sgn}(h(x)+1), & x \in X, \ h(x) \neq -1; \\ 0, & x \in X, \ h(x) = -1 \end{cases}$$

or

$$M(y) = y^{1/c} \ \text{for} \ y \in [0, \infty),$$

$$f(x) = \begin{cases} (h(x)+1)^c, & x \in X, \ h(x) > -1; \\ 0, & x \in X, \ h(x) \leq -1. \end{cases}$$

Observe that using the method from [15] we can prove a theorem which is analogous to Theorem 7.24; the most important change in the proof is to replace:

- [15, Lemma 6] by Theorem 7.4,
- [15, Lemma 7] by Lemma 7.2,
- [15, Lemma 8] by Theorem 7.23.

Then we obtain the following theorem.

Theorem 7.25 *Let* $f : X \to \mathbb{R}$, $M : \mathbb{R} \to \mathbb{R}$ *and* $|f(D)| \subset (0, a)$ *for a positive number* a *and a Borel set* $D \notin \mathcal{HM}$. *Then functions* f *and* M *satisfy Equation* (7.1) *if and only if one of the conditions (i)–(iii) of Theorem 7.24 holds.*

7.5 Modified Darji's and Christensen's Definitions

Doležal, Rmoutil, Vejnar and Vlasák [10] modified Darji's notion of meagerness in the following way.

Definition 7.4 A set $A \subset X$ is *naively Haar meager* if there is a compact metric space K and a continuous function $f : K \to X$ such that

$$f^{-1}(x + A) \ \text{is meager in} \ K \ \text{for every} \ x \in X.$$

They also have proved the next theorem.

Theorem 7.26 ([10, Theorem 16]) *If X is uncountable, then there exists a naively Haar meager subset of X, which is not Haar meager.*

In a similar way Elekes and Vindyánszky [12] have defined naively Haar null sets and showed a result analogous to Theorem 7.26.

Definition 7.5 A set A is called *naively Haar null* if there is a Borel probability measure μ on X such that

$$\mu(x + A) = 0 \quad \text{for all} \quad x \in X.$$

Theorem 7.27 ([12, Theorem 1.3]) *If X is uncountable, then there exists a naively Haar null subset of X which is not Haar null.*

Moreover, in non-abelian Polish groups definitions of Haar meager sets and Haar null sets have been modified in the following way.

Definition 7.6 A subset A of a Polish group X is *Haar null* if there are a universally measurable set $B \subset X$ with $A \subset B$ and a Borel probability measure μ on X such that

$$\mu(x + B + y) = 0 \quad \text{for all} \quad x, y \in X.$$

Definition 7.7 A subset A of a Polish group X is *Haar meager* if there are a Borel set $B \subset X$ with $A \subset B$, a compact metric space K and a continuous function $f : K \to X$ such that

$$f^{-1}(x + B + y) \quad \text{is meager in} \quad K \quad \text{for every} \quad x, y \in X.$$

Then both families—of all Haar null sets and of all Haar meager sets in X—form σ-ideals (see [12] and [10, Theorem 3]).

References

1. Banakh, T., Karchevska, L., Ravsky, A.: The closed Steinhaus properties of σ-ideals on topological groups. 30 Sept 2015, arXiv:1509.09073v1 [math.GN]
2. Banakh, T., Głąb, S., Jabłońska, E., Swaczyna, J.: Steinhaus properties on σ-ideals on Polish groups (manuscript)
3. Christensen, J.P.R.: On sets of Haar measure zero in abelian Polish groups. Israel J. Math. **13**, 255–260 (1972)
4. Christensen, J.P.R., Fischer, P.: Small sets and a class of general functional equations. Aequationes Math. **33**, 18–22 (1987)
5. Darji, U.B.: On Haar meager sets. Topol. Appl. **160**, 2396–2400 (2013)
6. Dodos, P.: Dichotomies of the set of test measures of a Haar null set. Israel J. Math. **144**, 15–28 (2004)
7. Dodos, P.: On certain regularity properties of Haar-null sets. Fund. Math. **181**, 97–109 (2004)
8. Dodos, P.: The Steinhaus property and Haar-null sets. Bull. Lond. Math. Soc. **41**, 377–384 (2009)

9. Doležal, M., Vlásak, V.: Haar meager sets, their hulls, and relationship to compact sets. J. Math. Anal. Appl. **446**, 852–863 (2017)
10. Doležal, M., Rmoutil, M., Vejnar, B., Vlásak, V.: Haar meager sets revisited. J. Math. Anal. Appl. **440**, 922–939 (2016)
11. Elekes, M., Vidnyánszky, Z.: Haar null sets without G_δ hulls. Israel J. Math. **209**, 199–214 (2015)
12. Elekes, M., Vidnyánszky, Z.: Naively Haar null sets in Polish groups. J. Math Anal. Appl. **446**, 193–200 (2017)
13. Gajda, Z.: Christensen measurability of polynomial functions and convex functions of higher orders. Ann. Polon. Math. **47**, 25–40 (1986)
14. Jabłońska, E.: Jensen convex functions bounded above on nonzero Christensen measurable sets. Ann. Math. Sil. **23**, 53–55 (2009)
15. Jabłońska, E.: Christensen measurability and some functional equation. Aequationes Math. **81**, 155–165 (2011)
16. Jabłońska, E.: Some analogies between Haar meager sets and Haar null sets in abelian Polish groups. J. Math. Anal. Appl. **421**, 1479–1486 (2015)
17. Jabłońska, E.: A theorem of Piccard's type and its applications to polynomial functions and convex functions of higher orders. Topology Appl. **209**, 46–55 (2016)
18. Jabłońska, E.: A theorem of Piccard's type in abelian Polish groups. Anal. Math. **42**, 159–164 (2016)
19. Jabłońska, E.: $\mathscr{D}$-measurability and t-Wright convex functions. J. Math. Inequal. **10**, 213–217 (2016)
20. Matoušková, E.: Convexity and Haar null sets. Proc. Am. Math. Soc. **125**, 1793–1799 (1997)
21. Matoušková, E., Stegall, C.: A characterization of reflexive Banach spaces. Proc. Amer. Math. Soc. **124**, 1083–1090 (1996)
22. Matoušková, E., Zelený, M.: A note on intersections of non-Haar null sets. Colloq. Math. **96**, 1–4 (2003)
23. Olbryś, A.: On the boundedness, Christensen measurability and continuity of t-Wright convex functions. Acta Math. Hungar. **141**, 68–77 (2013)
24. Oxtoby, J.C.: Measure and Category. Springer, New York/Heidelberg/Berlin (1971)

Chapter 8
On Some Inequalities Inspired by the Stability of Dynamical System

Zenon Moszner

Abstract We approximate the solutions of the system of inequalities

$$\begin{cases} |H(t,H(s,x)) - H(t+s,x)| \le \delta, \ x \in I, t, s \in \mathbb{R} \\ |H'(0,x) - a| \le \delta, \qquad\qquad\qquad x \in I \end{cases}$$

(I is nondegenerated interval) for $a \neq 0$ by the dynamical system and we consider the different stabilities of this system for $\delta = 0$.

Keywords Generalized dynamical system • Stability • b-Stability • Inverse stability • Inverse b-stability • Absolute stability

Mathematics Subject Classification (2010) Primary 39B82; Secondary 39B62

8.1 Introduction

The one-dimensional dynamical system is defined as the continuous function $F : \mathbb{R} \times I \to I$, where I is nondegenerated interval, for which

$$F(t, F(s,x)) = F(t+s,x), \qquad x \in I, t, s \in \mathbb{R} \tag{8.1}$$

(the translation equation) and

$$F(0,x) = x, \qquad x \in I. \tag{8.2}$$

Z. Moszner (✉)
Institute of Mathematics, Pedagogical University, Podchorążych 2, 30-084 Kraków, Poland
e-mail: zmoszner@up.krakow.pl

J. Brzdęk et al. (eds.), *Developments in Functional Equations and Related Topics*, Springer Optimization and Its Applications 124,
DOI 10.1007/978-3-319-61732-9_8

The conjunction of the above conditions is equivalent to the conjunction of conditions (8.1) and

$$F'(0,x) = a, \qquad x \in I, \tag{8.3}$$

(where $a \in \mathbb{R}$ and $F'(0,x)$ means the derivative of the function $F(0,.) : I \to I$ at the point x) only for $a = 1$.

The system (8.1) and (8.3) does not have the solution for $0 \neq a \neq 1$, and for $a = 0$ it has only the constant solution $F(t,x) = c \in I$.[1] Indeed, by (8.3) we have $F(0,x) = ax + b$ and by (8.1) we obtain $a(ax+b)+b = ax+b$, thus "$a = 0$ and b arbitrary" or "$a = 1$ and $b = 0$". For $a = 0$ we have $F(0,x) = b$ for a $b \in I$ and $F(t,x) = F(F(t,x),0) = b$ too.

The system (8.1) and (8.3) is said to be *Ulam–Hyers stable* (in short *stable*) if for every $\epsilon > 0$ there exists a $\delta > 0$ such that for every function $H : \mathbb{R} \times I \to I$ if

$$\begin{cases} |H(t,H(s,x)) - H(t+s,x)| \leq \delta, & x \in I, t,s \in \mathbb{R} \\ |H'(0,x) - a| \leq \delta, & x \in I, \end{cases} \tag{8.4}$$

then there exists a solution F of this system (8.1) and (8.3) for which

$$|F(t,x) - H(t,x)| \leq \varepsilon, \qquad x \in I, t \in \mathbb{R}.$$

8.2 Stability

Theorem 8.1 *If a function $H : \mathbb{R} \times I \to I$ satisfies the conditions* (8.4)

(i) for some $a > 0$ and positive $\delta \leq \frac{a}{2a+2}$ and it is continuous with each variable, then there exists a dynamical system $F^\star$ such that

$$|F^\star(t,x) - H(t,x)| \leq (20 + \frac{a+2}{a})\delta, \qquad x \in I, t \in \mathbb{R},$$

(ii) for some $a < 0$ and positive $\delta \leq \frac{a-a^2}{3a-2}$, then there exists a dynamical system $F^\star$ such that

$$|F^\star(t,x) - H(t,x)| \leq 4\frac{a-1}{a}\delta, \qquad x \in I, t \in \mathbb{R},$$

(iii) for $a = 0$ and positive $\delta \leq \frac{1}{2}$, then there exists a solution $F^\star$ of the system (8.1) *and* (8.3) *such that*

$$|F^\star(t,x) - H(t,x)| \leq 2\delta, \qquad x \in I, t \in \mathbb{R}.$$

---

[1] Moreover, for the solution F of (8.1) if $F'(0,x)$ exists (not necessarily constant), then $F(t,x) = c$ or $F(0,x) = x$ (see [1]).

Lemma 8.1 *Let a function $h : I \to I$ be such that $|h(h(x)) - h(x)| \le \delta$ and $|h'(x) - a| \le \delta$ for every $x \in I$ and for some $\delta > 0$ and $a \in \mathbb{R}$.*

(i) If $a > 0$ and $\delta \le \frac{a}{2a+2}$, then

$$|h(x) - x| \le \frac{a+2}{a}\delta, \qquad x \in I. \tag{8.5}$$

(ii) If $a < 0$ and $\delta \le \frac{a-a^2}{3a-2}$, then

$$|h(x) - x| \le 2\frac{a-1}{a}\delta, \qquad x \in I.$$

Proof (i) (a simple modification of the proof of Corollary 3.8 in [1]). We have

$$|h(h(x)) - h(x)| \le \delta \le \frac{a+2}{a}\delta, \qquad x \in I,$$

thus (8.5) is satisfied for $x \in h(I)$. Since $|h'(x) - a| \le \delta$, we see that

$$0 < a - \frac{a}{2a+2} \le a - \delta \le h'(x) \le a + \delta \le a + \frac{a}{2a+2}. \tag{8.6}$$

From here the function h is increasing.

Let $y_1 = \inf I, y_2 = \sup I, x_1 = \inf h(I), x_2 = \sup h(I)$. We consider two cases.

(1) For $y_1 > -\infty$ and $y_2 = +\infty$ the function h is unbounded, since, in the contrary case, we would have

$$\frac{h(n) - h(y_1 + 1)}{n - (y_1 + 1)} = h'(\theta(n)) \to 0 \quad \text{as} \quad n \to +\infty,$$

which is a contradiction with (8.6).

(a) If $x_1 = y_1$, then $h(I) = I$ and condition (8.5) is satisfied.

(b) If $x_1 > y_1$ and $y_1 \in I$, then we have $h(y_1) = x_1$ and $|h(x_1) - x_1| \le \delta$, and since

$$h(x_1) - x_1 = h(x_1) - h(y_1) = h'(\theta)(x_1 - y_1)$$

for a θ, we get

$$(a - \delta)(x_1 - y_1) \le h'(\theta)(x_1 - y_1) = |h(x_1) - x_1| \le \delta,$$

and consequently $x_1 - y_1 \le \frac{\delta}{a-\delta}$. If $x \in [y_1, x_1)$, then by (8.6) and $\delta \le \frac{a}{2a+2}$ we obtain

$$|h(x)-x| \le |h(x)-x_1| + |x_1-x| \le h'(\theta)(x-y_1)+(x_1-x)$$

$$\le \frac{\delta}{a-\delta}[h'(\theta)+1] \le \frac{\delta+a+1}{a-\delta}\delta \le \frac{a+2}{a}\delta.$$

Therefore (8.5) is satisfied for $x \in I$, because $h(I) = [x_1, +\infty)$.

(2) If $y_1 = -\infty$ and $y_2 = +\infty$, then by (8.6) the function h is unbounded from above and from below, thus $h(I) = \mathbb{R}$ and (8.5) is satisfied.
The proof in the other cases is analogous.

(ii) The proof is analogous as above with x_1 replaced by x_2, since the function h is decreasing in this case. □

Proof of Theorem 8.1 Put $h(x) = H(0,x)$ and $V = H(\mathbb{R} \times I)$.

(i) If $|V| \le 2\delta$, then by Lemma 8.1 for the dynamical system $F^\star(t,x) = x$ for $(t,x) \in \mathbb{R} \times I$ we have

$$|H(t,x) - F^\star(t,x)| \le |H(t,x) - H(0,x)| + |H(0,x) - x|$$

$$< [2 + \frac{a+2}{a}]\delta \le [20 + \frac{a+2}{a}]\delta.$$

By Theorem 1.1 in [2] there exists a continuous solution F of translation equation (8.1) for which

$$|F(t,x) - H(t,x)| \le 10\delta, \qquad x \in I, t \in \mathbb{R}.$$

Let $z_1 = \inf V$ and $z_2 = \sup V$. If $|V| > 2\delta$, then from the proof of Theorem 1.1 in [2] we see that

$$f(x) := F(0,x) = \begin{cases} x & \text{if } x \in [h(z_1), h(z_2)] \cap I, \\ h(z_1) & \text{if } x \in [z_1, h(z_1))] \cap I, \\ h(z_2) & \text{if } x \in [h(z_2), z_2] \cap I, \\ h(x) & \text{if } x \in I \setminus V \text{ and } h(x) \in [h(z_1), h(z_2)], \\ h(z_1) & \text{if } x \in I \setminus V \text{ and } h(x) \in [z_1, h(z_1)], \\ h(z_2) & \text{if } x \in I \setminus V \text{ and } h(x) \in [h(z_2), z_2], \end{cases}$$

where $h(z_1) := z_1$ if $z_1 \notin I$ and $h(z_2) := z_2$ if $z_2 \notin I$, and

$$F(t,x) = \begin{cases} h_n^{-1}(h_n(f(x)) + t) & \text{if } f(x) \in B_n \text{ and } x \in I, t \in \mathbb{R}, n \in M \subset \mathbb{N}, \\ f(x) & \text{for the other } x \in I \text{ and } t \in \mathbb{R}, \end{cases}$$

where $B_n \subset f(I)$ are open disjoint intervals and h_n is a homeomorphism from B_n onto $\mathbb{R}$.

Since f is increasing, by (8.4) we obtain

$$f(x) = \begin{cases} h(z_1) & \text{if } x < h(z_1) \text{ and } x \in I, \\ x & \text{if } x \in [h(z_1), h(z_2)], \\ h(z_2) & \text{if } x > h(z_2) \text{ and } x \in I, \end{cases}$$

and therefore $f(I) = [h(z_1), h(z_2)]$. Moreover, $B_n \subset (h(z_1), h(z_2))$ because B_n is an open interval and $B_n \subset f(I)$. This together with $f(x) \in B_n$ implies $x \in (h(z_1), h(z_2))$, and consequently $f(x) = x$ for $f(x) \in B_n$. Therefore the function F has the form

$$F(t,x) = \begin{cases} h_n^{-1}(h_n(x)+t) & \text{if } x \in B_n, t \in \mathbb{R}, n \in M \subset \mathbb{N}, \\ f(x) & \text{for the other } x \in I \text{ and } t \in \mathbb{R}. \end{cases}$$

The function

$$F^\star(t,x) = \begin{cases} h_n^{-1}(h_n(x)+t) & \text{if } x \in B_n, t \in \mathbb{R}, n \in M \subset \mathbb{N}, \\ x & \text{for the other } x \in I \text{ and } t \in \mathbb{R} \end{cases}$$

is a dynamical system. By the Lemma 8.1 we have

$$\begin{aligned} |H(t,x) - F^\star(t,x)| &= |H(t,x) - F(t,x)| + |F(t,x) - F^\star(t,x)| \\ &\le 10\delta + |f(x) - x| \le 10\delta + |f(x) - h(x)| + |h(x) - x| \\ &\le 10\delta + 10\delta + \frac{a+2}{a}\delta = \left(20 + \frac{a+2}{a}\right)\delta, \end{aligned}$$

because

$$|f(x) - h(x)| = |F(0,x) - H(0,x)| \le 10\delta.$$

(ii) The proof is as the proof of part (b) of Theorem 3.1 in [1]. However, we present this short proof for the reader's convenience. Since the function h is decreasing, the interval has to be bounded (otherwise we would have $\lim_{x\to+\infty} |h(x)-x| = +\infty$ or $\lim_{x\to-\infty} |h(x) - x| = +\infty$, which is impossible). Putting $h(y_1) = \lim_{x\to y_1^+} h(x)$ if $y_1 \notin I$ and $h(y_2) = \lim_{x\to y_2^-} h(x)$ if $y_2 \notin I$ we have

$$\begin{aligned} y_2 - y_1 &= y_2 - h(y_2) + h(y_2) - h(y_1) + h(y_1) - y_1 \\ &\le 2\frac{a-1}{a}\delta + (h(y_2) - h(y_1)) + 2\frac{a-1}{a}\delta \le 4\frac{a-1}{a}\delta, \end{aligned}$$

because $y_1 \le y_2$. Hence the function $F^\star(t,x) = x$ is a dynamical system for which

$$|F^\star(t,x) - H(t,x)| \le 4\frac{a-1}{a}\delta.$$

(iii) We proceed with the following remark. Let α, β be real numbers for which $|\alpha - \beta| \leq 2\delta$ and $|\frac{\alpha}{\beta}| \leq \delta$ for a positive $\delta \leq \frac{1}{2}$. Then we have $|\alpha(\frac{\alpha}{\beta} - 1)| \leq 2\delta^2$ and $0 < 1 - \delta \leq 1 - \frac{\alpha}{\beta}$, and thus $|\alpha| \leq 2\frac{\delta}{1-\delta}\delta \leq 2\delta$.

Since $|H(0, H(t,x)) - H(t,x)| \leq \delta$, we have $|h(x) - x| \leq \delta$ and $|h(y) - y| \leq \delta$ for $x, y \in h(V)$, where $V = h(H(\mathbb{R} \times I))$. Thus $|[h(x) - h(y)] - (x - y)| \leq 2\delta$ and

$$\left|\frac{h(x) - h(y)}{x - y}\right| = |h'(\theta)| \leq \delta, \qquad x \neq y.$$

By the above remark for $\alpha = h(x) - h(y)$ and $\beta = x - y$ we have $|h(x) - h(y)| \leq 2\delta$ for $x, y \in h(V)$, and therefore $\sup h(V) - \inf h(V) \leq 2\delta$. Let $c = \inf h(V)$ and $d = \sup h(V)$. Since $|H(0, H(t,x)) - H(t,x)| \leq \delta$, we see that

$$c - \delta \leq h(H(t,x)) - \delta \leq H(t,x) \leq h(H(t,x)) + \delta \leq d + \delta.$$

Hence $H(t,x) \in [c-\delta, d+\delta]$ and for the solution $F^\star(t,x) = \frac{c+d}{2}$ of system (8.1) and (8.3) we have

$$|F^\star(t,x) - H(t,x)| \leq \frac{d + \delta - c + \delta}{2} \leq \frac{4\delta}{2} = 2\delta, \qquad t \in \mathbb{R}, x \in I.$$

□

Corollary 8.1 *For* $0 \neq a \neq 1$ *there exists a* $\delta > 0$ *for which system* (8.4) *has a solution and there exists a* $\delta > 0$ *for which it does not have a solution.*

Proof For $\delta = |a|$ every constant function $H : \mathbb{R} \times I \to I$ is a solution of (8.4).

Assume that $0 \neq a \neq 1$ and that for every $\delta > 0$ system (8.4) has a solution H. Then for $h(x) := H(0, x)$ we have

$$|h'(x) - a| = |h'(x) - 1 - (a - 1)| \leq \delta$$

and

$$|a - 1| - \delta \leq |h'(x) - 1|.$$

By Lemma 8.1 there exist positive constants $A(a)$ and $B(a)$ such that $|h(x) - x| \leq B(a)\delta$ for $0 < \delta \leq A(a)$ and $x \in I$. Since we also have $|h(y) - y| \leq B(a)\delta$, thus for a θ we get

$$|h(x) - h(y) - (x - y)| = |h'(\theta)(x - y) - (x - y)| = |[h'(\theta) - 1](x - y)| \leq 2B(a)\delta.$$

Hence $(|a - 1| - \delta)|x - y| \leq 2B(a)\delta$, and if $\delta < |a - 1|$, then

$$|x - y| \leq 2B(a)\delta(|a - 1| - \delta)^{-1}, \qquad x, y \in I.$$

If the interval I is unbounded, then we have a contradiction. On the other hand, if it is bounded, then $0 < |I| \le 2B(a)\delta(|a-1|-\delta)^{-1}$, where $|I|$ is the length of I. Since $\lim_{\delta\to 0}\delta(|a-1|-\delta)^{-1} = 0$, we have a contradiction too. □

Problem The set $S_1(a)$ ($S_2(a)$) of $\delta > 0$ for which system (8.4) does not have a solution (has a solution) is an interval and $S_1 \cup S_2 = (0, +\infty)$. For $a = 0$ and $a = 1$ we have $S_1(0) = S_1(1) = \emptyset$ and $S_2(0) = S_2(1) = (0, +\infty)$. For $0 \neq a \neq 1$ give $\delta(a) := \sup S_1(a)$ (its existence follows from Corollary 8.1). Does $\delta(a)$ belong to S_1 (S_2)?

Corollary 8.2 *The system* (8.1) *and* (8.3) *is stable for* $0 \neq a \neq 1$ *as well as for* $a = 1$ *in the class of functions which are continuous in each variable with* $\delta = \frac{\varepsilon}{23}$, *and for* $a = 0$ *with* $\delta \le \min\{\frac{\varepsilon}{2}, \frac{1}{2}\}$.

Proof It is trivial for $0 \neq a \neq 1$, because δ for which system (8.4) does not have a solution is "good" for every $\varepsilon > 0$. For $a = 0$ and $a = 1$ the stability of (8.1) and (8.3) is a consequence of Theorem 8.1. □

8.2.1 Remarks

Remark 8.1 A continuous solution of (8.4) for $0 \neq a \neq 1$ is approximated by a dynamical system (Theorem 8.1 (i) and (ii)) and it is not approximated by a solution of (8.1) and (8.3), since system (8.1) and (8.3) does not have a solution.

Remark 8.2 The solution $F^\star$ of system (8.1) and (8.3) in Theorem 8.1(iii) cannot be replaced by a dynamical system. Indeed, if the interval I is unbounded, then the function $H(t,x) = c \in I$ satisfies condition (8.4) for $a = 0$ and for every dynamical system $F^\star$ the function $|F^\star(0,x) - H(0,x)| = |x - c|$ is unbounded. Now, let the interval I be bounded and nondegenerated and assume that for $\varepsilon = \frac{|I|}{4}$ there exists a $\delta > 0$ such that for every function H satisfying (8.4) for $a = 0$ there exists a dynamical system $F^\star$ for which $|F^\star(t,x) - H(t,x)| \le \varepsilon$. The function $H(t,x) = \frac{\inf I + \sup I}{2}$ satisfies (8.4) for every $\delta > 0$ and

$$\frac{|I|}{2} = \sup_{x\in I}\left|x - \frac{\inf I + \sup I}{2}\right| = \sup_{x\in I}|F^\star(0,x) - H(0,x)| \le \varepsilon = \frac{|I|}{4},$$

a contradiction.

Remark 8.3 The system (8.1) and (8.3) is equivalent to the equation

$$|F(t, F(s,x)) - F(t+s,x)| + |F'(0,x) - a| = 0. \tag{8.7}$$

Since for any $b, c \in \mathbb{R}$ we have

$$\Big(|b|+|c| \le \delta \Rightarrow (|b| \le \delta \wedge |c| \le \delta)\Big) \quad \text{and} \quad \Big((|b| \le \delta \wedge |c| \le \delta) \Rightarrow |b|+|c| \le 2\delta\Big),$$

Equation (8.7) is stable by Corollaries 8.1 and 8.2. This equation for $a = 1$ is equivalent to the equation

$$|F(t, F(s, x)) - F(t + s, x)| + |F(0, x) - x| = 0, \tag{8.8}$$

which is equivalent to system (8.1) and (8.2). This system is stable only for $I = \mathbb{R}$ (see [2]), thus Equation (8.8) is stable only for $I = \mathbb{R}$ too.

Remark 8.4 The stability of system (8.1) and (8.3) for $a = 1$ is proved in [1] with $\delta = \min\{\frac{\varepsilon}{10}, \frac{2}{5}\}$ by a complicated and longer proof than here.

Remark 8.5 A non-constant solution F of translation equation (8.1) for which $F'(0, x)$ exists is a dynamical system. The translation equation is unstable for every interval I in the class of functions $H : \mathbb{R} \times I \to I$ for which $H'(0, x)$ exists (see [1]).

8.3 *b*-Stability

The system (8.1) and (8.3) is said to be *b-stable* if for every function $H : \mathbb{R} \times I \to I$ the boundedness of the function

$$|H(t, H(s, x)) - H(t + s, x)| + |H'(0, x) - a| \tag{8.9}$$

implies the boundedness of the function $|F(t, x) - H(t, x)|$ for a solution F of the system (8.1) and (8.3).

Theorem 8.2 *The system* (8.1) *and* (8.3) *is*

(i) not b-stable for $0 \neq a \neq 1$,
(ii) b-stable both for $a = 1$ *and for* $a = 0$ *only if* I *is bounded.*

Proof (i) It is known (see Corollary 8.1) that for $0 \neq a \neq 1$ there exists a $\delta > 0$ such that the system (8.4) of the inequalities has a solution. Since the system (8.1) and (8.3) does not have the solutions, this system is not *b*-stable.

(ii) Let $a = 1$. The function (8.9) is bounded for the function $H(t, x) = c \in I$. Assume that there exists a solution F of (8.1) and (8.3) such that $|F(t, x) - H(t, x)|$ is bounded. Thus $|F(0, x) - H(0, x)| = |x - c|$ is bounded and if I is unbounded we have a contradiction. If I is bounded the function $|F(t, x) - H(t, x)|$ is bounded by the length of I for any functions $F, H : \mathbb{R} \times I \to I$, thus the system is *b*-stable in this case.

We have the same situation for $a = 0$: $H(t, x) = x$ is the solution of (8.4) for $\delta = 1$ and $F(t, x) = c \in I$ is the only solution of the system (8.1) and (8.3) in this case. □

8.4 Inverse Stability

The system (8.1) and (8.3) is said to be *inversely stable* if for every $\epsilon > 0$ there exists a $\delta > 0$ such that for every function $H : \mathbb{R} \times I \to I$ such that

$$|H(t,x) - F(t,x)| \leq \delta, \qquad (t,x) \in \mathbb{R} \times I$$

for some solution F of (8.1) and (8.3) we have

$$|H(t, H(s,x)) - H(t+s,x)| \leq \epsilon, \quad |H'(0,x) - a| \leq \epsilon, \qquad x \in I, t, s \in \mathbb{R}.$$

Theorem 8.3 *The system* (8.1) *and* (8.3) *is inversely stable for* $0 \neq a \neq 1$, *and it is not inversely stable for* $a = 0$ *and* $a = 1$.

Proof The system (8.1) and (8.3) does not have the solution for $0 \neq a \neq 1$, thus it is inversely stable.

For $a = 0$ ($a = 1$) assume that this system is inversely stable and for every $\delta > 0$ let $f : I \to I$ be a differentiable function for which $|f(x)| \leq \delta$ ($|f(x) - x| \leq \delta$) for $x \in I$ and there exists an x_0 such that $|f'(x_0)| > 1$ ($|f'(x_0) - 1| > 1$). For the function $H(t,x) = f(x)$ we have a contradiction (see the proof of Theorem 4.4 in [1]). □

8.5 Inverse b-Stability

The system (8.1) and (8.3) is said to be *inversely b-stable* if for every function $H : \mathbb{R} \times I \to I$ the boundedness of the function $|F(t,x) - H(t,x)|$ for some solution F of the system (8.1) and (8.3) implies the boundedness of the function (8.9).

Here the situation is the same as in Theorem 8.3.

8.6 Absolute Stability

The system (8.1) and (8.3) is stable and inversely stable (i.e. *absolutely stable*) only for $0 \neq a \neq 1$.

Acknowledgements The author would like to thank the referee for a number of valuable remarks and suggestions.

References

1. Moszner, Z., Przebieracz, B.: Is the dynamical system stable? Aequationes Math. **89**, 279–296 (2015)
2. Przebieracz, B.: On the stability of the translation equation and dynamical systems. Nonlinear Anal. **75**, 1980–1988 (2012)

Chapter 9
Homomorphisms from Functional Equations in Probability

Adam J. Ostaszewski

"Niedaleko jabłko spada od jabłoni"
"The apple never falls far from the tree"

Abstract We showcase the significance to probability theory of homomorphisms and their simplifying rôle by reference to the Goldie functional equation (*GFE*), an equation at the heart of regular variation theory (RV) encoding asymptotic flows, but with an apparent lack of symmetry. Like the Gołąb–Schinzel equation (*GS*), of which it is a disguised equivalent, it and its Pexiderized form can be transmuted into homomorphy under a 'generalized circle product' due to Popa, conformally with the *Pompeiu equation*. This not only forges a specific direct connection to Beurling's Tauberian Theorem, but also generally both helps simplify classical RV-analysis, lending it a flow-type intuition as a guide, and elevates it to unfamiliar contexts. This is illustrated by a new approach to the one-dimensional random walks with stable laws.

We review some new literature, offer some new insights and, in Sections 9.4 and 9.5, some new contributions; possible generalizations are indicated in Section 9.6.

Keywords Random walks • Stable laws • Goldie equation • Gołąb–Schinzel equation • Regular variation • Circle groups • Hypergroups

Mathematics Subject Classification (2010) Primary 26A03; Secondary 26E07

A.J. Ostaszewski (✉)
Department of Mathematics, London School of Economics, London WC2A 2AE, UK
e-mail: a.j.ostaszewski@lse.ac.uk

J. Brzdęk et al. (eds.), *Developments in Functional Equations and Related Topics*, Springer Optimization and Its Applications 124,
DOI 10.1007/978-3-319-61732-9_9

9.1 Introduction

The leitmotif of this article is the critical rôle of homomorphisms underlying several of the functional equations arising in probability theory. When homomorphy is patently present in a functional equation, then it surely guides the work of extending classical theorems to a wider context. As for the converse: if absent, one seeks out any latent structures capable of expressing homomorphy, and so of bearing the fruits of unity and clarity—through closeness to a paradigm, as in the introductory motto. We offer several examples, both old and new.

Generally speaking, functional equations, more properly their (continuous) solutions, play a significant rôle in the asymptotic analysis needed to elicit the characterization of various laws in probability theory (see [58] for the origins of such a programme). Below we meet familiar examples of functional equations in such situations.

The classical context of $\mathbb{R}$ generalizes naturally to the metric-group frameworks of harmonic analysis: a general locally compact group G, alternatively a linear space—indeed a Hilbert space H. A remarkable instance of generalization is to be seen in the characterization of infinitely divisible laws, which on $\mathbb{R}$ goes back to Lévy and Khintchine; here the most basic is the *Cauchy functional equation (CFE)* in the general form of a *homomorphy equation* between groups:

$$\chi(xy) = \chi(x)\chi(y)\,, \qquad (CFE)$$

its (continuous) solutions termed *characters*, and the symmetric bi-homomorphy variant:

$$\Psi(xy, z) = \Psi(x, z)\Psi(y, z) \qquad \text{with } \Psi(x, y) = \Psi(y, x)\,.$$

In the bi-additive case $\Psi : G^2 \to \mathbb{R}$, putting

$$\psi(x) := \Psi(x, x)$$

yields the important associated *quadratic form* $\psi : G \to \mathbb{R}$, which may be equivalently defined (as in [74, Section 6, (6.1)], or with more explicit details as in [50, L. 5.2.4]) by the *Apollonius* or *quadratic functional equation*:

$$\psi(xy) + \psi(xy^{-1}) = 2(\psi(x) + \psi(y))\,;$$

see also [2, Section 11.1; cf. Chapter 8, the related d'Alembert equation], [84, Chapter 13], and [86, Section 2.2], the latter in connection with the Chebyshev 'polynomial hypergroup'—for which see [22], Section 9.6 (and presently below). Their continuous solutions are critical in establishing the characterization of a Gaussian measure μ [27] either on a locally compact abelian G, or in Hilbert space H, along the following lines.

The first of the three equations above introduces duality considerations into a locally compact abelian G, employing the group $\hat{G}$ of continuous unitary characters $\chi : G \to \mathbb{T}$, with $\mathbb{T}$ the unit circle group in $\mathbb{C}$, and draws on the Pontryagin structure theorem for G. That and the third equation, with $\hat{G}$ replacing G, yields a functional characterization of a Gaussian measure μ via its Fourier transform $\hat{\mu}$: for some $g \in G$,

$$\hat{\mu}(\chi) = \chi(g)\exp(-\psi(\chi)) \qquad (\chi \in \hat{G}).$$

For details see [74, IV Theorem 6.1], or [50, Section 5.2], [85, Section 3.2]; for an example see Section 9.3.3 below. A similar result holds in Hilbert space, which is of course self-dual, so H replaces both G and $\hat{G}$ above—see [74, VI Theorem 4.9].

Noteworthy is that the last formula speaks *entirely* in the language of homomorphy.

Indeed, also the Fourier transformation taking μ to its 'characteristic function' (which uniquely determines the measure):

$$\hat{\mu}(\chi) := \int_G \chi(g)d\mu(g), \tag{$\dagger$}$$

is itself both an additive and multiplicative homomorphism (on the measures on G, which form a semigroup under convolution).

A further ubiquitous functional equation is the *Gołąb–Schinzel* equation [43], cf. [30]:

$$\eta(v + u\eta(v)) = \eta(u)\eta(v) \qquad (u, v \in \mathbb{R}), \tag{GS}$$

whose continuous solutions that are *positive on* $\mathbb{R}_+$ (briefly: *positive*) satisfy for some $\rho \geq 0$

$$\eta(t) \equiv \eta_\rho(t) := 1 + \rho t \qquad (t \in \mathbb{R}_+).$$

For a new approach to the proof see Section 9.5. We write $\eta \in GS$ to mean that η satisfies (GS). Equation (GS) is the focus for much of the text below, for good reason: indeed, for three reasons.

The classical theory of regular variation, RV for short, introduced by Karamata, studies for $f : \mathbb{R}_+ \to \mathbb{R}_+ := (0, \infty)$ the limit function

$$\kappa(t) = \kappa_f(t) := \lim_{x\to\infty} \frac{f(tx)}{f(x)} \qquad (t \in \mathbb{A}),$$

or *Karamata kernel*, with domain $\mathbb{A} \subseteq \mathbb{R}_+$; if $\mathbb{A} = \mathbb{R}_+$, f is called *regularly varying*. This is the *multiplicative* formulation, thematic here and of practical significance; for the additive variant, more convenient in theoretical considerations (for instance, in Section 9.3.1), see Section 9.7(1). The standard text for RV is [21], BGT below.

(There is also an associated notion of regularly varying measures: see [53], or [79], and Section 9.7(4) below.) In his seminal text on probability Feller laid claim to RV as an important tool: the opening second paragraph of [42, VIII.8], motivating the significance of RV to probability theory, highlights the *quantifier weakening* aspect (visited below) of being prepared to work on the premise of good limiting behaviour (as above) but initially on only a *dense* set $\mathbb{A}$ (cf. Section 9.2.1).

Above, if $\mathbb{A} = \mathbb{R}_+$ and if $\kappa \equiv 1$, then f is called *slowly varying*. In general, however, as κ satisfies the multiplicative Cauchy equation:

$$\kappa(st) = \kappa(s)\kappa(t)\,,$$

a regularly varying function that is measurable/Baire (i.e. with the Baire property) has a natural characterization as the product of a power function with a slowly varying factor.

For the purposes of extending the Wiener Tauberian theorem (Section 9.2.6 below) to encompass the Borel summability method (cf. [16, Section 1]), Beurling introduced what we now know as *Beurling slow variation*, BSV, employing functions $\varphi : \mathbb{R}_+ \to \mathbb{R}_+$ satisfying

$$\frac{\varphi(x + t\varphi(x))}{\varphi(x)} \to 1,$$

with $\varphi(x) = o(x)$ as $x \to \infty$. This includes the case, significant to the Borel and Valiron summability method, of

$$\varphi(x) := \sqrt{x}.$$

Such functions are called *self-neglecting,* $\varphi \in SN$, provided a further technical condition holds, that the convergence is locally uniform in t. Conditions implying self-neglect are studied in [16], where for $\varphi \in SN$ a more comprehensive theory of φ-regular variation is established by studying limit functions

$$g(t) := \lim_{x\to\infty} \frac{f(x + t\varphi(x))}{f(x)} \qquad (t \in \mathbb{A})\,.$$

It also emerged in [16, Section 10.3] (a matter followed through in [18]) that an even more satisfactory development may be had by going beyond BSV to obtain the even more comprehensive notion of *Beurling regular variation*, BRV, which encompasses both the Karamata theory and the related Bojanić-Karamata/de Haan theory (cf. BGT Chapter 3). BRV is built around functions φ that are *self-equivarying*, as in [71]; for these functions a limit value more general than the '1' above is permitted, so that

$$\frac{\varphi(x + t\varphi(x))}{\varphi(x)} \to \eta^{\varphi}(t) \qquad (t \in \mathbb{A})\,, \qquad (SE_{\mathbb{A}})$$

here with $\mathbb{A} = \mathbb{R}_+$ (but see Proposition 1 in Section 9.2.3), and this convergence demands a side-condition of local uniformity as in (*SN*) above (and $\varphi(x) = O(x)$). For $\eta \equiv 1$, these specialize to the *self-neglecting* functions of Beurling, as above. The key result from [71] is that the limit function $\eta^{\varphi}(t)$ satisfies (*GS*), and this is the *first reason* for interest in (*GS*) in RV.

The *second reason* is that (*GS*) may be 'converted' very simply into an expression of homomorphism and so throws much light on an alternative form of the equation occurring in RV, 'disguised from birth' in [11] (cf. BGT Lemma 3.2.1), now known as the *Goldie equation*. The latter contains a further *auxiliary function* ψ and takes the form:

$$K(x+y) - K(y) = \psi(y)K(x) \qquad (x, y \in \mathbb{R}_+). \qquad (GFE)$$

In the functional equations literature this is a special case of the *Levi-Civita* functional equation, albeit a *conditioned* one, as the quantifiers are bounded: quantifying over $\mathbb{R}_+$—cf. [84, Section 5.4]. However, tracing the direct connection of (*GFE*) to (*GS*), and so to homomorphy, brings untold benefits: see Section 9.2.4, as already mentioned.

The 'algebraicization' needed to release these benefits originates with a largely forgotten contribution, due first to Popa [76] and later Javor [57], based on the binary operation, generated from an arbitrary $\eta : \mathbb{R} \to \mathbb{R}$:

$$u \circ_\eta v := u + v\eta(u),$$

for which see Section 9.2.3 below (cf. [28]). This may be traced back to the 'circle product' of ring theory:

$$x \circ y := x + y + xy;$$

indeed, $\circ_\eta$ reduces to just that for

$$\eta(x) = 1 + x.$$

(For historical background see [72, Section 2.1].) This binary operation re-expresses (*GS*) as homomorphy:

$$\eta(u \circ_\eta v) = \eta(u)\eta(v),$$

where the right-hand side may be interpreted in various group structures (e.g. the *Pompeiu equation* of [84, Example 3.24], where the original circle product $\circ$ appears on both sides).

The *third reason* can now be declared as the benefit of homomorphy: homomorphism into $(\mathbb{R}_+, \times)$ lessens the burdens of proof in the Beurling theory of regular variation: the algebra becomes virtually identical to that of the $\mathbb{R}_+$ classical theory, leaving only the analysis of local uniformity to be undertaken (cf. Theorem 6 below).

We therefore advocate a more systematic use of the tool of homomorphy, as a unifier and clarifier.

The bulk of the material below falls naturally into two parts: first Sections 9.2 and 9.3, and then Sections 9.4 and 9.5, as follows.

In the first part, Section 9.2 discusses (*GFE*), indicating its relation to (*GS*), and considers the Popa operation $\circ_\eta$. We then describe the connection with the *Beurling Tauberian Theorem*, a proper extension of the celebrated *Wiener Tauberian Theorem*. In passing, we indicate briefly how to solve (*GFE*) using integration, which also permits a side glance at the rôle of flows—a natural consequence of the presence of a *group action*. In Section 9.3 we pass beyond Karamata kernels to the *Beurling kernels* of BRV, and as an application sketch how (*GFE*) helps to deduce very directly the form of *stable laws* associated with one-dimensional *random walks* (i.e. walks on the additive group $\mathbb{R}$—see [9] for an a very informative survey of the theory and application of random walks). The starting point is their *characteristic functional equation* (*ChFE*), which is briefly deduced ab initio and then reduced after some work to (*GFE*)—see Section 9.3.3 below. We also indicate further literature.

The second part, comprising Sections 9.4 and 9.5, contains new contributions as supporting material: a new theorem about (*ChFE*) and novel approaches to solutions of (*GS*) that are positive (on $\mathbb{R}_+$). The latter functions play a significant rôle in RV, so direct proofs are of interest.

We complete the circle of ideas in Section 9.6, ending as we began: with the theme of homomorphy—noting how the characteristic functions of random walks on some other groups give rise to an *integrated functional equation* (IFE)—for background here see [78], inspired by the work of Choquet and Deny [32]. However, the more natural setting for these is that of a *hypergroup* structure (sketchily reproduced below) with binary operation $\star$ and involution, within which these particular IFEs again reduce to a homomorphy:

$$K(x \star y) = K(x)K(y)\,. \tag{$\star$}$$

In brief, cf. [22], or [86, 87]: a *hypergroup* has as underlying domain a topological space X (possibly a topological group). The topology may be discrete. Upon this space is imposed (axiomatically) both a measure-theoretic and a group-like structure: first, the points x of X are identified with probability measures δ_x degenerate at the points of X; then a binary operation $\star$ is introduced on these (later extended to a wider domain of measures), and is interpreted much as convolution, so as to yield a probability measure with compact support (continuously mapped to the hyperspace $\mathscr{K}(X)$ of (nonempty) compact subsets of X, the latter equipped with the topology inherited from the *Vietoris* topology [41, 2.7.20] on the (nonempty) closed subsets, known also as the *Michael* topology, in view of the contribution [67]); and lastly, an involution operation is provided on the point-masses.

This allows a very broad algebraicization of random 'dynamics', generated by $\star$, within which measures describe the location of 'random points' of X. Sometimes the hypergroup is not much more than a group, as when

$$\delta_x \star \delta_y := \delta_{xy}.$$

But often the introduction of $\star$ calls for some quite intriguing ingenuity—as the two examples of Section 9.6 show.

We close in Section 9.7 with complements, including in Section 9.7(4) indications of some generalizations.

9.2 From Beurling via Goldie to Gołąb–Schinzel

We begin with a discussion of Equation (*GFE*) introduced in Section 9.1.

9.2.1 The Goldie Equation

In RV Equation (*GFE*) emerges from asymptotic analysis (see Section 9.3.1) and is initially valid on a *subset* of $\mathbb{R}$ (as the domain of convergence of a limit operation), so it is natural to formalize this phenomenon by *weakening the quantifiers*, as indicated in Section 9.1, allowing the free variables to range over a set $\mathbb{A}$ smaller than $\mathbb{R}$, which typically will be a subgroup that is dense. (There is an implicit appeal to Kronecker's density theorem here and the presence of two incommensurable elements in $\mathbb{A}$.) The functional equation in the result below, denoted by $(G_{\mathbb{A}})$, is thus a second form of the Goldie functional equation. As we see in Theorem 1 below, the two coincide in the principal case of interest—compare the insightful Footnote 3 of [26]. The notation H_γ below (originating in [26]) is from BGT Sections 3.1.7 and 3.2.1, implying

$$H_0(t) \equiv t.$$

Equation $(G_{\mathbb{A}})$ below when $\mathbb{A} = \mathbb{R}$ is a special case of a generalized Pexider equation studied by Aczél [1]. In Theorem 1 (*CEE*) is the *Cauchy exponential equation*. Versions of the specific result here, taken from [17, Theorem 1] (where the proof—based on the *Cauchy nucleus* of K [63, Section 18.5]—may be consulted), also appear elsewhere in the literature.

Theorem 1 ([26, (2.2)], BGT Lemma 3.2.1; cf. [3], [84, Proposition 5.8]) *For* ψ *with* $\psi(0) = 1$, *if* $K \not\equiv 0$ *satisfies*

$$K(u + v) = \psi(v)K(u) + K(v) \qquad (u, v \in \mathbb{A}), \qquad (G_{\mathbb{A}})$$

with $\mathbb{A}$ *a dense subgroup, then:*

(i) *the following is an additive subgroup on which* K *is additive:*

$$\mathbb{A}_\psi := \{u \in \mathbb{A} : \ \psi(u) = 1\};$$

(ii) *if* $\mathbb{A}_\psi \neq \mathbb{A}$ *and* $K \not\equiv 0$, *there is a constant* $\kappa \neq 0$ *with*

$$K(t) \equiv \kappa(\psi(t) - 1) \qquad (t \in \mathbb{A}), \qquad (*)$$

and ψ satisfies

$$\psi(u+v) = \psi(v)\psi(u) \qquad (u, v \in \mathbb{A}). \qquad (CEE)$$

(iii) *So for $\mathbb{A} = \mathbb{R}$ and ψ locally bounded at* 0 *with $\psi \neq 1$ except at* 0 :

$$\psi(x) \equiv e^{-\gamma x},$$

for some constant $\gamma \neq 0$, and so $K(t) \equiv cH_\gamma(t)$ for some constant c, where

$$H_\gamma(t) := (1 - e^{-\gamma t})/\gamma .$$

For the needs of Section 9.5 below, we note briefly that the proof rests on symmetry in the equation:

$$\begin{aligned} \psi(v)K(u) + K(v) &= K(u+v) = K(v+u) \\ &= \psi(u)K(v) + K(u). \end{aligned}$$

So, for u, v not in $\{x : \psi(x) = 1\}$, an additive subgroup,

$$K(u)[\psi(v) - 1] = K(v)[\psi(u) - 1],$$

$$\frac{K(u)}{\psi(u) - 1} = \frac{K(v)}{\psi(v) - 1} = \text{const.} = \kappa,$$

as in BGT Lemma 3.2.1. If $K(\cdot)$ is to satisfy (*GFE*), $\psi(\cdot)$ needs to satisfy (*CEE*).

9.2.2 The Disguised GS

By Theorem 1, assuming its local boundedness, the *auxiliary* function ψ of (*GFE*) is exponential; with this in mind, we can trace the connection to (*GS*) as follows.

Recall from Section 9.1 that a function is *positive* if it takes positive values on $\mathbb{R}_+$.

Recall also that the positive (and likewise, ultimately, the continuous) solutions of (*GS*) take the form

$$\eta \equiv \eta_\rho(x) := 1 + \rho x,$$

with $\rho > 0$, for $x > \rho^* := -\rho^{-1}$—see Section 9.5. Writing (*GS*) in the form

$$\eta(a + \eta(a)b) = \eta(a)\eta(b),$$

put

$$A := \eta(a) > 0, \quad B := \eta(b) > 0,$$

and take $f := \eta_\rho^{-1}$ (which exists to the right of ρ^*); then $a = f(A), b = f(B)$. Applying f to (GS) yields

$$a + Ab = f(AB) : \qquad f(A) + Af(B) = f(AB)\,.$$

Apply the logarithmic transformation: $u = \log A$, $v = \log B$, set $K(x) := f(e^x)$; then

$$f(e^u) + e^u f(e^v) = f(e^{u+v}) : \qquad K(u) + e^u K(v) = K(u+v)\,.$$

The reverse direction can be effected for non-trivial (i.e. invertible) solutions K of this last equation – see [20, §7].

9.2.3 *Popa (Circle) Operation: Basics*

The operation

$$x \circ_\eta y := x + y\eta(x)\,,$$

with $\eta : \mathbb{R} \to \mathbb{R}$ arbitrary, was introduced in 1965 for the study of Equation (GS) by Popa [76], and later Javor [57] (in the broader context of $\eta : \mathbb{E} \to \mathbb{F}$, with $\mathbb{E}$ a vector space over a commutative field $\mathbb{F}$), who observed that this equation is *equivalent to the operation* $\circ_\eta$ *being associative* on $\mathbb{R}$, and that then $\circ_\eta$ confers a group structure on $\mathbb{G}_\eta := \{g \in \mathbb{R} : \eta(g) \neq 0\}$—see [76, Proposition 2], [57, Lemma 1.2]. We term this a *Popa circle group*, or *Popa group* for short, as the case

$$\eta_1(x) = 1 + x$$

(i.e. for $\rho = 1$ above, so a translation) yields precisely the circle group of the ring $\mathbb{R}$, as noted in Section 9.1.

The operation $\circ_\eta$ turns η into a homomorphism from

$$\mathbb{G}_\eta^+ = \{g \in \mathbb{G}_\eta : \eta(g) > 0\}$$

to $(\mathbb{R}_+, \times)$. For $\eta = \eta^\varphi$, arising from $\varphi \in SE$ as in ($SE_{\mathbb{A}}$) with natural domain $\mathbb{A} = \mathbb{R}_+$, one may in fact extend the definition of η^φ from $\mathbb{R}_+$ to $\mathbb{G}_\eta^+$ preserving homomorphy, as we see presently (Proposition 1). Below, when

$$\eta(t) = 1 + \rho t,$$

we use the variants $(\mathbb{G}_\eta, \circ_\eta)$ and $(\mathbb{G}_\rho, \circ_\rho)$ interchangeably and call $\rho^* := -\rho^{-1}$ the *Popa centre* of $\mathbb{G}_\rho$. Other notation associated with $\mathbb{G}_\eta$ includes 1_η for the neutral element, and t_η^{-1} for the inverse of t, and obvious variants of these.

Proposition 1 (Non-zero Uniform Involutive Extension, [18, L.1]) *For $\varphi \in SE$, $\circ = \circ_\rho$ with $\rho = \rho_\varphi > 0$, put*

$$\eta^\varphi(t_\circ^{-1}) = \eta^\varphi(-t/\eta^\varphi(t)) := 1/\eta^\varphi(t) \qquad (t > 0)\,;$$

then $(SE_{\mathbb{A}})$ holds for $\mathbb{A} = \mathbb{G}_+^\rho = (\rho^*, \infty)$. Moreover, this is a maximal non-vanishing extension: for each $s < \rho^*$, assuming $\varphi(x + s\varphi(x)) > 0$ is defined for all large x,

$$\lim_{x\to\infty} \eta_x^\varphi(s) = \lim_{x\to\infty} \varphi(x + s\varphi(x))/\varphi(x) = 0 = \eta(\rho^*)\,.$$

Here we see the critical rôle of the Popa origin $\rho^* = -\rho^{-1}$: the domain of the limit operation

$$\lim_{x\to\infty} \eta_x^\varphi(s),$$

used to extend η^φ, is $\mathbb{G}_+^\rho$. So the argument s here has to take values to the right of the Popa origin. As $\rho \to 0+$ the Popa centre recedes to $-\infty$ and this extension falls into line with the natural extension to $\mathbb{R}_-$ (taken for granted) in the Karamata theory: see BGT (2.11.2).

With this much isomorphy in place (in fact conjugacy with $\mathbb{R}$), it is natural to seek further group structures in order to allow (*GFE*), as a statement about K, to assert homomorphism between Popa groups:

$$K(x \circ_\eta y) = K(y) \circ_\sigma K(x) \text{ for some } \sigma \in GS\,, \qquad (GBE)$$

with the side-condition

$$\sigma(K(y)) \equiv \psi(y).$$

We term the above the *Goldie–Beurling equation* (*GBE*), acknowledging the Beurling connection via η; it is a natural extension of the *Pompeiu equation* to which it reduces when $\eta \equiv \sigma \equiv \eta_1$ [84, Example 3.24], and so links with results not only of Aczél, but also of Chudziak [33–35], and Jabłońska [55], concerned with the equation

$$f(x \circ_g y) = f(x) \circ f(y) \qquad (ChE)$$

with $f : \mathbb{R} \to (S, \circ)$ for $(S, \circ)$ some group or semigroup, and $g : \mathbb{R} \to \mathbb{R}$ continuous, or locally bounded above.

Javor's observation regarding associativity has interesting corollaries. (Recall that *positive* means positive on $\mathbb{R}_+$.)

Lemma$_{\mathbf{com}}$ ([72]) *If (GBE) holds for some injective K, σ with $\circ_\sigma$ commutative, and $\eta : \mathbb{R}_+ \to \mathbb{R}$, then*

$$\eta(u) \equiv 1 + \rho u,$$

for some constant ρ.

Proof Here

$$K(u + v\eta(u)) = K(u) \circ_\sigma K(v) = K(v) \circ_\sigma K(u) = K(v + u\eta(v)),$$

as $\circ_\sigma$ is commutative. By injectivity, for all $u, v \geq 0$,

$$u + v\eta(u) = v + u\eta(v): \quad u(1 - \eta(v)) = v(1 - \eta(u)),$$

so, as in Theorem 1,

$$(\eta(u) - 1)/u \equiv \rho = \text{const.},$$

for $u > 0$; taking $v = 1$ above,

$$\eta(u) \equiv 1 + \rho u,$$

for all $u \geq 0$. □

Lemma$_{\mathbf{assoc}}$ **([72])** *If* (*GBE*) *holds for some injective* K, σ *with* $\circ_\sigma$ *associative, and a positive continuous* $\eta : \mathbb{R} \to \mathbb{R}$, *then*

$$\eta(u) = 1 + \rho u \quad (u \geq 0),$$

for some constant ρ.

Proof This follows, e.g., from Javor's observation above connecting associativity with (*GS*) [57, p. 235].

9.2.4 Creating Homomorphisms

In this section we demonstrate how to convert two functional equations into expressions of homomorphy. The immediate use this serves is to enable the solutions to be 'read back' from those of the Cauchy functional equation (*CFE*), as in Theorem 5 below. This process is captured in the following routine result concerning (*GBE*). For

$$\circ_\eta = \circ_0 \quad \text{and} \quad \circ_\sigma = \circ_\infty,$$

the equation reduces to the exponential format of (*CFE*) ([63, Section 13.1]; cf. [54]). The critical case for Beurling regular variation is for $\rho \in (0, \infty)$, with positive continuous solutions described in the table below; the four corner formulas correspond to classical variants of (*CFE*). The proof, which we omit, proceeds by a straightforward reduction to a classical variant of (*CFE*) by an appropriate shift and rescaling.

Proposition 2 ([72, Proposition A]; cf. [33]) *For*

$$\circ_\eta = \circ_r, \quad \circ_\sigma = \circ_s,$$

and K Baire/measurable satisfying (*GBE*), *there is* $\gamma \in \mathbb{R}$ *so that* $K(t)$ *is given by:*

Popa parameter	$s = 0$	$s \in (0, \infty)$	$s = \infty$
$r = 0$	γt	$(e^{\gamma t} - 1)/s$	$e^{\gamma t}$
$r \in (0, \infty)$	$\gamma \log(1 + rt)$	$[(1 + rt)^\gamma - 1]/s$	$(1 + rt)^\gamma$
$r = \infty$	$\gamma \log t$	$(t^\gamma - 1)/s$	t^γ

Below and elsewhere a function K is *non-trivial* if $K \not\equiv 0$ and $K \not\equiv 1$.

Theorem 2 (Conversion to Homomorphy, [72, Theorem 1]) *For* $\eta \in GS$ *in the setting above,* (*GBE*) *holds for positive* ψ *in the side-condition and a non-trivial K iff*

(i) *K is injective;*
(ii) $\sigma =: \psi K^{-1} \in GS$, *equivalently, either* $\psi \equiv 1$, *or, for some* $s > 0$,

$$K(u) \equiv (\psi(u) - 1)/s \text{ and } \psi(0) = 1, \text{ so } K(0) = 0;$$

(iii)

$$K(x \circ_\eta y) = K(x) \circ_\sigma K(y)\,. \qquad \text{(Hom-1)}$$

Then
(iv) *for some constants* c, γ,

$$K(x) \equiv c \cdot [(1 + \rho x)^\gamma - 1]/\rho\gamma, \text{ or } \qquad K(t) \equiv \gamma \log(1 + \rho t)$$

$$(\rho = \rho_\eta > 0),$$

$$\text{or} \qquad K(x) \equiv c \cdot (e^{\gamma x} - 1)/\gamma \qquad (\rho_\eta = 0).$$

A related functional equation replaces one instance of K on the right of (*GBE*) by a further unknown function κ multiplying ψ, yielding a 'Pexiderized' generalization[1]

$$K(x + y\eta(x)) - K(y) = \psi(y)\kappa(x) \qquad (x, y \in \mathbb{R})\,, \qquad (GBE\text{-}P)$$

[1] Acknowledging the connection, the qualifier *P* in (*GBE-P*) is for 'Pexiderized' Goldie–Beurling equation—referring to Pexider's equation: $f(xy) = g(x) + h(y)$ and its generalizations—cf. [29, 30], and the recent [54].

considered also in [36]. Passage to this more general form enables the inclusion of (*GS*) as the case

$$K \equiv \psi \equiv \eta$$

with $\kappa \equiv \eta - 1$.

To apply the earlier argument here, an extension of the Popa binary operation suggests itself; put

$$u \circ v = u \circ_{\alpha\beta} v := \alpha(u) + v\beta(u),$$

with α, β continuous and α invertible; this seems reminiscent of [3].

Proposition 3 ([72, Proposition A]) *The operation* $\circ$ *is a group operation on* $\mathbb{A} \subseteq \mathbb{R}$ *with* $0 \in \mathbb{A}$ *iff* $\mathbb{A}$ *is closed under* $\circ$ *and for some constants* b, c *with* $bc = 0$

$$\alpha(x) \equiv x + b \text{ and } \beta(x) \equiv 1 + cx.$$

That is:

$$\alpha(x) \equiv x \text{ and } \beta(x) \equiv 1 + cx, \text{ OR } \alpha(x) \equiv x + b \text{ and } \beta(x) \equiv 1.$$

So this is either a Popa group with

$$x \circ y = x \circ_c y := x + y(1 + cx),$$

or the b-shifted additive reals with the operation

$$x +_b y := x + y + b.$$

Remark For the b-shifted additive reals, the neutral element is $e := -b$ and $x^{-1} = -x - 2b$.

Applying Proposition 3, we deduce the circumstances when (*GBE-P*) may be transformed to a homomorphism. Here we see that

$$K(x) \equiv (\psi(y) - 1)/s$$

only in the cases (i) and (iii), but not in (ii)—compare Theorem 2. Note that in all cases κ is a homomorphism between Popa groups.

Theorem 2′ (Conversion to Homomorphy, [72, Theorem 1′]) *If* (*GBE-P*) *is solved by* K *for* ψ *positive,* κ *positive and invertible,* $\eta(x) \equiv 1 + \rho x$ *(with* $\rho \geq 0$*), then in the equation below* $\circ$ *is a group operation and* K^{-1} *is a homomorphism under* $\circ$*:*

$$K^{-1}(u \circ_\sigma v) = K^{-1}(u) \circ K^{-1}(v) \qquad (u, v \in \mathbb{R}), \tag{Hom-2}$$

iff $\sigma := \psi K^{-1} \in GS$ *and one of the following three conditions holds:*

(i) $\rho = 0$, $\circ = \circ_0$ *and* $\circ_\sigma = \circ_s$ *for some* $s > 0$*; then, for some* $\gamma \in \mathbb{R}$,

$$K(t) \equiv \kappa(t) \equiv (e^{\gamma t} - 1)/s, \qquad \psi(t) \equiv e^{\gamma t};$$

(ii) $\rho = 0$, $\circ_\sigma = \circ_0$ *and* $\circ = +_b$ *for some* $b \in \mathbb{R}$*; then*

$$K(t) \equiv \kappa(t + b) = \kappa(t) + \kappa(b), \qquad \psi(t) \equiv 1 \qquad (t \in \mathbb{R}),$$

and $\kappa : \mathbb{G}_0 \to \mathbb{G}_0$ *is linear;*

(iii) $\rho > 0$, $\circ = \circ_\rho$ *and* $\circ_\sigma = \circ_s$ *for some* $s \geq 0$*; then, for some* $\gamma \in \mathbb{R}$,

$$K(t) \equiv \kappa(t) \equiv [(1 + \rho t)^\gamma - 1]/s, \quad (s > 0), \quad \text{or} \quad \gamma \log(1 + rt) \quad (s = 0),$$
$$\psi(t) \equiv (1 + \rho t)^\gamma \quad (s > 0), \quad \text{or} \quad \psi(t) \equiv 1 \quad (s = 0).$$

This recovers results in [33].

9.2.5 Beck Sequences, Integration, and Flows

Assuming continuity, we show in this section how to use integration to find the non-trivial solutions of the following variant of (*GFE*):

$$K(x + y\eta(x)) - K(y) = \psi(y)K(x).$$

A key tool here, and also in later sections, is an appropriate partitioning of any interval (range of integration); for this we refer to what we term the *Beck* φ*-sequence* $t_m = t_m(u)$, defined recursively for $u > 0$ and φ a solution of (*GS*) by

$$t_{m+1} = t_m \circ_\varphi u = t_m + u\varphi(t_m) \text{ with } t_0 = 0.$$

Albeit present in [43], the systematic use of such iterations seems to stem from Beck's oeuvre on continuous flows in the plane—[5, L. 1.6.4]. The Popa notation inserted above clarifies that this is the sequence of *Popa powers* of u under $\circ_\varphi$ and so may also be written u_φ^m. So, from the group perspective, this is the natural *discretization* with 'mesh' size u for the purposes of integration. As φ is a homomorphism,

$$\varphi(u_\varphi^{m+1}) = \varphi(u)\varphi(u_\varphi^m) = \varphi(u)^{m+1}\varphi(0).$$

So ([17, Theorem 5], or Theorem 8 below) the sequence t_m is divergent, since either $\varphi(u) = 1$ and $t_m = mu$ (directly, from the inductive definition), or else $\varphi(u) \neq 1$

and

$$t_m = u\frac{\varphi(u)^m - 1}{\varphi(u) - 1} = (\varphi(u)^m - 1) \bigg/ \frac{\varphi(u) - 1}{u} \tag{**}$$

—see, e.g., by Ostaszewski [71, L. 4] (cf. a lemma of Bloom: BGT Lemma 2.11.2). In either case, for $u, t > 0$ there exists a unique integer $m = m_t(u)$, the *jump index of t*, satisfying

$$t_m \le t < t_{m+1}\,.$$

Application: Solutions by Integration To solve the equation above for K, ψ (and η) continuous, note that if K is non-trivial with $K(0) = 0$, then for all small enough $u > 0$ we have $K(u)$ non-zero; otherwise the $\circ_\eta$-subgroup[2]

$$\{u : K(u) = 0\}$$

accumulates at the origin, and so is dense in $\mathbb{R}_+$ (forcing K into triviality). Now proceed as follows. Fix $x_0, x_1 > 0$, and denote the corresponding jump indices $i_0 = i_0(u)$ and $i_1 = i_1(u)$: so for $j \in \{0, 1\}$

$$t_{i_j} \le x_j < t_{i_j+1}\,.$$

Now, for the Beck η-sequence $t_m = u_\eta^m$,

$$K(t_{m+1}) - K(t_m) = K(u)\psi(t_m)\,.$$

Summing, and setting

$$h(t) := \psi(t)/\eta(t) \ge 0 \qquad (t \in \mathbb{R}_+)$$

(valid as η is positive),

$$K(t_m) = K(t_m) - K(t_0) = K(u)\sum_{n=0}^{m-1}\psi(t_n) = \frac{K(u)}{u}\sum_{n=0}^{m-1}u\eta(t_n)h(t_n)\,,$$

since $t_0 = 0$.

As above, $K(u) > 0$ for small enough $u > 0$, so we may write with the obvious notation

---

2

$$0 = K(0) = K(1_\eta) = K(u \circ_\eta u_\eta^{-1}) = \psi(u_\eta^{-1})K(u) + K(u_\eta^{-1}) = K(u_\eta^{-1}).$$

$$\frac{K(t_{i_1})}{K(t_{i_0})} = \frac{K(u)\sum_{n=0}^{i_1-1}\psi(t_n)}{K(u)\sum_{n=0}^{i_0-1}\psi(t_n)} = \frac{\sum_{n=0}^{i_1-1}u\eta(t_n)h(t_n)}{\sum_{n=0}^{i_0-1}u\eta(t_n)h(t_n)}$$
$$= \frac{\sum_{n=0}^{i_1-1}(t_{n+1}-t_n)h(t_n)}{\sum_{n=0}^{i_0-1}(t_{n+1}-t_n)h(t_n)} \to \frac{\int_0^{x_1}h(t)dt}{\int_0^{x_0}h(t)dt} = \frac{H(x_1)}{H(x_0)}\,.$$

Here passage to the limit in the rightmost terms is as $u \downarrow 0$. Above we assume without loss of generality that $H(x_0) > 0$. (Otherwise $\psi \equiv 0$ on $[0,\infty)$, implying that K is constant and yielding the trivial case $K \equiv 0$.) Passing to the limit as $u \downarrow 0$ in the leftmost term above, by continuity of K, as $t_{i_j} \to x_j$

$$K(x_1)/K(x_0) = H(x_1)/H(x_0)\,.$$

Put

$$c := K(x_0)/H(x_0)\,;$$

then, with x for x_1,

$$K(x) = cH(x) := c\int_0^x h(t)\,\mathrm{d}t\,,$$

valid for $x \geq 0$, as $K(0) = 0$.

Remark When

$$\eta(t) \equiv 1\,, \qquad \psi(t) \equiv e^{\gamma t}\,, \qquad h(t) \equiv e^{\gamma t}\,,$$

the analysis above lends new clarification, via the language of homomorphisms, to the 'classical relation' in RV that

$$K = c(\psi - 1)\,,$$

connecting K and the auxiliary function ψ, as in Theorem 1.

Flows ('Translation Equation') Subject to $K(0) = 0$, assuming positivity of K (i.e. to the right of 0), and continuity and positivity of ψ, we have just seen that the solution K satisfies, for some $c \geq 0$,

$$K(x) = c \cdot \tau_f(x)\,,$$

for

$$\tau_f(x) := \int_0^x \mathrm{d}u/f(u), \text{ with } f := \eta/\psi\,.$$

Inspired by Beck [5, 5.25], we may interpret τ_f as the *occupation time measure* (of $[0,x]$) of the continuous f-flow: $dx/dt = f(x)$, where f as above measures the relative velocity of η and ψ. Furthermore, interpreting $\circ_\eta$ as a *flow* or *group action* (yielding the *translation equation*, cf. [69], [77]) it emerges surprisingly that the underlying homomorphy is now expressed not by K, but by the *relative flow-velocity* f: under mild regularity assumptions, if K solves (*GBE-P*), then f satisfies

$$f(x \circ_\eta y) = f(x)f(y) \qquad (x, y \in \mathbb{R}_+)\,.$$

There is a converse for $\psi := \eta/f$: see [17, 72].

9.2.6 Beurling's Tauberian Theorem

For $\varphi : \mathbb{R} \to \mathbb{R}_+$ introduce the following 'Beurling convolution':

$$\begin{aligned} F *_\varphi H(x) &:= \int F\Big(\frac{x-u}{\varphi(x)}\Big)H(u)\frac{\mathrm{d}u}{\varphi(x)} \\ &= \int F(-t)H(x + t\varphi(x))\,\mathrm{d}t\,, \end{aligned}$$

reducing for $\varphi \equiv 1$ to the classical counterpart

$$F * H(x) = \int F(x-t)H(t)\,\mathrm{d}t\,.$$

See [18] for background. Substitution of $t = (u-x)/\varphi(x)$ yields

$$u = u_x(t) := x + t\varphi(x),$$

so that $t \mapsto u_x(t)$ is a 'speeded-up' version of the x-shift $t \mapsto x+t$. This includes for

$$H = (1/a)\mathbf{1}_{[0,a]}$$

and

$$G(x) := \sum_{n<x} g_n$$

the *moving average* 'speeded up' by φ, introducing alternative summability methods:

$$MA_a^\varphi(x) = G *_\varphi H(x) = \frac{1}{a}\int_x^{x+a\varphi(x)} G(u)du = \frac{1}{a}\sum_x^{x+a\varphi(x)} g_n\,.$$

Theorem BT (Beurling's Tauberian Theorem) *For $K \in L_1(\mathbb{R})$ with $\hat{K}$ non-zero on $\mathbb{R}$, and φ Beurling slowly varying, i.e. with*

$$\varphi(x + t\varphi(x))/\varphi(x) \to 1, \qquad (x \to \infty) \qquad (t \geq 0): \qquad (BSV)$$

if H is bounded, and the following holds for some $c \in \mathbb{R}$

$$K *_{\varphi} H(x) \to c \int K(y)\,\mathrm{d}y\,, \qquad (K *_{\varphi} H)$$

then for all $F \in L_1(\mathbb{R})$

$$F *_{\varphi} H(x) \to c \int F(y)\,\mathrm{d}y \qquad (x \to \infty)\,.$$

As a sample, we note that the Popa algebraicization enables the following generalization:

Theorem 3 (Extension to Beurling's Tauberian Theorem, [18, Theorem 2]) *Suppose that:*

(i) $\varphi \in SE$*, i.e. locally uniformly in* t

$$\varphi(x + t\varphi(x))/\varphi(x) \to \eta(t) \in GS \qquad (x \to \infty) \qquad (t \geq 0)\,,$$

(ii) $K \in L_1(\mathbb{R})$ *with* $\hat{K}$ *non-zero on* $\mathbb{R}$,
(iii) H *is bounded, and*
(iv) $(K *_{\varphi} H)$ *holds*
then for all $G \in L_1(\mathbb{R})$

$$G *_{\varphi} H(x) \to c \int G(y)\mathrm{d}y \qquad (x \to \infty)\,.$$

9.3 Beurling Kernels

We begin by describing the context in which Beurling kernels arise.

9.3.1 Asymptotics

We refer below again to the self-equivarying functions defined by (*SE*) of Section 9.1. We adopt the additive formulation here. At its simplest, a functional equation such as (*GFE*) arises when taking limits

$$K_F(t) := \lim_{x\to\infty}[F(x + t\varphi(x)) - F(x)] = \text{briefly, } \lim \Delta_t^{\varphi} F(x)\,, \qquad (BK)$$

for $\varphi \in SE$; then, with η the associated limit as in (*SE*) above, for s, t ranging over the set $\mathbb{A}$ on which the limit function K_F, the *Beurling kernel* of F, exists as a locally uniform limit:

$$K_F(s + t) = K_F(s/\eta(t)) + K_F(t) : \qquad K_F(t + s\eta(t)) = K_F(s) + K_F(t)\,.$$

So with $\circ_\eta$ in mind, both $\mathbb{A}$ and $K_F(\mathbb{A})$ carry group structures under which K_F is a homomorphism. Thus, even in the classical context, (*GS*) plays a significant rôle albeit disguised and previously unnoticed, despite its finger-print: namely, the terms $+1$ or -1, appearing in the formulas for K_F (as in Theorem 2).

The more general functional equation, arising in Beurling RV, is the *generalized Goldie–Beurling equation* on $\mathbb{R}_+ := [0, \infty)$, noted in Section 9.2.3:

$$K(x + y\eta(x)) - K(y) = \psi(y)K(x) \qquad (x, y \in \mathbb{R}_+) \qquad (GBE_\psi)$$

(in the two unknowns K and ψ), where $\eta(x) = \eta_\rho(x)$ for some $\rho \in \mathbb{R}_+$. This arises quite similarly to (*BK*) in the context

$$K(t) = \lim \Delta_t^\varphi F(x) / \Phi(x) \text{ with } \psi(t) := \lim \Phi(x + t\varphi(x))/\Phi(x)\,,$$

assuming these limits exist.

The classical Karamata case is $\rho = 0$ with $\mathbb{A} = \mathbb{R}$, and the general Beurling case $\rho > 0$ with $\mathbb{A} = \mathbb{G}_\rho^+$ (in which case $\Phi(x)$ is *Beurling φ-regularly varying*). In the RV literature this equation appears in [11], in work inspired by Bojanić and Karamata [26], and is due principally to Goldie. In both these cases the solution K to (*GBE*) describes a function derived from the limiting behaviour of some *regularly varying* function F for a suitable auxiliary Φ.

Example ([18, Corollary 2]) For $\varphi \in SE$, if U satisfies

$$\frac{U(x + t\varphi(x)) - U(x)}{\varphi(x)} \to c_U t \text{ as } x \to \infty, \text{ for all } t \geq 0\,, \qquad (BMA_\varphi)$$

and

$$K_V(u) := \lim_{x\to\infty} \Delta_u^\varphi F(x)/\varphi(x)\,,$$

for $V(\cdot) := U(\tau_\varphi^{-1}(\cdot))$ with τ_φ as in Section 9.2.5, then for $\rho = \rho_\varphi$

$$K_V(s + t) = K_V(s)e^{\rho t} + K_V(t)\,,$$

and so with the notation H_ρ of Section 9.2.1 above, for some c,

$$K_V(s) = cH_\rho(s)\,.$$

9.3.2 Some "Advanced" Popa Theory: Quantifier Weakening

We illustrate the usefulness of the Popa group structure by surveying some further results from the recent [18]. These culminate in a theorem on quantifier weakening (Theorem 5 below) in the demanding context of local uniformity; it in turn relies on the 'subgroup property' of the domain of definition of certain limit operations (the sets $\mathbb{A}^{\varphi}$ and $\mathbb{A}_u$ below). For additional motivation see Proposition 10 in Section 9.7(5).

The definition of *SE* in Section 9.1 demands locally uniform convergence: this motivates the introduction of the following weak notion of uniformity, which is key to Theorem 4 below. Say that $f_n \to f$ *uniformly near* t if for every $\varepsilon > 0$ there is $\delta > 0$ and $m \in \mathbb{N}$ such that

$$f(t) - \varepsilon < f_n(s) < f(t) + \varepsilon \text{ for } n > m \text{ and } s \in (t - \delta, t + \delta).$$

For instance, for $\varphi \in SN$, x_n divergent, and $f(s) \equiv 1$, if $f_n(s) := \varphi(x_n + s\varphi(x_n))/\varphi(x_n)$, then '$f_n \to f$ uniformly near t for all $t > 0$.'

The notion above is easier to satisfy than Hobson's 'uniform convergence *at* t' which replaces $f(t)$ above by $f(s)$ twice, [52, p. 110]; suffice it to refer to $f_n \equiv 0$, and f with $f(0) = 0$ and $f \equiv 1$ elsewhere. (See also Klippert and Williams [62], where though Hobson's condition is satisfied at all points of a set, the choice of δ cannot itself be uniform in t.)

The above notion of uniformity may be equivalently stated in limsup language, which presently (in Proposition 6) brings to the fore the underlying *uniform upper and lower semicontinuity*. We refer to [18, Section 5] for details.

For $\varphi \in SE$ we now introduce a further binary operation, one in which a point x appears as a parameter (we think of this as a *circle operation localized* to x):

$$s \circ_{\varphi x} t := s + t\eta_x^{\varphi}(s),$$

where

$$\eta_x^{\varphi}(s) := \varphi(x + s\varphi(x))/\varphi(x).$$

This notation neatly summarizes two frequently used facts in (Karamata/Beurling) regular variation: firstly,

$$x \circ_{\varphi} (b \circ_{\varphi x} a) = y \circ_{\varphi} a, \text{ for } y := x \circ_{\varphi} b = x + b\varphi(x)$$

(so an 'absorption' property), and secondly, as $x \to \infty$, locally uniformly in s, t :

$$s \circ_{\varphi x} t \to s \circ_{\eta} t, \text{ for } \eta := \lim_x \eta_x^{\varphi} \in GS.$$

Here η satisfies (*GS*), by Ostaszewski [71], so the localized operation $\circ_{\varphi x}$ is asymptotic to a Popa operation $\circ_{\eta}$. This is used in Proposition 8.

An important rôle is played by the corresponding *localized Beck* η_x^{φ}*-sequence* (or *iteration*):

$$a_{\varphi x}^{n+1} = a_{\varphi x}^{n} \circ_{\varphi x} a, \qquad a_{\varphi x}^{1} = a. \tag{η_x^{φ}}$$

Its properties are listed below; here to avoid excessive bracketing, the usual arithmetic operations bind more strongly than Popa operations.

Proposition 4 (Arithmetic of Popa Operations, [18, Proposition 2])

(i)	$a^0_{\varphi x} = 1_{\varphi x} = 0\,, \qquad a \circ_{\varphi x} a^{-1}_{\varphi x} = 0\,,$	*for* $a^{-1}_{\varphi x} := (-a)/\eta^{\varphi}_x(a)\,;$
(ii)	$x \circ_\varphi (b \circ_{\varphi x} a) = y \circ_\varphi a\,,$	*for* $y := x \circ_\varphi b\,;$
(iii)	$x \circ_\varphi (b \circ_\eta a) = y \circ_\varphi a\eta(b)/\eta^{\varphi}_x(b)\,,$	*for* $y := x \circ_\varphi b\,;$
(iv)	$x = y \circ_\varphi b^{-1}_{\varphi x}\,,$	*for* $y := x \circ_\varphi b\,;$
(v)	$\eta^{\varphi}_x(a^m_{\varphi x}) = \prod_{k=0}^{m-1} \eta^{\varphi}_{y_k}(a)\,,$	*for* $y_k = x \circ_\varphi a^k_{\varphi x}\,.$

Definitions Recalling from Section 9.3.1 that

$$\Delta^{\varphi}_t h(x) := h(x + t\varphi(x)) - h(x),$$

and, taking limits here and below as $x \to \infty$, as before (rather than sequentially as $n \to \infty$), put for $\varphi \in SE$ and with $\rho = \rho_\varphi$ and $\rho^* = -\rho^{-1}$

$$\mathbb{A}^{\varphi} := \{t > \rho^* : \Delta^{\varphi}_t h \text{ converges to a finite limit}\}\,,$$

$$\mathbb{A}_{\mathrm{u}} := \{t > \rho^* : \Delta^{\varphi}_t h \text{ converges to a finite limit locally uniformly near } t\}\,.$$

So

$$0 \in \mathbb{A}^{\varphi},$$

but we cannot yet assume either that $\mathbb{A}^{\varphi}$ is a subgroup, or that $0 \in \mathbb{A}_{\mathrm{u}}$, a critical point in Proposition 5 below. In the Karamata case $\varphi \equiv 1$, $\mathbb{A}^{\varphi} = \mathbb{A}^1$ is indeed a subgroup (see [20, Proposition 1] and Section 9.7(5) below).

For $t \in \mathbb{A}^{\varphi}$ put

$$K(t) := \lim_{x\to\infty} \Delta^{\varphi}_t h\,. \tag{K}$$

So $K(0) = 0$.

Proposition 5 ([18, Proposition 6]) *For $\varphi \in SE$, $\mathbb{A}_{\mathrm{u}}$ is a subgroup of $\mathbb{G}^{\rho}_{+}$ for $\rho = \rho_\varphi$ iff $0 \in \mathbb{A}_u$; then $K : (\mathbb{A}_u, \circ) \to (\mathbb{R}, +)$, defined by (K) above, is a homomorphism.*

Theorem 4 ([18, Theorem 4]) *If the pointwise convergence (K) holds on a co-meagre set in $\mathbb{G}^{\rho}_{+}$ with the limit function K upper semicontinuous also on a co-meagre set, and, furthermore, the one-sided condition*

$$K(t) = \lim_{\delta\downarrow 0} \limsup_{x\to\infty} \sup\{h(x + s\varphi(x)) - h(x) : s \in [t, t+\delta)\} \tag{UNIF$^+$}$$

holds at the origin, then two-sided limsup convergence holds everywhere:

$$\mathbb{A}^{\varphi} = \mathbb{A}_{\mathrm{u}} = \mathbb{G}_{+}^{\rho}.$$

This last result is based on the following monotone convergence theorem, akin to those of Dini and of Pólya-Szegő; the proof relies on the Baire category theorem.

Proposition 6 (Uniform Upper Semicontinuity, [18, Proposition 4]) *If quasi everywhere f_n converges pointwise to f, an upper semicontinuous limit satisfying quasi everywhere in its domain the one-sided condition*

$$f(t) = \lim_{\delta \downarrow 0} \limsup_{n} \sup\{f_n(s) : s \in [t, t+\delta)\},$$

then quasi everywhere f is uniformly upper semicontinuous:

$$f(t) = \lim_{\delta \downarrow 0} \limsup_{n} \sup\{f_n(s) : s \in (t-\delta, t+\delta)\}.$$

Definitions For $\varphi \in SE$ and $\rho = \rho_{\varphi}$, put

$$H^{\dagger}(t) := \lim_{\delta \downarrow 0} \limsup_{x \to \infty} \sup\left\{h(x \circ_{\varphi} s) - h(x) : s \in [t, t+\delta)\right\} \qquad (t > \rho^*),$$

$$\mathbb{A}_{\mathrm{u}}^{\dagger} := \{t > \rho^* : H^{\dagger}(t) < \infty\}.$$

So $\mathbb{A}_{\mathrm{u}} \subseteq \mathbb{A}_{\mathrm{u}}^{\dagger}$, as $H^{\dagger}(t) = K(t)$ on $\mathbb{A}_{\mathrm{u}}$.

The following result clarifies the rôle of uniformity in classical 'Heiberg–Seneta boundedness' terms (for background see BGT (3.2.4) and [17, Section 1,2]).

Proposition 7 ([18, Proposition 9]) *For $\varphi \in SE$, the following are equivalent:*

(i) $0 \in \mathbb{A}_u$ *(i.e. $\mathbb{A}_u \neq \emptyset$ and so a subgroup);*
(ii) $\lim_{x \to \infty}[h(x + u\varphi(x)) - h(x)] = 0$ *uniformly near $u = 0$;*
(iii) $H^{\dagger}(t)$ *satisfies the two-sided Heiberg–Seneta condition:*

$$\limsup_{u \to 0} H^{\dagger}(u) \leq 0. \qquad (HS_{\pm}(H^{\dagger}))$$

Theorem 5 (Quantifier Weakening from Uniformity, [18, Theorem 6]) *If $\mathbb{A}_u$ is dense in $\mathbb{G}_{+}^{\rho}$ and $H^{\dagger}(t) = K(t)$ on $\mathbb{A}_u$—i.e. $H^{\dagger} : (\mathbb{A}_u, \circ_{\rho}) \to (\mathbb{R}, +)$ is a homomorphism, then $\mathbb{A}_u = \mathbb{G}_{+}^{\rho}$ and for some $c \in \mathbb{R}$:*

$$H^{\dagger}(t) = c \log(1 + \rho t) \quad (t > \rho^*).$$

This uses Proposition 2. Below, again working additively, we put for $\varphi \in SE$

$$H^*(t) := \limsup_{x \to \infty} h(x \circ_{\varphi} t) - h(x) \qquad (t > \rho_{\varphi}^*),$$

$$H_*(t) := \liminf_{x\to\infty} h(x \circ_\varphi t) - h(x) \qquad (t > \rho_\varphi^*).$$

Theorem 6 ([18, Theorem 10]) *In the setting of Theorem 5, for $\varphi \in SE$, if the set S on which $H^*(t)$ and $H_*(t)$ are both finite contains a half-interval $[a, \infty)$ for some $a > 0$, then there is a constant $K > 0$ such that for all large enough x and u*

$$h(u\varphi(x) + x) - h(x) \le K \log u\,.$$

The proof parallels a classical result (that of BGT Theorem 2.0.1), but with the usual powers a^n replaced by the (localized) Beck η_x^φ-iterates, as in Equation (η_x^φ) above. But there is heavy reliance on the estimation results below for $a_{\varphi x}^m$ that are uniform in m (this only needs $\eta_x^\varphi \to \eta_\rho$ pointwise):

Proposition 8 ([18, Proposition 11]) *If $\varphi \in SE$ with $\rho = \rho_\varphi > 0$, then for any $a > 1$ and $0 < \varepsilon < 1$:*

(i) *($a_{\varphi x}^m$-estimates under η_x^φ) for all large enough x,*

$$(1-\varepsilon) \le \eta_x^\varphi(a_{\varphi x}^m)^{1/m}/\eta_\rho(a) \le (1+\varepsilon) \qquad (m \in \mathbb{N})\,,$$

(ii) *($a_{\varphi x}^m$-*estimates under *η_ρ) for all large enough x,*

$$\frac{\eta_\rho(a(1-\varepsilon))^m}{1-\varepsilon} - \frac{\varepsilon}{1-\varepsilon} \le \eta_\rho(a_{\varphi x}^m) \le \frac{\eta_\rho(a(1+\varepsilon))^m}{1+\varepsilon} + \frac{\varepsilon}{1+\varepsilon} \qquad (m \in \mathbb{N})\,,$$

(iii) *$a_{\varphi x}^m \to \infty$, and*

(iv) *there are $C_\pm = C_\pm(\rho, a, \varepsilon) > 0$ such that, for all large enough x and u,*

$$a_{\varphi x}^m \le u < a_{\varphi x}^{m+1} \Longrightarrow mC_- \le \log u \le (m+1)C_+\,.$$

9.3.3 Random Walks with Stable Laws: GFE Again

A random variable X has a *stable law* if the probability law (measure) μ of the random walk $S_n := X_1 + \ldots + X_n$, in which the steps are executed on the group of additive reals $\mathbb{R}$ independently and with identical law, is again of the same *type*. The latter means that the distribution function

$$F(x) = \mathrm{Prob}^\mu[X \le x]$$

of X and that of each S_n should be equal up to a change of 'scale and location':

$$S_n \stackrel{D}{=} a_n X + b_n\,, \qquad \text{(D)}$$

for some (real) norming constants a_n, b_n with $a_n > 0$. Here $\stackrel{D}{=}$ denotes equality of distributions. Such a law may be exactly characterized by its *characteristic functional equation,* Equation (*ChFE*) below, obtained from (D) on taking its Fourier transform (using the linearity and multiplicativity features of the transform). Since the characteristic function here is

$$\varphi(t) = \mathbb{E}[\exp(\mathrm{i}tX)] = \int_{\mathbb{R}} \exp(\mathrm{i}tx)\,\mathrm{d}F(x)$$

(identifying the characters as

$$\chi_t(x) = e^{\mathrm{i}tx}$$

—cf. [84, Example 3.7]), (D) above yields

$$\varphi(t)^n = \varphi(a_n t)\exp(\mathrm{i}b_n t) \qquad (n \in \mathbb{N}). \qquad (ChFE)$$

In what follows we restrict attention to $t \geq 0$, without loss of generality (as $\varphi(-t)$ may be reconstructed via complex conjugation). The standard way of solving (*ChFE*) is to derive from it the equations satisfied by the functions $a : n \mapsto a_n$ and $b : n \mapsto a_n$. A direct approach to the characterization of the laws was recently demonstrated in Pitman and Pitman [75], who proceed by proving the map *a injective,* extending both of the maps a and b to $\mathbb{R}_+$, and exploiting the classical Cauchy functional equation (*CFE*) in both cases. For a background textbook account see [58] and for subsequent developments, based on the Choquet–Deny Theorem [45]; the stable laws are given a sketchy account in [78, Chapter 3], and more recent studies include [46] and [47].

Here, however, we indicate why (*ChFE*) can be re-configured to (*GFE*), so that (*GFE*) may be used just once, thereby simplifying the Pitman approach and yielding an even more direct approach. Though we adopt a somewhat cavalier fashion here, the procedure is made entirely rigorous in [73], and we comment below on the underlying justification. Take logarithms (trickery!—see below) and, adjusting notation, pass first to the form

$$f(g(n)t) = nf(t) + h(n)t \qquad (n \in \mathbb{N}, t \in \mathbb{R}_+),$$

where now $\mathbb{R}_+ := (0,\infty)$. Suppose both that g is injective and that one may pass to continuous arguments, in the manner of Kendall's Theorem, for which see Section 9.7(4) (for the double trickery involved here—again see below); then, taking $s = g(n)$, this is

$$f(st) = g^{-1}(s)f(t) + h(g^{-1}(s))t \qquad (s,t \in \mathbb{R}_+),$$

or with $F(t) := f(t)/t$, $G(s) := g^{-1}(s)/s$, $H(s) := hg^{-1}(s)/s$, by symmetry:

$$F(st) = F(t)G(s) + H(s) = F(s)G(t) + H(t).$$

There are now two cases to consider, both leading to the multiplicative form of (*GFE*):

Case (i). If $F(1) = 0$, then taking $s = 1$ yields $F(t) = H(t)$, and so

$$F(st) - F(s) = F(t)G(s)\,.$$

So $\kappa = F$ indeed satisfies the *multiplicative* form of the Goldie equation.

Case (ii). On the other hand, if $F(1) \neq 0$, then passing from F to $F/F(1)$ and from H to $H/F(1)$ we may assume without loss of generality that $F(1) = 1$ (i.e. $f(1) = 1$); then, taking $t = 1$,

$$F(s) = G(s) + H(s).$$

Eliminating H gives

$$F(st) - F(s) = (F(st) - 1) - (F(s) - 1) = (F(t) - 1)G(s)\,,$$

so $\kappa = F - 1$ now satisfies the multiplicative form of the Goldie equation.

Either way, putting $s = e^u$ and $t = e^v$, and $K(u) = \kappa(e^u)$ and $\psi(u) = G(e^u)$, we obtain the additive form:

$$K(u + v) - K(u) = K(v)\psi(u)\,.$$

So (*ChFE*) is (*GFE*), again in disguise!

As to the trickery above: application of the logarithm and the passage from discrete to continuous in the transformation of (*ChFE*) into (*GFE*) is justified in [73] from knowledge of the norming constants, that $a_n = n^k$ for some $k \neq 0$ (as then a extends to an injective function g, and the values a_m/a_n form a dense set). That is an acceptable way to proceed for probabilists, by virtue of an elementary probabilistic proof identifying the norming constants (cf. [42, VI.1, Theorem 1], [75, Lemma 5.3]); the next section (Section 9.4) rids us of this dependence on 'outside material'.

The first trick above (taking logarithms) is justified by Lemma 1 below; the subsequent trick relies on continuity of K and on reference to a dense subset of $\mathbb{R}$, via the simple Corollary below, the routine proof of which we omit: it is similar in spirit to the proof of Lemma 1. (Unlike for the constants a_n, an explicit form for the b_n is not needed.)

Lemma 1 ([73, L. 1]) *For continuous $\varphi \not\equiv 0$ satisfying (ChFE) with $a_n = n^k$ $(k \neq 0)$, φ has no zeros on $\mathbb{R}_+$.*

Proof If $\varphi(\tau) = 0$ for some $\tau > 0$, then $\varphi(a_m\tau) = 0$ for all m, by (*ChFE*). Again by (*ChFE*),

$$|\varphi(\tau a_m/a_n)|^n = |\varphi(a_m\tau)| = 0,$$

so φ is zero on the dense subset of points $\tau a_m/a_n$; then, by continuity, $\varphi \equiv 0$ on $\mathbb{R}_+$, a contradiction. □

Corollary *([73, C. 1]) Equation (ChFE) with $a_n = n^k$ $(k \neq 0)$ holds on the dense subgroup*

$$\mathbb{A}_{\mathbb{Q}} := \{a_m/a_n : m, n \in \mathbb{N}\} :$$

there are constants $\{b_{m/n}\}_{m,n \in \mathbb{N}}$ with

$$\varphi(t)^{m/n} = \varphi(ta_m/a_n)\exp(ib_{m/n}t) \qquad (t \geq 0)\,.$$

Reference to case (ii) in the reduction to (*GFE*) above and to the known continuous solutions of (*GFE*) yields the form of the (non-degenerate) stable law: for some $\gamma \in \mathbb{R}$, $\kappa \in \mathbb{C}$ and with $A := \kappa/\gamma$ and $B := 1 - A$ (for $\gamma \neq 0$),

$$f(t) = \log\varphi(t) = \begin{cases} f(1)(At^{\gamma+1} + Bt), & \text{for } \gamma \neq 0, \\ f(1)(t + \kappa t \log t), & \text{for } \gamma = 0, \end{cases} \qquad (t > 0)\,. \tag{‡}$$

Here $\alpha := \gamma + 1$ is called the *characteristic exponent.*

Remark The form (‡) here takes no account of a further probabilistic ingredient: restrictions on the two parameters γ and κ (equivalently α and κ). Such restrictions follow from the asymptotic analysis of the 'initial' behaviour of the characteristic function φ (i.e. near the origin). This is equivalent to the 'final' or tail behaviour (i.e. at infinity) of the corresponding distribution function, and relates to its *skewness*, i.e. its 'tail balance' ratio—the asymptotic ratio of the distribution's tail difference to its tail sums; for the details see [75, Section 8].

9.4 The Stable Laws Equation on $\mathbb{R}$

Treating the stable laws equation (*ChFE*) *purely* as a functional equation for determining continuous solutions calls for the removal of spurious probabilistic assumptions. It emerges that knowledge of a_n may be deduced from (*ChFE*) provided the continuous solution φ is to be *non-trivial,* i.e. neither $|\varphi| \equiv 0$ nor $|\varphi| \equiv 1$ holds on $[0, \infty)$. That is: the explicit form of a_n may be deduced without assuming that φ is the characteristic function of a (non-degenerate) distribution, as we now show.

Theorem 7 *If φ is a non-trivial continuous function and satisfies (ChFE) for some sequence $a_n \geq 0$, then $a_n = n^k$ for some $k \neq 0$.*

We will first need to establish a further lemma and proposition.

Lemma 2 *If (ChFE) is satisfied by a continuous and non-trivial function φ, then the sequence a_n is either convergent to 0, or divergent ('convergent to $+\infty$').*

Proof Suppose otherwise. Assume first that, as $a_n \geq 0$, for some infinite $\mathbb{M} \subseteq \mathbb{N}$, and $a > 0$,

$$a_m \to a \text{ through } \mathbb{M}.$$

Without loss of generality $\mathbb{M} = \mathbb{N}$, otherwise interpret m below as restricted to $\mathbb{M}$. For any fixed t, $a_m t \to at$, so

$$K_t := \sup_m\{|\varphi(a_m t)|\}$$

is finite by the continuity of φ. Then, for all m,

$$|\varphi(t)|^m = |\varphi(a_m t)| \leq K_t,$$

and so $|\varphi(t)| \leq 1$, for each t. Then, by continuity,

$$|\varphi(at)| = \lim_m |\varphi(a_m t)| = \lim_m |\varphi(t)|^m = 0 \text{ or } 1\,.$$

So, setting $N_k := \{t : |\varphi(at)| = k\}$,

$$\mathbb{R}_+ = N_0 \cup N_1\,.$$

By the connectedness of $\mathbb{R}_+$, one of N_0, N_1 is empty, as the disjoint sets N_k are closed; so respectively $|\varphi| \equiv 0$ or $|\varphi| \equiv 1$, contradicting non-triviality.

To complete the proof, suppose there exist $\mathbb{M} \subseteq \mathbb{N}$ and $\mathbb{M}' \subseteq \mathbb{N}$ such that $\lim_{m\in\mathbb{M}} a_m = \infty$ and $\lim_{m\in\mathbb{M}'} a_m = 0$. The former implies that $|\varphi(0)| = 1$: as φ is non-trivial, we may choose t with $|\varphi(t)| \neq 0$; then, by continuity at 0,

$$|\varphi(0)| = \lim_{m\in\mathbb{M}} |\varphi(t/a_m)| = \lim_{n\in\mathbb{M}} \exp\left(\frac{1}{m}\log|\varphi(t)|\right) = 1.$$

But, again by continuity at 0, for each t,

$$\lim_{m\in\mathbb{M}'} |\varphi(t)|^m = \lim_{m\in\mathbb{M}'} |\varphi(a_m t)| = |\varphi(0)| = 1,$$

and so $|\varphi(t)| = 1$ for all t, contradicting non-triviality. □

The next result essentially contains [75, Lemma 5.2]; the latter relies on $|\varphi(0)| = 1$, the continuity of φ, and the existence of some t with $\varphi(t) < 1$ (guaranteed below by the non-triviality of φ). We assume less here, and so must also consider the possibility that $|\varphi(0)| = 0$ (automatically excluded if φ is the characteristic function of a distribution [42, Chapter XV, Lemma 1]).

Proposition 9 *If (ChFE) is satisfied by a continuous and non-trivial function φ and for some $c > 0$, $|\varphi(t)| = |\varphi(ct)|$ for all $t > 0$, then $c = 1$.*

Proof Note first that $a_n > 0$ for all n; indeed, otherwise, $a_k = 0$ for some $k \geq 1$ and

$$|\varphi(t)|^k = |\varphi(0)| \qquad (t \geq 0)\,.$$

Assume first that $k > 1$; taking $t = 0$ yields $|\varphi(0)| = 0$ or 1, which as in Lemma 2 implies $|\varphi| \equiv 0$ or $|\varphi| \equiv 1$. If $k = 1$, then $|\varphi(t)| = |\varphi(0)|$, and for all $n > 1$,

$$|\varphi(0)|^n = |\varphi(0)|;$$

so again $|\varphi(0)| = 0$ or 1, which again implies $|\varphi| \equiv 0$ or $|\varphi| \equiv 1$.

Applying Lemma 2, the sequence a_n converges either to 0 or to ∞.

We consider these two cases separately.

(i) Suppose that $a_n \to 0$. Then, as above (referring again to K_t), we obtain

$$|\varphi(t)| \leq 1,$$

for all t. Now, since

$$|\varphi(0)| = \lim_n |\varphi(a_n t)| = \lim_n |\varphi(t)|^n\,,$$

if $|\varphi(t)| = 1$ for *some* t, then $|\varphi(0)| = 1$, and that in turn yields, for the very same reason, that

$$|\varphi(t)| \equiv 1$$

for *all* t, a trivial solution, which is ruled out. So in fact $|\varphi(t)| < 1$ for *all* t, and so $|\varphi(0)| = 0$.

Now suppose that for some $c > 0$, $|\varphi(t)| = |\varphi(ct)|$ for all $t > 0$. We show that $c = 1$. If not, without loss of generality $c < 1$, (otherwise replace c by c^{-1} and so, by hypothesis, $|\varphi(t/c)| = |\varphi(ct/c)| = |\varphi(t)|$); then

$$0 = |\varphi(0)| = \lim_n |\varphi(c^n t)| = |\varphi(t)|, \text{ for } t > 0\,,$$

and also for $t = 0$; so φ is trivial, a contradiction. So indeed $c = 1$ in this case.

(ii) Suppose now that $a_n \to \infty$. Choose s with $\varphi(s) \neq 0$; then, by (*ChFE*),

$$|\varphi(0)| = \lim_n |\varphi(s/a_n)| = \lim_n \exp\left(\frac{1}{n}\log|\varphi(s)|\right) = 1\,,$$

i.e. $|\varphi(0)| = 1$. Again as in case (i) above, suppose that for some $c > 0$,

$$|\varphi(t)| = |\varphi(ct)|$$

for all $t > 0$. To show that $c = 1$, suppose again without loss of generality that $c < 1$; then

$$1 = |\varphi(0)| = \lim_n |\varphi(c^n t)| = |\varphi(t)| \text{ for } t > 0\,,$$

and so $|\varphi(t)| \equiv 1$, for $t \geq 0$, again a trivial solution. So again $c = 1$. □

Proof of the Theorem 7 (*ChFE*) implies that

$$|\varphi(a_{mn}t)| = |\varphi(t)|^{mn} = |\varphi(a_m t)|^n = |\varphi(a_m a_n t)| \qquad (t \geq 0)\,.$$

By Proposition 9, a_n satisfies the discrete version of the Cauchy equation

$$a_{mn} = a_m a_n \qquad (m, n \in \mathbb{N})\,,$$

whose solution is known to take the form n^k, since $a_n > 0$ (as at the start of the proof of Proposition 9). If $a_n = 1$ for some $n > 1$, then, for each $t > 0$, $|\varphi(t)| = 0$ or 1 (as $|\varphi(t)| = |\varphi(t)|^n$) and so again, by continuity as in Lemma 2, φ is trivial. So $k \neq 0$. □

Remark Continuity is essential to the theorem: take $a_n \equiv 1$, then a Borel function φ may take the values 0 and 1 arbitrarily.

9.5 Positive Solutions of *GS*

In this section we include various new arguments providing information on the positive solutions of (*GS*) by way of fairly direct links to the equation. Theorem BM, with a family resemblance to Theorem 1, is derived here more directly than if we were to specialize results from Brzdęk [29] and Brzdęk-Mureńko [31]. Theorem B, which follows it and in combination yields *the dichotomy*: f is either never 1 or always 1 on $\mathbb{R}_+ := (0, \infty)$, is taken from these papers, but again the proof here is more direct, and shorter. The final result is Theorem 9, suggested by the recent [71, Theorem 6].

For completeness, as it is needed in Theorem B (and obliquely referred to in Section 9.2.5 above), we begin with the following, which we quote verbatim, as it is short.

Theorem 8 (From [17, Theorem 5]) *If $\varphi : \mathbb{R}_+ \to \mathbb{R}_+$ satisfies (GS), then $\varphi(x) \geq 1$ for all $x > 0$.*

Proof Suppose that $\varphi(u) < 1$ for some $u > 0$; then $v := u/(1 - \varphi(u)) > 0$ and so, since $v = u + v\varphi(u)$,

$$0 < \varphi(v) = \varphi\,(u + v\varphi(u)) = \varphi(u)\varphi(v)\,.$$

So, cancelling by $\varphi(v) > 0$, one has $\varphi(u) = 1$, a contradiction. □

In Theorems BM and B below we use f rather than φ for ease of comparison with [31].

Theorem BM ([31, Lemma 7]) *For $f > 0$ on $\mathbb{R}_+$ a solution of (GS), if $f \neq 1$ at all points, then $f(x) = 1 + cx$ $(x > 0)$ for some $c > 0$.*

Proof By symmetry, for any $x, y > 0$

$$f(x + yf(x)) = f(x)f(y) = f(y + xf(y))\,.$$

Fix x and y and put $u := x + yf(x)$ and $v := y + xf(y)$. If these are unequal, without loss of generality suppose that $v > u$. Then $(v - u)/f(u) > 0$, so

$$0 < f(u) = f(v) = f(u + f(u)(u - v)/f(u)) = f(u)f((u - v)/f(u))\,.$$

Cancelling by $f(u) > 0$ gives $f((u - v)/f(u)) = 1$, contradicting the hypothesis that f is never 1. So $u = v$: that is, for all $x, y > 0$

$$x + yf(x) = y + xf(y)\,;$$

equivalently, for all $x, y > 0$

$$x/(1 - f(x)) = y/(1 - f(y)) = \text{ const. } = c\,,$$

say. Then $f(x) = 1 + cx$ for all $x > 0$. So $c > 0$. □

Below we suppose that $f(a) = 1$, for some fixed $a > 0$. Note that $t_n := na$ is a *Beck* sequence under $\circ_f$ with step size a; so $f(na) = 1$, since $f(t_n) = f(t_1)^n$ (see Section 9.2.5).

For f a positive solution of (GS), we denote here the *positive range* of f by

$$R_f := \{w : (\exists x > 0) w = f(x)\}.$$

If $f \equiv 1$, then $R_f = \{1\}$.

Lemma B ([29, Corollary 1], cf. [31, Lemmas 1,2]) *If the value 1 is achieved at $a > 0$ by a solution $f > 0$ on $\mathbb{R}_+$ of (GS), then*

(i) *the range set R_f is a multiplicative subgroup;*
(ii) *$f(x + a) = f(x)$ for all $x > 0$;*
(iii) *$f(wa) = 1$ for $w \in R_f$.*

Proof For (i), (GS) itself implies that R_f is a semigroup. We only need to find the inverse of $w := f(x)$ with $x > 0$. Choose $n \in \mathbb{N}$ with $na > x$. Put $y = (na - x)/f(x)$; then $y > 0$ and

$$f(x)f(y) = f(x + yf(x)) = f(na) = 1\,.$$

So $f(y) \in R_f$. For (ii), note that, as $f(a) = 1$,

$$f(x) = f(x)f(a) = f(a + xf(a)) = f(x + a).$$

For (iii), since (i) holds, this time write $w = 1/f(x)$ for some $x > 0$; then by (ii)

$$f(x) = f(x + a) = f(x + f(x)a/f(x)) = f(x)f(aw),$$

and cancelling by $f(x) > 0$ gives $f(aw) = 1$. □

Theorem B ([29, Theorem 3]) *For f a positive solution of (GS), if $1 \in R_f$, then $f \equiv 1$.*

Proof Suppose otherwise; then, by Theorem 8 above, $f(u) > 1$, for some $u > 0$. Choose $a > 0$ with $f(a) = 1$ and $n \in \mathbb{N}$ with

$$na > u/(f(u) - 1) > 0.$$

Put

$$v := na + u/(1 - f(u)) > 0; \quad v + naf(u) = u + vf(u) + na.$$

So, since $f(u) \in R_f$, applying Lemma B (first (iii) with $f(u)$ in place of w giving $f(af(u)) = 1$, then (ii) repeatedly, but with $af(u)$ in place of a, and then again (ii) repeatedly, but this time with a)

$$\begin{aligned} 0 < f(v) &= f(v + naf(u)) = f(u + vf(u) + na) \\ &= f(u + vf(u)) = f(u)f(v), \end{aligned}$$

yielding the contradiction $f(u) = 1$. Hence $f(x) = 1$ for all x. □

We now revert to the φ notation. In Section 9.2.5 above, $K(u) > 0$ was posited for $u > 0$ near 0. Below a similar assumption, justified by Theorem 8 above, is made for $K := \varphi - 1$. For $\varphi : [0, \infty) \to \mathbb{R}$, denote its level set above unity by:

$$L_+(\varphi) := \{t \in \mathbb{R}_+ : \varphi(t) > 1\}.$$

Theorem 9 *If the continuous solution φ of (GS) with $\varphi(0) = 1$ has a nonempty level set $L_+(\varphi)$ containing an interval $(0, \delta)$ for some $\delta > 0$, then φ is differentiable and for some $\rho > 0$*

$$\varphi(t) = 1 + \rho t.$$

Proof For $T \in L_+ := L_+(\varphi)$ and $u > 0$, write $m(u) = m_T(u)$ for the jump index of T for the Beck sequence $t_m(u)$, as in Section 9.2.5 above; then

$$t_{m(u)}(u) \le T < t_{m(u)+1}(u).$$

By (**) of Section 9.2.5 (with $m = m(u)$) and continuity at 0 of φ,

$$\begin{aligned}\Delta_{m(u)}(u) &:= t_{m(u)+1}(u) - t_{m(u)}(u) = u\varphi(u)^{m(u)} \\ &\leq T(\varphi(u) - 1) + u \to 0 \quad \text{as } u \to 0\,,\end{aligned}$$

for $u \in L_+$ *uniformly* in $T > 0$ on compacts. Likewise for $u \notin L_+$, as then

$$\Delta_{m(u)}(u) = u.$$

Consider any null sequence $u_n \to 0$ with $u_n > 0$. We will show that

$$\{(\varphi(u_n) - 1)\,/u_n\}$$

is convergent, by showing that down every subsequence $\{(\varphi(u_n) - 1)\,/u_n\}_{n\in\mathbb{M}}$ there is a convergent sub-subsequence with limit independent of $\mathbb{M}$.

Without loss of generality we take $0 < u_n \in L_+$ for all n (so $u_n < \delta$). Now consider an arbitrary $T \in L_+$. Passing, if necessary, to a subsequence (dependent on T) of $\{(\varphi(u_n) - 1)\,/u_n\}_{n\in\mathbb{M}}$, we may suppose, for $k(n) := m_T(u_n)$, that

$$\Delta_{k(n)}(u_n) \to 0\,;$$

then along $\mathbb{M}$

$$|T - t_{m(u_n)}(u_n)| \leq \Delta_{m(u_n)}(u_n)\,,$$

and so

$$t_{k(n)}(u_n) = t_{m(u_n)}(u_n) \to T.$$

Again by (**) and continuity at T of φ, putting $\rho := (\varphi(T) - 1)/T > 0$,

$$\frac{\varphi(u_n) - 1}{u_n} = \frac{\varphi(u_n)^{m(u_n)} - 1}{t_{m(u_n)}(u_n)} = \frac{\varphi(t_{m(n)}(u_n)) - 1}{t_{m(u_n)}(u_n)} \to \frac{\varphi(T) - 1}{T} = \rho\,,$$

along $\mathbb{M}$ to a limit ρ dependent only on T (and not on $\mathbb{M}$). So $\{(\varphi(u_n) - 1)\,/u_n\}$ is itself convergent to ρ. But this holds for any null sequence $\{u_n\}$ in $\mathbb{R}_+$, so the function φ is differentiable at 0, and so is right-differentiable everywhere in L_+ (see [71, Lemma 3]). It is also left-differentiable at any $x > 0$, as follows. For y with $0 < y < x$, put

$$t := (x - y)/\varphi(y) > 0.$$

Then $x = y + t\varphi(y)$, so

$$\frac{\varphi(x)-\varphi(y)}{x-y}=\frac{\varphi(y+t\varphi(y))-\varphi(y)}{x-y}=\frac{[\varphi(t)-1]\varphi(y)}{x-y}=\frac{\varphi(t)-1}{t}.$$

But $t\downarrow 0$ as $y\uparrow x$ (by continuity of φ at x), and

$$(\varphi(t)-1)/t\to\varphi'(0).$$

So φ is left-differentiable at x and so differentiable; from here

$$\varphi'(x)=\varphi'(0).$$

Integration then yields the form of $\varphi(x)$; also, since T above was arbitrary, for any $T\in L_+$ and with $u_n\to 0$ as above,

$$\rho=\lim_{n\in\mathbb{N}}\{(\varphi(u_n)-1)/u_n\}=\varphi'(0)=(\varphi(T)-1)/T:$$

$$\varphi(x)=1+\rho x\quad(x\in\mathbb{R}_+).$$

□

9.6 Two Random Walks in $\mathbb{R}^3$

We close by taking note of two higher-dimensional analogues of the random walk of Section 9.3.2, one unbounded, the other not. These are random walks involving independence both of the step size and of the direction, the latter with (directional) symmetry, i.e. its probability law is invariant under rotation; the object of study is the distribution of the distance from a designated starting point o. The unbounded, locally compact, case is a motion in space starting from the origin with spherical symmetry (which can thus be described by the distribution of its radial component), the other, compact, case a motion on the sphere with starting point o at its north pole (yielding angular, or great circle, distance from o). The correspondingly radial or angular-wise characteristic function satisfies a functional equation involving an 'averaging homomorphy':

$$K(x)K(y)=\int_{-1}^{1}K(x\circ_\lambda y)\,\mathrm{d}\,\psi(\lambda)\,,\qquad (AH)$$

with the auxiliary function ψ a direction-cosine distribution, and two corresponding commutative binary operations with real parameter λ:

$$x\circ_\lambda y=(x^2+y^2+2\lambda xy)^{1/2}\,,$$

$$x\circ_\lambda y=xy+\lambda\sqrt{1-x^2}\sqrt{1-y^2}\,.$$

These expressions arise from the cosine rules for Euclidean and spherical Triangles, respectively. The first of the two gives the radial distance generated by the two step lengths x, y with λ the direction-cosine of the angle between them (note the relation to the *Gauss functional equation* [2, Chapter 3, Example 6]); similarly, the second measures angular distance. As the action which generates motion is not associative in the usual sense, associativity has to be replaced by a probabilistic variant. Replacing the step-length realizations by random variables, the usual associativity property is re-interpreted modulo 'equality in distribution' (cf. $\stackrel{D}{=}$ in Section 9.3.2) for the corresponding random outcomes '$(X \circ_\psi Y) \circ_\psi Z$' and '$X \circ_\psi (Y \circ_\psi Z)$' (with ψ denoting the law of λ). The two kinds of motion were studied, respectively, first by Kingman [61] and next by Bingham [8]. They were very much driven by the work of Bochner, especially [23–25]; indeed, on the basis of this link, one may regard Bochner as the forerunner to/founding father of hypergroups.

The Kingman non-degenerate case finds that probabilistic associativity holds iff the direction-determining *auxiliary function* is ψ_σ with

$$d\psi_\sigma(\lambda) \propto (1-\lambda^2)^{\sigma-1/2} d\lambda$$

(for a parameter $\sigma > -1/2$), a matter earlier recognized by Haldane [48]; the (radial) characteristic function of the walk is then

$$K(u) = \int_{-1}^{1} e^{\mathrm{i}u\lambda} \mathrm{d}\,\psi_\sigma(\lambda) \equiv \Lambda_\sigma(u)\,,$$

where the *lambda Bessel function* is defined by

$$\Lambda_\sigma(t) := (t/2)^{-\sigma} J_\sigma(t) \Gamma(\sigma+1)\,.$$

The Bingham *non-degenerate* case finds that probabilistic associativity holds iff the auxiliary function ψ again has the same ψ_σ form and, up to normalization, the corresponding (angular) characteristic functions K are the *Gegenbauer orthogonal polynomials* (ultraspherical polynomials): Gegenbauer's original analysis plays a rôle in both random walks.

The two *degenerate* cases of (*AH*) in the spherical case correspond to ψ representing either δ_0—a unit point-mass at 0, or $\frac{1}{2}(\delta_{-1} + \delta_{+1})$—two half-unit masses at ± 1. The former yields the Cauchy multiplicative equation on $[-1, 1]$, as may be expected, the latter the *cosine functional equation.*

The general framework for non-deterministic binary operations is provided by the theory of *hypergroups,* as noted in the introduction. Thus the two examples above yield *Kingman's Bessel hypergroups* [22, 3.5.68] (cf. [86, Section 4.1], [87]), and *Bingham's Gegenbauer polynomial hypergroups* [22, 3.4.23] (cf. [86, Chapter 2]). A few words may help to provide some context.

The latter 'polynomial hypergroup' is the easier to describe. Its underlying topological space is discrete: $\mathbb{N}$. Convolution is defined using a family of orthogonal polynomials $\{C_n(t)\}$ acting as a base in the linear space of all polynomials; the binary operation on the pair $k, l \in \mathbb{N}$ is computed from the product $C_k C_l$ via its 'linearization'—its orthogonal expansion. The indices n for C_n with non-zero coefficients in the expansion (the direction cosines) are the possible locations in $\mathbb{N}$, with the cosines prescribing the probability of random selection. This calculation is also at the heart of [8, Proposition 3b], which uses classical orthogonal polynomials with weight function ψ_σ.

The other example is a hypergroup on $\mathbb{R}_+ := [0, \infty)$ with Euclidean topology. The connection with Bessel's differential equation makes Kingman's random walks a canonical example of hypergroups generated by a standard Sturm–Liouville (S-L) differential operator

$$\mathscr{L}_x := -\partial_x^2 - \frac{p'(x)}{p(x)} \partial_x ,$$

where $p(x)$ denotes, as usual, the S-L coefficient function, so that the subscript x signifies the variable of differentiation (cf. [22, 3.5]). The convolution of two unit point-masses at x and y is determined by their action on a $C^\infty(\mathbb{R}_+)$ function f, which action maps f to the evaluation $u_f(x, y)$ at (x, y) of the unique function $u = u_f(.,.)$ defined on $\mathbb{R}_+^2$ and satisfying the p.d.e.

$$\mathscr{L}_x u(x, y) = \mathscr{L}_y u(x, y) ,$$

with boundary information along the axes $x = 0$ and $y = 0$ provided by f.

The upshot of this is to fulfil a like aim as in the earlier example: to define a binary operation $\star$. The continuous analogue, based on (*AH*) above, is

$$f(x \star y) = \int_{\mathbb{R}_+} f(t)(\delta_x \star \delta_y)(\mathrm{d}t) := \int_{-1}^{1} f(x \circ_\lambda y)\psi(\mathrm{d}\lambda) ,$$

where $f(x \star y)$ stands for $f(\delta_x \star \delta_y)$, and so is the mean value of f under the measure $\delta_x \star \delta_y$, and the function $u(x, y) := f(\delta_x \star \delta_y)$ is to satisfy the S-L p.d.e. as above. (This assumes f is integrable with respect to such measures.)

The characteristic function K now solves (*AH*) above iff it solves the functional equation

$$K(x \star y) = K(x)K(y) , \qquad (\star)$$

and now this again expresses homomorphy. In the Sturm–Liouville case, by dint of the construction of the hypergroup relying on the operator $\mathscr{L}_x$, Equation ($\star$) reduces (via separation of variables) to solving a Sturm–Liouville eigenvalue problem:

$$\mathscr{L}_x K(x) = \text{const.} ,$$

with

$$p(x)/p'(x) \equiv x\,,$$

which identifies that K is a lambda Bessel function [22, 3.5.23]. In the polynomial case, Equation ($\star$) reduces to a polynomial recurrence equation, with solution yielding the Gegenbauer polynomials.

Remarks We note two significant underlying features, correlated with the homomorphy asserted by ($\star$).

Firstly, the r-th normalized coefficient (i.e. modulo division by the usual binomial coefficients) θ_r in any valid *finite* Taylor expansions of $\log K(t)$ is 'additive':

$$\theta_r(X \circ_\psi Y) = \theta_r(X) + \theta_r(Y)$$

(these are the Haldane 'cumulants')—see [61, Section 4]; here by a *valid* expansion is meant that the powers in the expansion corresponding to r are taken only as far as the finiteness of the corresponding moments allows.

Secondly, the radial characteristic function encodes homomorphy:

$$\mathbb{E}[K(tX)]\mathbb{E}[K(tY)] = \mathbb{E}[K(t(X \circ_\psi Y))]\,.$$

9.7 Complements

1. Additive Versus Multiplicative, and Double Sweep The definition of a regularly varying f defined on $\mathbb{R}_+$ is usually given in multiplicative form, as that is generally found most useful in applications; the definition immediately suggests a connection with *scaling phenomena*, as in the *Fechner theorem* in physics—see [10]. One is tempted to interpret these phenomena as *functional equations of absent scaling*: to solve $f(x) = \varphi(g(x))$ in the absence of any natural scaling effect between f and g. This is solved on the assumption of *asymptotic* scale independence of f from g :

$$f(\lambda x) \sim \psi(\lambda)f(x)$$

for some ψ, i.e. on the assumption that f is regularly varying. [10] is a very illuminating survey of the applications of RV also in other fields.

The theoretical work in RV, on the other hand, prefers the equivalent additive form of regular variation (as in Section 9.3.1), with f defined on $\mathbb{R}$ satisfying

$$f(x+t) - f(x) \to k(t)\,,$$

so that k will satisfy the additive Cauchy equation. This limit function k may be regarded as the first-order derivative of f 'at infinity'. Of interest is then a second-order asymptotic form arising from the divided difference:

$$[f(x+t) - f(x)] / g(x)$$

(comparing growth rates) studied in the Bojanić-Karamata/de Haan theory, BGT Chapter 3. The general denominator yields the advantages of 'double sweep' (BGT 3.13.1) by capturing both first- and second-order at once (setting $g \equiv 1$ in the former case). Consequently, the Beurling divided difference story of BRV captures the best of both worlds and encompasses all the forms of RV see especially [20, §7].

2. Automatic Continuity In the presence of even the merest hint of additional good behaviour, an additive function is beautifully well-behaved—it is (continuous, and hence) linear. The general context for results like this is that of *automatic continuity*, studied, e.g., by us [12, 13, 15, 17] for real analysis, Hoffmann-Jørgensen in [80, Part 3, Section 2], [83] and [82] for groups, etc. For Banach algebras and Gelfand theory, see, e.g., Dales [37, 38], Helson [49, p. 51], [39, 40], and the recent [60, esp. Corollary 16.7]. The pathology of discontinuity in the absence of good behaviour here is tied to set-theoretic axioms (cf. the foundational discussion in [19, Appendix 1]).

For a study of these features and the *up-grade phenomenon* (as in Theorem 9), that continuity implies differentiability, see [44] and the textbook [56].

3. Generalized Quantifiers Relevant for us are weakenings of the universal quantifier, along such lines as 'for quasi all x', i.e. for x off a negligible set (and elsewhere 'there exist an infinite subset of $\mathbb{N}$' [20]). Mostowski [68] was the first in modern times to begin a study of generalized quantifiers, followed by Lindström [66] (for a textbook treatment see [6, Chapter 13]), and most notably Barwise [4]—see [89] for an account of this important development, and e.g. [65] for some recent developments in this field. Van Lambalgen [88] traces connections here with the conditional expectation of probability theory.

4. Sequential Limits The quantifier weakening here has been concerned with thinning as much as possible the set of λ occurring in $\lambda + x$ or λx. Related, and equally important, is the question of thinning the set of x here—that is, in letting $x \to \infty$ through not all the reals, but some thinned subset. The most familiar case is taking limits *sequentially*, as in *Kendall's theorem* (BGT, Theorem 1.9.2; cf. [10] and Section 9.3.3): for any sequence $\{x_n\}$ with $\lim\sup x_n = \infty$ and $\lim\sup x_{n+1}/x_n \to 1$ (for instance, $x_n = n$), if f is smooth enough (e.g. continuous) and

$$a_n f(\lambda x_n) \to g(\lambda) \in (0, \infty) \quad \forall \lambda \in I$$

for some finite interval $I \subseteq (0, \infty)$ and some sequence $a_n \to \infty$, then f is regularly varying. (Here a_n regularly varying follows from smoothness of f.) The question arises of simultaneous thinning of λ and x together. Another case here is regular variation—in many dimensions, or of measures:

$$n\mathbb{P}(a_n^{-1}\mathbf{x} \in .) \to \nu(.) \qquad (n \to \infty),$$

(here regular variation of $a_n \to \infty$ is assumed) and the limit (spectral) measure ν is on the unit sphere **S**; see, e.g., Hult et al. [53] or [79, Chapter 6] for background. Now thinning is to be done on subsets of **S** on which convergence is assumed. For *convergence-determining classes* here, see, e.g., Billingsley [7, Section 1.2], Landers [64], Rogge [81].

5. Regular Variation Without Limits In the absence of limit functions one studies the 'limsup' variants. As these are subadditive, one asks when does this subadditivity lead to additivity. The following identifies where naturally to apply quantifier weakening; Theorem 5 of Section 9.3.2 yields a sample answer: see also [18, 20].

Proposition 10 (Additive Kernel, [20, Proposition 1]) *For* $F : \mathbb{R} \to \mathbb{R}$ *put*

$$\mathbb{A}_F := \{u : \lim_{x\to\infty} [F(u+x) - F(x)] \text{ exists and is finite}\},$$

and, for $a \in \mathbb{A}_F$, *put* $G(a) := \lim_{x\to\infty}[F(a+x) - F(x)]$. *For* $u \in \mathbb{R}$ *define*

$$F^*(u) := \limsup_{x\to\infty}[F(u+x) - F(x)].$$

Then:

(i) $\mathbb{A}_F$ *is an additive subgroup;*
(ii) G *is an additive function on* $\mathbb{A}_F$*;*
(iii) $F^* : \mathbb{R} \to \mathbb{R} \cup \{-\infty, +\infty\}$ *is a subadditive extension of* G*;*
(iv) F^* *is finite-valued and additive iff* $\mathbb{A}_F = \mathbb{R}$ *and* $F^*(u) = G(u)$ *for all* u.

This directly connects to Theorem 1 in Section 9.2, as the identity

$$uv - u - v + 1 \equiv (1-u)(1-v)$$

gives that $(1-e^{-\gamma x})/\gamma$ is *subadditive* on $\mathbb{R}_+ := (0,\infty)$ for $\gamma \geq 0$, and *superadditive* on $\mathbb{R}_+$ for $\gamma \leq 0$.

6. Functional Equations of Associativity The equivalence noticed by Javor of (GS) with the associativity of $\circ_\eta$ has further analogues in connecting functional equations with the associativity of binary operations. For example, one may consider the operations

$$x *_\lambda y := xy \pm \lambda^2 p(x)p(y)$$

with p either involutary or skew-involutary. These are associative iff $g(x) := \lambda p(x)/x$ solves the equation

$$g(x *_\lambda y) = \frac{g(x) + g(y)}{1 \mp g(x)g(y)/\lambda^2};$$

converting g into a homomorphism calls for the right-hand side to be interpreted as the combination of the elements $u = g(x)$ and $v = g(y)$ by means of a group operation on the interval $(-\lambda, \lambda)$, $\circ_\lambda$ say, given by

$$u \circ_\lambda v = \frac{u+v}{1 \pm uv/\lambda^2} \,.$$

Then g is seen to satisfy the *functional equation of competition* introduced recently by Kahlig and Matkowski [59]; cf. the hyperbolic semi-group of [51, 8.3]. As there, the choice of sign '−' or '+' yields the familiar tangent or hyperbolic tangent addition formulas. In the skew case the operations $*_\lambda$ include both

$$xy \pm \lambda^2(1-x)(1-y)$$

and the 'cosine formula', similarly as in Section 9.6:

$$xy \pm \lambda^2\sqrt{1-x^2}\sqrt{1-y^2} \,.$$

The operation

$$x * y = xy + p(x) + p(y) \,,$$

with $p(0) = 0$, is associative only for $p(x) \equiv 0$ and $p(x) \equiv x$.

7. The Cocycle Equation The cocycle functional equation

$$F(st, x) = F(s, tx)F(t, x)$$

for $F : G \times X \to G$ may be regarded as an entry-point into RV, using flow language, as in [70, Section 4] and [14]; indeed, if F is to be a h-coboundary for some continuous h, then

$$h(tx) = F(t, x)h(x) \,.$$

References

1. Aczél, J.: Extension of a generalized Pexider equation. Proc. Am. Math. Soc. **133**, 3227–3233 (2005)
2. Aczél, J., Dhombres, J.: Functional Equations in Several Variables. With Applications to Mathematics, Information Theory and to the Natural and Social Sciences. Encyclopedia of Mathematics and Its Applications, vol. 31. Cambridge University Press, Cambridge (1989)
3. Aczél, J., Gołąb, S.: Remarks on one-parameter subsemigroups of the affine group and their homo- and isomorphisms. Aequationes Math. **4**, 1–10 (1970)
4. Barwise, J.: Axioms for abstract model theory. Ann. Math. Logic **7**, 221–265 (1974)

5. Beck, A.: Continuous Flows on the Plane. Grundl. math. Wiss., vol. 201. Springer, Berlin (1974)
6. Bell, J.L., Slomson, A.B.: Models and Ultraproducts: An Introduction. Springer, Berlin (1969)
7. Billingsley, P.: Convergence of Probability Measures. Wiley, New York (1999)
8. Bingham, N.H.: Random walk on spheres. Z. Wahrscheinlichkeitstheorie und Verw. Gebiete **22**, 169–192 (1972)
9. Bingham, N.H.: Random walk and fluctuation theory. In: Shanbhag, D.N., Rao, C.R. (eds.) Handbook of Statistics, vol. 19, pp. 171–213. Elsevier, Amsterdam (2001)
10. Bingham, N.H.: Scaling and regular variation. Publ. Inst. Math. Beograd **97**(111), 161–174 (2015)
11. Bingham, N.H., Goldie, C.M.: Extensions of regular variation. I. Uniformity and quantifiers. Proc. Lond. Math. Soc. (3) **44**, 473–496 (1982); Extensions of regular variation. II. Representations and indices. Proc. London Math. Soc. (3) **44**, 497–534 (1982)
12. Bingham, N.H., Ostaszewski, A.J.: Automatic continuity: subadditivity, convexity, uniformity. Aequationes Math. **78**, 257–270 (2009)
13. Bingham, N.H., Ostaszewski, A.J.: Infinite combinatorics in function spaces: category methods. Publ. Inst. Math. Beograd **86** (100), 55–73 (2009)
14. Bingham, N.H., Ostaszewski, A.J.: The index theorem of topological regular variation and its applications. J. Math. Anal. Appl. **358**, 238–248 (2009)
15. Bingham, N.H., Ostaszewski, A.J.: Automatic continuity via analytic thinning. Proc. Am. Math. Soc. **138**, 907–919 (2010)
16. Bingham, N.H., Ostaszewski, A.J.: Beurling slow and regular variation. Trans. Lond. Math. Soc. **1**, 29–56 (2014) (fuller versions: arXiv 1301.5894 and 1307.5305)
17. Bingham, N.H., Ostaszewski, A.J.: Cauchy's functional equation and extensions: Goldie's equation and inequality, the Gołąb-Schinzel equation and Beurling's equation. Aequationes Math. **89**, 1293–1310 (2015)
18. Bingham, N.H., Ostaszewski, A.J.: Beurling moving averages and approximate homomorphisms. Indag. Math. **27**, 601–633 (2016) (fuller version: arXiv 1407.4093)
19. Bingham, N.H., Ostaszewski, A.J.: Category-measure duality: convexity, mid-point convexity and Berz sublinearity. Aequationes Math. (to appear) (fuller version: arXiv 1607.05750)
20. Bingham, N.H., Ostaszewski, A.J.: Additivity, subadditivity and linearity: automatic continuity and quantifier weakening. arXiv 1405.3948v4
21. Bingham, N.H., Goldie, C.M., Teugels, J.L.: Regular Variation. Cambridge University Press, Cambridge (1989)
22. Bloom, W.R., Heyer, H.: Harmonic Analysis of Probability Measures on Hypergroups. De Gruyter Studies in Mathematics, vol. 20. Walter de Gruyter, Berlin (1995)
23. Bochner, S.: Positive zonal functions on spheres. Proc. Natl. Acad. Sci. **40**, 1141–1147 (1954)
24. Bochner, S.: Sturm-Liouville and heat equations whose eigenfunctions are ultraspherical polynomials or associated Bessel functions. In: Proceedings of the conference on Differential Equations (dedicated to A. Weinstein), pp. 23–48. University of Maryland Book Store, College Park, MD (1956)
25. Bochner, S.: Positivity of the heat kernel for ultraspherical polynomials and similar functions. Arch. Ration. Mech. Anal. **70**, 211–217 (1979)
26. Bojanić, R., Karamata, J.: On a class of functions of regular asymptotic behavior. Mathematics Research Center Technical Report 436, Madison, Wisconsin 1963; reprinted in: Marić, V. (ed.) Selected papers of Jovan Karamata, pp. 545–569, Zevod Udžbenika, Beograd (2009)
27. Bogachev, V.I.: Gaussian Measures. Mathematical Surveys & Monographs, vol. 62. American Mathematical Society, Providence, RI (1998)
28. Brillouët, N., Dhombres, J.: Equations fonctionelles et recherche de sous-groupes. Aequationes Math. **31**, 253–293 (1986)
29. Brzdęk, J.: Subgroups of the group $\mathbb{Z}_n$ and a generalization of the Gołąb-Schinzel functional equation. Aequationes Math. **43**, 59–71 (1992)
30. Brzdęk, J.: The Gołąb-Schinzel equation and its generalizations. Aequationes Math. **70**, 14–24 (2005)

31. Brzdęk, J., Mureńko, A.: On a conditional Gołąb-Schinzel equation. Arch. Math. (Basel) **84**, 503–511 (2005)
32. Choquet, G., Deny, J.: Sur l'équation de convolution $\mu = \mu * \sigma$. C. R. Acad. Sci. Paris **250**, 799–801 (1960)
33. Chudziak, J.: Semigroup-valued solutions of the Gołąb-Schinzel type functional equation. Abh. Math. Semin. Univ. Hambg. **76**, 91–98 (2006)
34. Chudziak, J.: Semigroup-valued solutions of some composite equations. Aequationes Math. **88**, 183–198 (2014)
35. Chudziak, J., Kočan, Z.: Continuous solutions of conditional composite type functional equations. Results Math. **66**, 199–211 (2014)
36. Chudziak, J., Tabor, J.: Generalized Pexider equation on a restricted domain. J. Math. Psychol. **52**, 389–392 (2008)
37. Dales, H.G.: Automatic continuity: a survey. Bull. Lond. Math. Soc. **10**, 129–183 (1978)
38. Dales, H.G.: Banach Algebras and Automatic Continuity. London Math. Soc. Monographs **24**, Oxford University Press (2000)
39. Dales, H.G., Woodin, W.H.: An Introduction to Independence for Analysts. London Mathematical Society Lecture Note Series, vol. 115. Cambridge University Press, Cambridge (1987)
40. Dales, H.G., Woodin, W.H.: Super-Real Fields. Totally Ordered Fields with Additional Structure. London Mathematical Society Monographs, vol. 14. Oxford Science Publications. Oxford University Press, Oxford (1996)
41. Engelking, R.: General topology. Heldermann Verlag, Berlin (1989)
42. Feller, W.: An Introduction to Probability Theory and Its Applications, vol. 2. Wiley, New York (1971)
43. Gołąb, S., Schinzel, A.: Sur l'équation fonctionelle $f(x+yf(x)) = f(x)f(y)$. Publ. Math. Debr. **6**, 113–125 (1959)
44. Grossse-Erdmann, K.-G.: Regularity properties of functional equations and inequalities. Aequationes Math. **37**, 233–251 (1989)
45. Gu, H.-M., Lau, H.-S.: Integrated Cauchy functional equation with an error term and the exponential law. Sankhya, Indian J. Stat. **46**, 339–354 (1984)
46. Gupta, A.K., Nguyen, T.T., Zeng, W.-B.: Characterization of multivariate distributions through a functional equation of their characteristic functions. J. Stat. Plann. Inference **63**, 187–201 (1997)
47. Gupta, A.K., Jagannanthan, K., Nguyen, T.T., Shanbhag, D.: Characterization of stable laws via functional equations. Math. Nachr. **279**, 571–580 (2006)
48. Haldane, J.B.S.: The addition of random vectors. Sankhya, Indian J. Stat. **22**, 213–220 (1960)
49. Helson, H.: Harmonic Analysis. Hindustan Book Agency, New Delhi (1995)
50. Heyer, H.: Probability Measures on Locally Compact Groups. Ergebnisse Math. Grenzgebiete, vol. 94. Springer, Berlin (1977)
51. Hille, E., Phillips, R.S.: Functional analysis and semi-groups. Am. Math. Soc. Coll. Publ. **31**, Rev. Ed. (1957)
52. Hobson, E.W.: The Theory of Functions of a Real Variable and the Theory of Fourier's Series, vol. 2. Cambridge University Press, Cambridge (1926)
53. Hult, H., Lindskog, F., Mikosch, T., Samorodnitsky, G.: Functional large deviations for multivariate regularly varying random walks. Ann. Appl. Probab. **15**, 2651–2680 (2005)
54. Jabłońska, E.: On solutions of some generalizations of the Gołąb-Schinzel equation. In: Rassias, Th.M., Brzdęk, J. (eds.) Functional Equations in Mathematical Analysis, pp. 509–521. Springer, New York (2012)
55. Jabłońska, E.: On locally bounded above solutions of an equation of the Gołąb-Schinzel type. Aequationes Math. **87**, 125–133 (2014)
56. Járai, A.: Regularity Properties of Functional Equations in Several Variables. Springer, New York (2005)
57. Javor, P.: On the general solution of the functional equation $f(x + yf(x)) = f(x)f(y)$. Aequationes Math. **1**, 235–238 (1968)

58. Kagan, A.M., Linnik, Yu.V., Rao, C.R.: Characterization Problems in Mathematical Statistics. Wiley, New York (1973)
59. Kahlig, P., Matkowski, J.: On a functional equation related to competition. Aequationes Math. **87**, 301–308 (2014)
60. Kąkol, J., Kubiś, W., López-Pellicer, M.: Descriptive Topology in Selected Topics of Functional Analysis. Developments in Mathematics, vol. 24. Springer, New York (2011)
61. Kingman, J.F.C.: Random walks with spherical symmetry. Acta Math. **109**, 11–53 (1963)
62. Klippert, J., Williams, G.: Uniform convergence of a sequence of functions at a point. Int. J. Math. Educ. Sci. Technol. **33**(1), 51–58 (2002)
63. Kuczma, M.: An Introduction to the Theory of Functional Equations and Inequalities. Cauchy's Equation and Jensen's Inequality. Birkhäuser, Basel (2009)
64. Landers, D.: The convergence-determining class of connected open sets on product spaces. Proc. Am. Math. Soc. **33**, 529–533 (1972)
65. Lapierre, S.: New results in the analysis of some conditional quantifiers and their logics. Log. Anal. (N.S.) **34**, 105–120 (1991)
66. Lindström, P.: First-order predicate logic with generalized quantifiers. Theoria **32**, 186–195 (1966)
67. Michael, E.: Topologies on the spaces of subsets. Trans. Am. Math. Soc. **71**, 152–182 (1951)
68. Mostowski, A.: On a generalization of quantifiers. Fund. Math. **44**, 12–36 (1957)
69. Moszner, Z.: General theory of the translation equation. Aequationes Math. **50**, 17–37 (1995)
70. Ostaszewski, A.J.: Regular variation, topological dynamics, and the uniform boundedness theorem. Topology Proc. **36**, 305–336 (2010)
71. Ostaszewski, A.J.: Beurling regular variation, Bloom dichotomy, and the Gołąb-Schinzel functional equation. Aequationes Math. **89**, 725–744 (2015)
72. Ostaszewski, A.J.: Homomorphisms from functional equations: the Goldie equation. Aequationes Math. **90**, 427–448 (2016)
73. Ostaszewski, A.J.: Stable laws and Beurling kernels. Adv. Appl. Probab. **48A** (N. H. Bingham Festschrift), 239–248 (2016)
74. Parthasarathy, K.R.: Probability Measures on Metric Spaces. AMS Chelsea Publishing, Providence, RI (2005); Reprint of the 1967 original
75. Pitman, E.J.G., Pitman, J.: A direct approach to the stable distributions. Adv. Appl. Probab. **48A** (N. H. Bingham Festschrift), 261–282 (2016)
76. Popa, C.G.: Sur l'équation fonctionelle $f(x + yf(x)) = f(x)f(y)$. Ann. Polon. Math. **17**, 193–198 (1965)
77. Przebieracz, B.: Recent developments in the translation equation and its stability. In: Brzdęk, J., et al. (eds.) Developments in Functional Equations and Related Topics. Springer Optimization and Its Applications, vol. 124. Springer International Publishing, Cham (2017), doi:10.1007/978-3-319-61732-9_10
78. Ramachandran, B., Lau, K.-S.: Functional Equations in Probability Theory. Academic Press, New York (1991)
79. Resnick, S.I.: Heavy-Tail Phenomena. Springer, New York (2007)
80. Rogers, C.A., Jayne, J., Dellacherie, C., Topsøe, F., Hoffmann-Jørgensen, J., Martin, D.A., Kechris, A.S., Stone, A.H.: Analytic Sets. Academic Press, New York (1980)
81. Rogge, L.: The convergence-determining class of regular open sets. Proc. Am. Math. Soc. **37**, 581–585 (1973)
82. Rosendal, C., Solecki, S.: Automatic continuity of homomorphisms and fixed points on metric compacta. Isr. J. Math. **162**, 349–371 (2007)
83. Solecki, S., Srivastava, S.M.: Automatic continuity of group operations. Topology Appl. **77**, 65–75 (1997)
84. Stetkær, H.: Functional Equations on Groups. World Scientific, Singapore (2013)
85. Stroock, D.W.: Probability Theory: An Analytic View. Cambridge University Press, Cambridge (1999)
86. Székelyhidi, L.: Functional Equations on Hypergroups. World Scientific, Singapore (2013)

87. Székelyhidi, L.: Functional equations and stability problems on hypergroups. In: Brzdęk, J., Ciepliński, K., Rassias, T.M. (eds.) Developments in Functional Equations and Related Topics. Springer Optimization and Its Applications, vol. 124. Springer, Cham (2017). doi:10.1007/978-3-319-61732-9
88. van Lambalgen, M.: Conditional quantification, or poor man's probability. J. Log. Comput. **11**, 295–335 (2001)
89. Väänänen, J.: Barwise: Abstract model theory and generalized quantifiers. Bull. Symb. Logic **10**, 37–53 (2004)

Chapter 10
Recent Developments in the Translation Equation and Its Stability

Barbara Przebieracz

Abstract The aim of this chapter is to present some of the recent results concerning the theory of the translation equation and its stability.

Keywords Translation equation • Stability of functional equations • Iterative roots • Embeddability in iteration groups

Mathematics Subject Classification (2010) Primary 39B12; Secondary 39B82, 26A18, 37E10, 37C15, 37E05, 39B22

10.1 Introduction

The functional equation

$$F(s, F(t, x)) = F(s \cdot t, x), \quad s, t \in G, x \in X,$$

where $F: G \times X \to X$, G is a set with binary operation $\cdot$, and X is an arbitrary set, is called *the translation equation*. Here, we gather only a personal choice of results concerning the translation equation and its stability published in recent years. We focus in a more detailed way only on these results, which are not discussed in the previous survey papers.[1] We refer the reader to the earlier survey papers on this topic:

[1]Especially the recent ones: [6, 25, 26] are published as open access papers, and [13] is also free available.

B. Przebieracz (✉)
Institute of Mathematics, University of Silesia, Bankowa 14, 40-007 Katowice, Poland
e-mail: barbara.przebieracz@us.edu.pl

J. Brzdęk et al. (eds.), *Developments in Functional Equations and Related Topics*, Springer Optimization and Its Applications 124,
DOI 10.1007/978-3-319-61732-9_10

1. In paper [11], Moszner listed several mathematical domains in which the translation equation appears. It includes among others abstract geometric and algebraic objects, groups of transformations, iterations, and dynamical systems. The author then presents many results concerning the solutions of the translation equation, including his own construction for the general solution on some domain. Continuity problems are also discussed.
2. In [13], Moszner continues the presentation of achievements in the theory. Further results on structure of solutions are listed. This paper also contains survey on regular (continuous, differentiable, analytic, and monotonic) solutions, problem of extendability of solutions to bigger domain, and papers of Smajdor on set-valued iteration semigroups.
3. The survey [14] covers among others the results on stability of the translation equation obtained by Mach and Moszner.
4. Zdun and Solarz [26] is an extensive survey on iteration theory. Here, we consider G an additive subgroup or subsemigroup of $\mathbb{R}$ or $\mathbb{C}$; in case $G = \mathbb{R}$ or $G = \mathbb{R}_+$, we say about iteration group (flow) or semigroup (semiflow), respectively. We usually write $F^t(x)$ instead of $F(t, x)$, hence, the translation equation takes the form

$$F^t \circ F^s(x) = F^{s+t}(x).$$

The origin of the notion of iteration group is extending the iterates F^n, $n \in \mathbb{N}$, of a given $F: X \to X$, to "real" iterates F^t, $t \in \mathbb{R}$. We often interpret $F^t(x)$ as the state of a point (object) x at the time t.

Topics covered in this paper (quite in detail)[2]:

- Measurable iteration semigroups: results of Baron, Chojnacki, Jarczyk, and Zdun on the problem *under what condition the measurability of iteration group/semigroup implies its continuity*;
- Embeddability of f into iteration groups or semigroups: *when for a given f there exists $\{F^t\}$ such that $F^1 = f$*, moreover, we can demand that iteration group or semigroup is of suitable regularity. This issue was examined for diffeomorphisms in $\mathbb{R}^N$, Brouwer homeomorphisms on the plane (mainly Leśniak's results), and interval homeomorphisms (mainly the result obtained by Zdun, Krassowska, and Zhang);
- When two commuting functions (f and g defined on an open interval, without fixed points) can be embeddable in the same iteration group (i.e. $f, g \in \{F^t\}$) (mainly the results of Zdun, Krassowska, and Ciepliński);
- Problem of existence of iterative roots (φ is an iterative root of order n of a given f, if $\varphi^n = f$, where φ^n denotes n-th iterate of φ) of piecewise monotonic functions, homeomorphism of the circle, and homeomorphisms of the plane (Zhang, Liu, Li, Yang, Jarczyk, Jarczyk, Zdun, and Solarz);

[2] Here, we signal them only, and mention some main authors; for detailed references, we refer the reader to [26].

- The structure of iteration groups of homeomorphisms of an interval, and of homeomorphisms of the circle (Zdun and Ciepliński);
- Different notions of "near" embeddability into iteration semigroup and characterization of such functions (Jarczyk and Przebieracz);
- A few problems concerning set-valued iteration semigroups (existence of iteration semigroup of single valued functions which is a selection of a given set-valued iteration group, and existence of majorizing iteration semigroups (Smajdor, Olko, Piszczek, and Łydzińska);
- Theorems of Matkowski and Jarczyk on iterates of mean-type mappings; and
- Stability of the translation equation (Moszner, Mach, Chudziak, Przebieracz, Reich, and Jabłoński).

5. The readers interested in the topic of iterative roots should read [6], where many results (recent and older) are presented in detail, also some open problems are listed. Here (in Section 10.3), we develop only the topic of conjugacy between F and its iterative root, for piecewise monotonic F.
6. In [25], Zdun discussed the existence of embeddings of given mappings in real iteration groups with suitable regularity, the conditions which imply the uniqueness of embeddings, and the formulas expressing the above embeddings or their general constructions. Here, in the next section, we refine some new approach to this subject proposed in [7].

10.2 Recent Advances in the Problem of Embeddability in Iteration Groups: Embeddability of Homeomorphisms of the Circle in Set-Valued Iteration Groups

Let $\mathbb{S}^1 = \{z \in \mathbb{C} : |z| = 1\}$ be the unit circle with positive orientation, $cc[\mathbb{S}^1]$ be the family of all non-empty convex and compact subsets of $\mathbb{S}^1$ (that is, the family of closed arcs and points of $\mathbb{S}^1$). Let $F\colon \mathbb{S}^1 \to \mathbb{S}^1$ be a homeomorphism without periodic points (its rotation number ρ is irrational). Let L_F be the set of all limit points of orbits of F (it is known that L_F is either equal to $\mathbb{S}^1$ or is a nowhere dense perfect set [4]). Moreover, F is embeddable in continuous iteration group[3] if and only if $L_F = \mathbb{S}^1$; in such a case, the continuous embedding is unique up to a constant [24]. Necessary and sufficient conditions for embeddability in the discontinuous iteration groups were given in [2] (in this case, F has infinitely many nonmeasurable embeddings).

In the paper [7], authors proposed a new approach to the problem of embeddability. They constructed some substitution of an iteration group in which F can be embedded.

[3]That is, there exists an iteration group $\{f^t\colon \mathbb{S}^1 \to \mathbb{S}^1;\, t \in \mathbb{R}\}$, such that $F = f^1$ and for every z the orbits $t \mapsto f^t(z)$ are continuous.

Before formulating main theorems from that paper, let us fix some notation. We assume that $L_F \neq \mathbb{S}^1$. In this case, the set $\mathbb{S}^1 \setminus L_F$ is a countable sum of pairwise disjoint open arcs, let $\mathscr{A}$ be a family of these arcs, $\alpha(I)$ be the middle point of the arc I, $M := \{\alpha(I) : I \in \mathscr{A}\}$, and $I_p := \alpha^{-1}(p)$ for $p \in M$. Hence, $\bigcup_{p\in M} I_p$ is a decomposition of $\mathbb{S}^1 \setminus L_F$ into open pairwise disjoint arcs. Let $L^* := \mathbb{S}^1 \setminus \bigcup_{p\in M} \operatorname{cl} I_p$. There exists exactly one continuous solution Φ of equation

$$\Phi(F(z)) = e^{2\pi i\rho}\Phi(z), \quad z \in \mathbb{S}^1,$$

such that $\Phi(1) = 1$. This solution is surjective and increasing (see [3, 23]). Define $F^t(z)$ as preimages of singletons:

$$F^t(z) := \Phi^{-1}[\{e^{2\pi it\rho}\Phi(z)\}], \quad t \in \mathbb{R},\ z \in \mathbb{S}^1.$$

The family $\{F^t\colon \mathbb{S}^1 \to cc[\mathbb{S}^1];\ t \in \mathbb{R}\}$ is an iteration group such that $F(z) \in F^1(z)$ for $z \in \mathbb{S}^1$. It will be called *the main set-valued embedding* of F. It has the following properties:

(A1) $\forall_{t\in\mathbb{R},\, z\in\mathbb{S}^1}$ $F^t(z)$ is either a closed arc $\operatorname{cl} I_p$ for some $p \in M$ or a singleton belonging to L^*;
(A2) $\forall_{t\in\mathbb{R}}$ the function $z \mapsto F^t(z)$ is increasing and constant on the arcs $\operatorname{cl} I_p$, $p \in M$;
(A3) $\forall_{z\in\mathbb{S}^1}$ the function $t \mapsto F^t(z)$ is periodic with the period $\frac{1}{\rho}$ and strictly increasing on the arcs $\operatorname{cl} I_p$, $p \in M$;
(A4) if $F^u(z) \cap F^v(z) \neq \emptyset$, then $u = v + \frac{k}{\rho}$ for a $k \in \mathbb{Z}$;
(A5) $\forall_{p\in M}$ F^0 is constant on $\operatorname{cl} I_p$, $F^0[\operatorname{cl} I_p] = \operatorname{cl} I_p$; $F^0(z) = z$ for $z \in L^*$;
(A6) $\forall_{z\in\mathbb{S}^1}$ $\bigcup_{t\in\mathbb{R}} F^t(z) = \mathbb{S}^1$; and
(A7) $\forall_{z\in\mathbb{S}^1}$ $\exists_{t_1,t_2\in\mathbb{R}}$ $F^{t_1}(z)$ is an arc, $F^{t_2}(z)$ is a singleton.

Some of the above properties characterize the main set-valued embeddings of F, namely, if a set-valued group $\{F^t\colon \mathbb{S}^1 \to cc[\mathbb{S}^1];\ t \in \mathbb{R}\}$ fulfills conditions (A1), (A3), and (A6) (only for one point $z_0 \in \mathbb{S}^1$, not necessarily for all $z \in \mathbb{S}^1$) and $F(z) \in F^1(z)$, then it is the main set-valued embedding of F.

Moreover, the set

$$T := \{t \in \mathbb{R};\ \Phi[\mathbb{S}^1 \setminus L_F] = e^{2\pi it\rho}\Phi[\mathbb{S}^1 \setminus L_F]\},$$

is an additive, countable, and dense subgroup of $\mathbb{R}$ and $1 \in T$. It will be called *the supporting group of F*.

Let $\mathscr{F} := \{F^t\colon \mathbb{S}^1 \to cc[\mathbb{S}^1];\ t \in \mathbb{R}\}$ be the main set-valued embedding of F. It turns out that for every $t \in \mathbb{R}$ and every $p \in M$ the function F^t is constant on $\operatorname{cl} I_p$, whence, for every $z \in I_p$, $F^t(z) = F^t[\operatorname{cl} I_p]$. The set $F^t[\operatorname{cl} I_p]$ is either an arc or a point. Similarly, if $z \in L^*$, then $F^t(z)$ is either an arc or a point. Group T characterizes these indices for which F^t maps arcs $\operatorname{cl} I_p$ onto arcs and points from L^* onto points from L^*.

The subgroup $\{F^t: \mathbb{S}^1 \to cc[\mathbb{S}^1]; t \in T\}$ of $\mathcal{F}$ is said to be the *refinement set-valued embedding of* F. It possesses a piecewise linear selection $\{v^t: \mathbb{S}^1 \to \mathbb{S}^1; t \in T\}$ of homeomorphisms. Moreover, $H \in \mathcal{F}$ has a continuous and injective selection if and only if H belongs to the refinement set-valued embedding of F.

10.3 Recent Advances in the Subject of Iterative Roots: Conjugacy Between Piecewise Monotonic Functions and Their Iterative Roots

First, we set some notations in order to formulate theorems in this section in a more concise way.

Let $I := [a, b]$ for $a < b < \infty$ and $F: I \to \mathbb{R}$ be a continuous function. A point $c \in (a, b)$ is called a *fort* of F if F is not strictly monotonic in any neighbourhood of c. We say that F is piecewise monotonic ($F \in \mathcal{PM}[I]$) if the number $N(F)$ of forts of F is finite.

We put $S(F) := \{c_1, c_2, \ldots, c_{N(F)}\}$ for the set of all forts of piecewise monotonic F. Additionally, put $c_0 = a$ and $c_{N(F)+1} = b$ and define $I_i := [c_i, c_{i+1}]$ for $i = 0, 1, \ldots, N(F)$. It is known that [27, 28] either there exists an integer $r \in \mathbb{N} \cup \{0\}$ such that

$$0 = N(F^0) < N(F) < N(F^2) < \ldots < N(F^r) = N(F^{r+1}) = N(F^{r+2}) = \ldots,$$

or for every $k \in \mathbb{N} \cup \{0\}$ we have $N(F^k) < N(F^{k+1})$. In the first case, we put $H(F) := r$, and in the second $H(F) := \infty$, where $H(F)$ is called the *non-monotonicity height* of piecewise monotonic F.

For $F \in \mathcal{PM}[I]$ with $H(F) = 1$, the maximal interval $K(F)$, containing $F[I]$ and such that F is monotonic on it, is called the *characteristic interval* of F [27, 28].

If f is a continuous iterative root of F of order n, then for every $i \in \{0, \ldots, N(F)\}$ there exists a positive integer $k \le \min\{n, N(F)\}$ and $i_1, \ldots i_{k-1} \in \{0, \ldots, N(F)\}$ such that

$$I_i \to I_{i_1} \to \ldots \to I_{i_{k-1}} \to K(F),$$

where by $I_{l_1} \to I_{l_2}$ we mean $f(I_{l_1}) \subset I_{l_2}$. Let $k_f(i)$ denote the number k described above. The *pace* ℓ, of iterative root f, is defined as $\max\{k_f(i); i \in \{0, 1, \ldots, N(F)\}\}$.

Every iterative root f of F can be extended from the characteristic interval $K(F)$ [9].

It turns out that all continuous monotonic functions are conjugate to their iterative roots [29] (we say that f is conjugate to g if there exists a homeomorphism Φ such that $\Phi \circ f = g \circ \Phi$). It enables us to understand the topological dynamics properties of iterative root f (explicit formulas can be complicated) having given $F = f^n$. In [8], authors gave examples of continuous piecewise monotonic but not monotonic functions, in order to prove that such functions:

- May have no iterative roots conjugate to them;
- May have some iterative roots not conjugate to them; and
- May have some iterative roots ($n \neq 1$) conjugate to them.

Moreover, they give necessary and sufficient conditions under which piecewise monotonic F is conjugate to its iterative root f.

Theorem 10.1 *Suppose that the mapping $F \in \mathscr{PM}[I]$ with $N(F) \geq 1$ and its continuous iterative root F having pace 1 are conjugate. Suppose that F is strictly increasing on its characteristic interval $K(F)$. Moreover, assume that $K(F) = \mathrm{Fix}(f) \cup J_1 \cup J_2 \cup \ldots \cup J_d$, where $\mathrm{Fix}(f)$ is the set of all fixed points of f and J_m's ($m = 1, 2, \ldots d$) are pairwise different intervals with endpoints being fixed points of f and interiors without fixed points. Then, f is strictly increasing on $K(F)$ and for each interval J_m, $m = 1, \ldots, d$, either*

(H1) $\{f(c_i);\ i = 1, 2 \ldots, N(F)\} \cap \operatorname{int} J_m = \emptyset$, *or*
(H2) There is a point $c^* \in \operatorname{int} J_m$ such that

$$\begin{cases} f(c_j) \in (f^2(c^*), f(c^*)], & \text{if } f(c^*) < c^* \text{ or} \\ f(c_j) \in (c^*, f(c^*)], & \text{if } f(c^*) > c^* \end{cases}$$

for all c_j's ($j = 1, 2, \ldots, N(F)$) satisfying $f(c_j) \in \operatorname{int} J_m$.

Also,

(H1') $\{F(c_i);\ i = 1, 2 \ldots, N(F)\} \cap \operatorname{int} J_m = \emptyset$, *or*
(H2') There is a point $c^* \in \operatorname{int} J_m$ such that

$$\begin{cases} F(c_j) \in (F \circ (f|_{K(F)})(c^*), F(c^*)], & \text{if } F(c^*) < c^* \text{ or} \\ F(c_j) \in (F \circ (f|_{K(F)})^{-1}(c^*), F(c^*)], & \text{if } F(c^*) > c^* \end{cases}$$

for all c_j's ($j = 1, 2, \ldots, N(F)$) satisfying $F(c_j) \in \operatorname{int} J_m$.

Theorem 10.2 *Suppose that the mapping $F \in \mathscr{PM}[I]$ with $N(F) \geq 1$ is strictly increasing on its characteristic interval $K(F)$. Assume that $K(F) = \mathrm{Fix}(F) \cup J_1 \cup J_2 \cup \ldots \cup J_d$, where $\mathrm{Fix}(F)$ is the set of all fixed points of F and J_m's ($m = 1, 2, \ldots d$) are pairwise different intervals with endpoints being fixed points of F and interiors without fixed points. Suppose that a continuous iterative root f of F is strictly increasing on $K(F)$. Moreover, let F and f satisfy either*

(H3) $\{F(c_i);\ i = 0, 1, 2 \ldots, N(F) + 1\} \cap \operatorname{int} J_m = \emptyset$, *or*
(H4) There is a point $c^* \in \operatorname{int} J_m$ such that

$$\begin{cases} F(c_j) \in (F \circ (f|_{K(F)})(c^*), F(c^*)], & \text{if } F(c^*) < c^* \text{ or} \\ F(c_j) \in (F \circ (f|_{K(F)})^{-1}(c^*), F(c^*)], & \text{if } F(c^*) > c^* \end{cases}$$

for all c_j's ($j = 0, 1, 2, \ldots, N(F) + 1$) satisfying $F(c_j) \in \operatorname{int} J_m$.

Then, F is conjugate to f.

10.4 Different Definitions of Stability of the Translation Equation

The question of Ulam, concerning the stability of group homomorphisms, posed in 1940, and the partial affirmative answer of Hyers [5] is often considered as the origin of the theory of stability of functional equations. But, even in these papers: [5, 21, 22], the precise formulation of what to understand as stability differs. Moszner devoted a few papers to define different kind of stabilities and examined the relations between them. See [10, 12, 14–17]. In this section, we present some of the results concerning the different stabilities of the translation equation and, in the next section, of the systems of functional equations defining (equivalently) dynamical systems (see [17, 18] and [16]).

In this section, let (S, d) be a metric space, $(G, \cdot)$ a groupoid. We start with reminding some definitions.

Definition 10.1 We say that the translation equation is *stable in the Hyers–Ulam sense* (shortly *stable*) if there exists a function $\Phi\colon (0,\infty) \to (0,\infty)$ (called *measure of stability*) such that for every $\varepsilon > 0$ and every function $H\colon G \times S \to S$, if

$$d(H(x, H(y, \alpha)), H(x \cdot y, \alpha)) \leq \Phi(\varepsilon), \quad \alpha \in S,\ x, y \in G,$$

then there exists a solution $F\colon G \times S \to S$ of the translation equation

$$F(x, F(y, \alpha)) = F(x \cdot y, \alpha) \tag{10.1}$$

such that

$$d(G(x, \alpha), F(x, \alpha)) \leq \varepsilon, \quad x \in G,\ \alpha \in S.$$

Moreover, if there exists such a function Φ which is unbounded, we say that Equation (10.1) is *normally stable.*

If there exists such Φ of the form $\Phi(\varepsilon) = K\varepsilon$, we say that Equation (10.1) is *strongly stable.*

Definition 10.2 We say that the translation equation is *uniformly b-stable* if there exists a function $\Psi\colon (0,\infty) \to (0,\infty)$ (called *measure of uniform b-stability*), such that for every $\delta > 0$ and every function $H\colon G \times S \to S$, if

$$d(H(x, H(y, \alpha)), H(x \cdot y, \alpha)) \leq \delta, \quad \alpha \in S,\ x, y \in G,$$

then there exists a solution $F\colon G \times S \to S$ of the translation equation (10.1) such that

$$d(H(x, \alpha), F(x, \alpha)) \leq \Psi(\delta), \quad \alpha \in S,\ x \in G.$$

Moreover, if there exists such a function Ψ which is unbounded, we say that the *uniform b-stability is normal.*

If there exists such Ψ of the form $\Psi(\delta) = k\delta$, we say that Equation (10.1) is *strongly b-stable.*

Definition 10.3 We say that the translation equation is *b-stable* if for every function $H: G \times S \to S$ such that

$$G \times G \times S \ni (x, y, \alpha) \mapsto d(H(x, H(y, \alpha)), H(x \cdot y, \alpha))$$

is bounded there exists a solution F of (10.1) such that

$$G \times S \ni (x, \alpha) \mapsto d(H(x, \alpha), F(x, \alpha))$$

is bounded.

Notice that uniform b-stability implies b-stability.

We have the following results concerning these notions.

Theorem 10.3 (1–4 in [17], 5 in [19] and [1])

1. *If the stability of (10.1) is normal, then this equation is uniformly b-stable.*
2. *Stable equation (10.1) does not need to be necessarily b-stable.*
3. *If the b-stability of (10.1) is uniform and normal, then this equation is normally stable.*
4. *Uniform b-stability of (10.1) does not necessarily imply stability.*
5. *The translation equation is normally stable with $\Phi(\varepsilon) = \varepsilon/10$ and normally uniformly b-stable with $\Psi(\delta) = 10\delta$, in the class of continuous functions with $(G, \cdot) = (\mathbb{R}, +)$ and S being a real interval.*

10.5 Stability of Dynamical Systems

In this section, we confine ourselves to continuous function $\mathbb{R} \times I \to I$, where $I \subset \mathbb{R}$ is nondegenerate interval. Such class of function is natural for consideration of dynamical systems.

Definition 10.4 The continuous function $F: \mathbb{R} \times I \to I$ is called *dynamical system* if F is a solution of the translation equation

$$F(s, F(t, x)) = F(s + t, x), \quad s, t \in \mathbb{R},\ x \in I, \tag{10.2}$$

and satisfies *one* or (equivalently, as it appears), *every*, of the following conditions:

1. $F(0, x) = x$, for $x \in I$,
2. $(F^0)'(x) = 1$, for $x \in I$, where $F^0 = F(0, \cdot)$,

3. $I \ni x \mapsto F(0,x)$ is strictly increasing,
4. $(F^0)'$ exists, and
5. F is a surjection.

Hence, we can consider stability problem for systems: translation equation and one of the equations appearing in the first two of the above conditions; and stability problem of the translation equation in the class of functions described by one of the last three of the above conditions. Full research on this topic can be found in [16–18]. Here, we present the selected results. The definitions from the previous section can be complemented by the notion of *restricted uniform b-stability* (definition almost the same as the definition of uniform b-stability, only the function Ψ is defined on some interval $(0, \delta_0)$ instead of on the whole positive halfline).

Theorem 10.4 *1. The translation equation is normally stable and normally uniformly b-stable in the class of surjective functions.*

2. *The translation equation is not stable in any of the classes: such F that F^0 is strictly increasing, and such F that the derivative of F^0 exists.*
3. *The translation equation is b-stable, uniformly b-stable, restrictedly uniformly b-stable, and normally uniformly b-stable only for I bounded , in both classes: such F that F^0 is strictly increasing, and such F that the derivative of F^0 exists.*
4. *The system of equations: "(10.2) & $(F^0)' \equiv 1$" is stable and restrictedly normally uniformly b-stable for every I; normally stable, normally uniformly b-stable, b-stable, and uniformly b-stable only for I bounded.*
5. *The system of equations: "(10.2) & $F^0 = \mathrm{id}$" is stable and normally stable only for $I = \mathbb{R}$; b-stable, uniformly b-stable, restrictedly uniformly b-stable, and normally uniformly b-stable only for I bounded and $I = \mathbb{R}$.*

10.6 Approximate Continuous Solutions of the Translation Equation

In this section, we concentrate only on a class of continuous function $\mathbb{R} \times I \to I$, where $I \subset \mathbb{R}$ is a nondegenerate interval.

In paper [20], there were listed some conditions which every approximate continuous solution of the translation equation, G, satisfies. These conditions show similarities between an exact solution and approximate solution of the translation equation. One of them is the existence of an exact solution of the translation equation in some neighbourhood of G. It is of interest that assuming only the existence of a solution of the translation equation in a neighbourhood of $G: \mathbb{R} \times I \to I$ does not suffice to obtain that G satisfies the translation equation approximately. More precisely, in paper [18] it was shown that

- The translation equation is not *inversely stable* (i.e. it is **not** true that

  for every $\varepsilon > 0$ *there exists a* $\delta > 0$ *such that for every continuous function* $H: \mathbb{R} \times I \to I$ *if there exists a continuous solution* F *of the translation equation such that*

$$|F(t,x) - H(t,x)| \leq \delta, \qquad t \in \mathbb{R},\ x \in I,$$

  then

$$|H(s, H(t,x)) - H(t+s,x)| \leq \varepsilon, \qquad s,t \in \mathbb{R},\ x \in I);$$

- The translation equation is not *inversely b-stable* for unbounded intervals I (i.e. it is **not** true that

  for every continuous $F, H: \mathbb{R} \times I \to I$ *if* F *is a solution of the translation equation and*

$$\mathbb{R} \times I \ni (t,x) \mapsto |F(t,x) - H(t,x)|$$

  is bounded, then

$$\mathbb{R} \times \mathbb{R} \times I \ni (t,s,x) \mapsto |H(s, H(t,x)) - H(t+s,x)|$$

  is bounded); *and*
- The translation equation is not *inversely uniformly b-stable* for unbounded intervals I (i.e. it is **not** true that

  for every $\delta > 0$ *there exists a* $\varepsilon > 0$ *such that for every continuous function* $H: \mathbb{R} \times I \to I$ *if there exists a continuous solution* F *of the translation equation such that*

$$|F(t,x) - H(t,x)| \leq \delta, \qquad t \in \mathbb{R},\ x \in I,$$

  then

$$|H(s, H(t,x)) - H(t+s,x)| \leq \varepsilon, \qquad s,t \in \mathbb{R},\ x \in I).$$

Now, we are going to remind the characterization of a continuous solution of the translation equation (Theorem 10.5). Next, we present the necessary (Theorem 10.6) and sufficient condition (Theorem 10.7) for satisfying the translation equation approximately.

Theorem 10.5 *Let* $F: \mathbb{R} \times I \to I$ *be a solution of the translation equation, i.e.*

$$F(s, F(t,x)) = F(s+t,x), \quad s,t \in \mathbb{R}, x \in I.$$

Put $V = H(\mathbb{R} \times I)$. Then, there exist open, disjoint, intervals $U_n \subset V$ and homeomorphisms $h_n: \mathbb{R} \to U_n$ such that for every $x \in U_n$

$$F(t,x) = h_n(h_n^{-1}(x) + t), \quad t \in \mathbb{R},$$

and

$$F(t,x) = x, \quad x \in V \setminus \bigcup_n U_n, \; t \in \mathbb{R}.$$

Moreover, there exists a continuous function $f: I \to V$, such that $f(x) = x$ for $x \in V$ and

$$F(t,x) = F(t,f(x)), \quad t \in \mathbb{R}, x \in I \setminus V.$$

Conversely, for every continuous $f: I \to I$ such that $f \circ f = f$, a family of open, disjoint intervals $\{U_n;\ n \in N \subset \mathbb{N}\}$ such that $U_n \subset f(I)$, and a family of homeomorphisms $h_n: \mathbb{R} \to U_n$, $n \in N$, every function of the form

$$F(t,x) = \begin{cases} h_n(h_n^{-1}(f(x)) + t), & \text{if } f(x) \in U_n,\ t \in \mathbb{R}; \\ f(x), & \text{if } f(x) \notin \bigcup_{n \in N} U_n,\ t \in \mathbb{R} \end{cases}$$

is a continuous solution of the translation equation.

Theorem 10.6 *Suppose that $H: \mathbb{R} \times I \to I$ is a continuous solution of*[4]

$$|H(s,H(t,x)) - H(s+t,x)| \le \delta, \quad x \in I,\ s,t \in \mathbb{R}.$$

Then,

(a) There exist open, disjoint intervals $U_n \subset I$, $n \in N$, of the length greater or equal to 6δ, homeomorphisms $h_n: \mathbb{R} \to U_n$, $n \in N$, and a continuous function $f: I \to I$, such that $f \circ f = f$, $U_n \subset f(I)$, $n \in N$,

$$|H(t,x) - f(x)| \le 10\delta, \qquad t \in \mathbb{R},\ f(x) \notin \bigcup_{n \in N} U_n,$$

$$|H(t,x) - h_n(h_n^{-1}(f(x)) + t)| \le 10\delta, \qquad t \in \mathbb{R},\ f(x) \in U_n,\ n \in N;$$

(b) $\forall_{(x \in I,\, n \in N)}\ (f(x) \in U_n \Rightarrow H(\mathbb{R},x) = U_n)$;
(c) $\forall_{(x \in I,\, n \in N)}\ (x \in U_n \Rightarrow f(x) = x)$;
(d) $\forall_{(x \in I,\, t \in \mathbb{R})}\ (|f(H(t,x)) - H(t,x)| \le 2\delta)$;

[4]The proof of this theorem can be found in [19] and [20], and the construction of homeomorphisms $h_n: \mathbb{R} \to U_n$ was done in [1].

(e) $\forall_{x\in I}\left(f(x)\notin\bigcup_{n\in N}U_n \Rightarrow \left(\forall_{t\in\mathbb{R}} f(H(t,x))\notin\bigcup_{n\in N}U_n\right)\right)$;

(f) $\forall_{x\in I}\left(f(x)\notin\bigcup_{n\in N}U_n \Rightarrow (\forall_{s_1,s_2\in\mathbb{R}}\,|H(s_1,x)-H(s_2,x)|\leq 6\delta)\right)$;

(g) *The set of values of function f, V_f, is contained in the set of values of function H, V_H, i.e. $V_f \subset V_H$;*

(h) *Every interval U_n is "invariant", more precisely*

$$H(\mathbb{R},x)=U_n, \qquad x\in U_n,\ n\in N,$$

and

$$H(t,U_n)=U_n, \qquad t\in\mathbb{R},\ n\in N;$$

(i) *For every $n\in N$, put $a_n := \inf U_n$, $b_n := \sup U_n$. Either h_n is an increasing homeomorphism,*

$$\lim_{t\to\infty}H(t,x)=b_n, \quad \lim_{t\to-\infty}H(t,x)=a_n, \qquad x\in U_n,$$

and $H(\cdot,x)$ "almost increases", i.e. for every $t\in\mathbb{R}$ we have $H(s,x)>H(t,x)-2\delta$ for $s>t$; or h_n is a decreasing homeomorphism,

$$\lim_{t\to\infty}H(t,x)=a_n, \quad \lim_{t\to-\infty}H(t,x)=b_n, \qquad x\in U_n,$$

and $H(\cdot,x)$ "almost decreases", i.e. for every $t\in\mathbb{R}$ we have $H(s,x)<H(t,x)+2\delta$ for $s>t$;

(j) *For every $n\in N$*

$$H(t,a_n)=a_n,\ H(t,b_n)=b_n, \qquad t,\in\mathbb{R},$$

whenever a_n, b_n are in I;

(k) *For every $x\in I$ such that $x\notin\bigcup_{n\in N}U_n$ but there are $n,m\in N$ with $b_n\leq x\leq a_m$, we have*

$$|H(t,x)-x|\leq 6\delta, \qquad t\in\mathbb{R};$$

(l)

$$|H(t,x)-H(t,f(x))|\leq 10\delta, \qquad t\in\mathbb{R},\ x\in I;\text{and}$$

(m) *Moreover, for every $n \in N$ there are two possibilities:*

- *Either there exists $\eta_n > 0$ such that*

$$|t_1 - t_2| \le \eta_n \Rightarrow |h_n(t_1) - h_n(t_2)| \le 21\delta, \quad t_1, t_2 \in \mathbb{R}, \tag{10.3}$$

for $\eta_n^* := \sup\{\eta_n > 0 : (10.3) \text{ holds}\} \in (0, \infty]$ we have[5]

$$h_n(t-\eta_n^*+h_n^{-1}(f(x))) \le H(t,x) \le h_n(t+\eta_n^*+h_n^{-1}(f(x))), \quad t \in \mathbb{R}, f(x) \in U_n,$$

if h_n is increasing,

$$h_n(t-\eta_n^*+h_n^{-1}(f(x))) \ge H(t,x) \ge h_n(t+\eta_n^*+h_n^{-1}(f(x))), \quad t \in \mathbb{R}, f(x) \in U_n,$$

if h_n is decreasing,

- *or such η_n, for which (10.3) holds, does not exist and*

$$H(t,x) = h_n(t + h_n^{-1}(f(x)), \quad t \in \mathbb{R}, f(x) \in U_n.$$

Theorem 10.7 *Let I be a nondegenerate real interval, $\delta, A_1, A_2, B, C, D > 0$, suppose that $H: \mathbb{R} \times I \to I$ is a continuous function. If*

(a) *There exist open, disjoint intervals $U_n \subset I$, $n \in N$, homeomorphisms $h_n: \mathbb{R} \to U_n$, $n \in N$, and a continuous function $f: I \to I$, such that $f \circ f = f$, $U_n \subset f(I)$, $n \in N$,*

$$|H(t,x) - f(x)| \le A_1\delta, \qquad t \in \mathbb{R}, f(x) \notin \bigcup_{n\in N} U_n,$$

$$|H(t,x) - h_n(h_n^{-1}(f(x)) + t)| \le A_2\delta, \qquad t \in \mathbb{R}, f(x) \in U_n, n \in N;$$

(b) $\forall_{(x\in I, n\in N)} \, (f(x) \in U_n \Rightarrow H(\mathbb{R}, x) \subset U_n)$;

(c) $\forall_{(x\in I, n\in N)} \, (x \in U_n \Rightarrow f(x) = x)$;

(d) $\forall_{(x\in I, t\in\mathbb{R})} \, (|f(H(t,x)) - H(t,x)| \le B\delta)$;

(e) $\forall_{x\in I} \left(f(x) \notin \bigcup_{n\in N} U_n \Rightarrow \left(\forall_{t\in\mathbb{R}} f(H(t,x)) \notin \bigcup_{n\in N} U_n \right)\right)$;

(f) $\forall_{x\in I} \left(f(x) \notin \bigcup_{n\in N} U_n \Rightarrow (\forall_{s_1,s_2\in\mathbb{R}} \, |H(s_1,x) - H(s_2,x)| \le C\delta) \right)$; *and*

(g) *Moreover, for every $n \in N$ there are two possibilities:*

- *Either there exists $\eta_n > 0$ such that*

$$|t_1 - t_2| \le \eta_n \Rightarrow |h_n(t_1) - h_n(t_2)| \le D\delta, \qquad t_1, t_2 \in \mathbb{R}, \tag{10.4}$$

[5] If $\eta_n^* = \infty$, then by $h_n(\pm\infty)$ we understand $\lim_{t\to\pm\infty} h_n(t)$.

for $\eta_n^* := \sup\{\eta_n > 0 :$ *(10.4) holds*$\} \in (0, \infty]$ *we have*[6]

$$h_n(t - \eta_n^* + h_n^{-1}(f(x))) \le H(t,x) \le h_n(t + \eta_n^* + h_n^{-1}(f(x))), \qquad t \in \mathbb{R}, f(x) \in U_n,$$

if h_n *is increasing,*

$$h_n(t - \eta_n^* + h_n^{-1}(f(x))) \ge H(t,x) \ge h_n(t + \eta_n^* + h_n^{-1}(f(x))), \qquad t \in \mathbb{R}, f(x) \in U_n,$$

if h_n *is decreasing,*

- *or such* η_n, *for which (10.4) holds, does not exist and*

$$H(t,x) = h_n(t + h_n^{-1}(f(x)), \qquad t \in \mathbb{R}, f(x) \in U_n,$$

then

$$|H(s, H(t,x)) - H(t+s,x)| \le E\delta, \qquad s,t \in \mathbb{R}, x \in I,$$

where $E := \max\{(2A_2 + D), \min\{3A_1 + B, A_1 + B + C\}\}$.

Acknowledgements This research is a part of *Iterative functional equations and real analysis* program (Institute of Mathematics, University of Silesia, Katowice, Poland).

References

1. Chudziak, J.: Approximate dynamical systems on interval. Appl. Math. Lett. **25**, 532–537 (2012)
2. Ciepliński, K.: On the embeddability of a homeomorphism of the unit circle in disjoint iteration groups. Publ. Math. Debrecen **55**, 363–383 (1999)
3. Ciepliński, K., Zdun, M.C.: On a system of Schr´oder equations on a circle. Int. J. Bifurcation Chaos Appl. Sci. Eng. **13**, 1883–1888 (2003)
4. Cornfeld, I.P., Fomin, S.V., Sinai, Ya.G.: Ergodic Theory. Springer, New York (1982)
5. Hyers, D.H.: On the stability of the linear functional equation. Proc. Natl. Acad. Sci. USA **27**, 222–224 (1941)
6. Jarczyk, W.: Recent results on iterative roots. In: Jarczyk, W., Fournier-Prunaret, D., Cabral, J.M.G. (eds.) ECIT 2012, 19th European Conference on Iteration Theory. ESAIM Proceedings of Surveys, vol. 46, pp. 47–62. EDP Science, Les Ulis (2014)
7. Krassowska, D., Zdun, M.C.: Embeddability of homeomorphisms of the circle in set-valued iteration groups. J. Math. Anal. Appl. **433**, 1647–1658 (2016)
8. Li, L., Zhang, W.M.: Conjugacy between piecewise monotonic functions and their iterative roots. Sci. China Math. **59**, 367–378 (2016)
9. Liu, L., Zhang, W.: Non-monotonic iterative roots extended from characteristic intervals. J. Math. Anal. Appl. **378**, 359–373 (2011)

[6] If $\eta_n^* = \infty$, then by $h_n(\pm\infty)$ we understand $\lim_{t \to \pm\infty} h_n(t)$.

10. Mach, A., Moszner, Z.: Unstable (stable) system of stable (unstable) functional equations. Ann. Univ. Paedagog. Crac. Stud. Math. **9**, 43–47 (2010)
11. Moszner, Z.: The translation equation and its application. Demonstratio Math. **6**, 309–327 (1973)
12. Moszner, Z.: Sur la définition de Hyers de la stabilité de l'équation fonctionnelle. Opuscula Math. **3**, 47–57 (1987)
13. Moszner, Z.: General theory of the translation equation. Aequationes Math. **50**, 17–37 (1995)
14. Moszner, Z.: On the stability of functional equations. Aequationes Math. **77**, 33–88 (2009)
15. Moszner, Z.: On the inverse stability of functional equations. In: Brzdęk, J., Chmieliński, J., Ciepliński, K., Ger, R., Páles, Z., Zdun, M.C. (eds.) Recent Developments in Functional Equations and Inequalities. Banach Center Publications, vol. 99, , pp. 111–121. Polish Academy of Sciences, Institute of Mathematics, Warsaw (2013)
16. Moszner, Z.: Stability has many names. Aequationes Math. **90**, 983–999 (2016)
17. Moszner, Z.: On the normal stability of functional equations. Ann. Math. Silesianae **30**, 111–128 (2016)
18. Moszner, Z., Przebieracz, B.: Is the dynamical system stable? Aequationes Math. **89**, 279–296 (2015)
19. Przebieracz, B.: On the stability of the translation equation and dynamical systems. Nonlinear Anal. **75**, 1980–1988 (2012)
20. Przebieracz, B.: A characterization of the approximate solutions of the translation equation. J. Differ. Equ. Appl. **21**, 1058–1067 (2015)
21. Ulam, S.M.: A Collection of Mathematical Problems. Interscience Publishers, New York (1960)
22. Ulam, S.M.: Problems in Modern Mathematics. Science Editions Wiley, New York (1964)
23. Walters, P.: An Introduction to Ergodic Theory. Springer, New York (1982)
24. Zdun, M.C.: On embedding of homeomorphism of the circle in continuous flow. In: Liedl, R., Reich, L., Targonski, G. (eds.) Iteration Theory and Its Functional Equations (Lochau, 1984), vol. 1163, pp. 218–231. Lecture Notes in Mathematics. Springer, Berlin (1985)
25. Zdun, M.C.: The embedding problem in iteration theory. In: Jarczyk, W., Fournier-Prunaret, D., Cabral, J.M.G. (eds.) ECIT 2012, 19th European Conference on Iteration Theory. ESAIM Proceedings Surveys, vol. 46, pp. 86–97. EDP Science, Les Ulis (2014)
26. Zdun, M.C., Solarz, P.: Recent results on iteration theory: iteration groups and semigroups in the real case. Aequationes Math. **87**, 201–245 (2014)
27. Zhang, W.: PM functions, their characteristic intervals and iterative roots. Ann. Polon. Math. **65**, 119–128 (1997)
28. Zhang, J.Z., Yang, L.: Iterative roots of a piecewise monotone continuous self-mapping. (Chinese) Acta Math. Sinica **26**, 398–412 (1983)
29. Zhang, J.Z., Yang, L., Zhang, W.: Some advances on functional equations. Adv. Math. (China) **24**, 385–405 (1995)

Chapter 11
On Some Recent Applications of Stochastic Convex Ordering Theorems to Some Functional Inequalities for Convex Functions: A Survey

Teresa Rajba

Abstract This is a survey paper concerning some theorems on stochastic convex ordering and their applications to functional inequalities for convex functions. We present the recent results on those subjects.

Keywords Convex functions • Higher-order convex functions • Hermite–Hadamard inequalities • Convex stochastic ordering

Mathematics Subject Classification (2010) Primary 26A51; Secondary 26D10, 39B62

11.1 Introduction

In the present paper, we look at Hermite–Hadamard type inequalities from the perspective provided by the stochastic convex order. This approach is mainly due to Cal and Cárcamo. In the paper [12], the Hermite–Hadamard type inequalities are interpreted in terms of the convex stochastic ordering between random variables. Recently, also in [19, 32, 35–38, 40–42], the Hermite–Hadamard inequalities are studied based on the convex ordering properties. Here, we want to attract the reader's attention to some selected topics by presenting some theorems on the convex ordering that can be useful in the study of the Hermite–Hadamard type inequalities.

The Ohlin lemma [31] on sufficient conditions for convex stochastic ordering was first used in [36], to get a simple proof of some known Hermite–Hadamard

---

T. Rajba (✉)
Department of Mathematics, University of Bielsko-Biala, Willowa 2,
43-309 Bielsko-Biala, Poland
e-mail: trajba@ath.bielsko.pl

J. Brzdęk et al. (eds.), *Developments in Functional Equations and Related Topics*, Springer Optimization and Its Applications 124,
DOI 10.1007/978-3-319-61732-9_11

type inequalities as well as to obtaining new Hermite–Hadamard type inequalities. In [32, 41, 42], the authors used the Levin–Stečkin theorem [25] to study Hermite–Hadamard type inequalities.

Many results on higher-order generalizations of the Hermite–Hadamard type inequality one can find, among others, in [1–5, 16, 36, 37]. In recent papers [36, 37], the theorem of Denuit, Lefèvre, and Shaked [13] was used to prove Hermite–Hadamard type inequalities for higher-order convex functions. The theorem of Denuit, Lefèvre, and Shaked [13] on sufficient conditions for s-convex ordering is a counterpart of the Ohlin lemma concerning convex ordering. A theorem on necessary and sufficient conditions for higher-order convex stochastic ordering, which is a counterpart of the Levin–Stečkin theorem [25] concerning convex stochastic ordering, is given in the paper [38]. Based on this theorem, useful criteria for the verification of higher-order convex stochastic ordering are given. These criteria can be useful in the study of Hermite–Hadamard type inequalities for higher-order convex functions, and in particular inequalities between the quadrature operators. They may be easier to verify the higher-order convex orders, than those given in [13, 22].

In Section 11.2, we give simple proofs of known as well as new Hermite–Hadamard type inequalities, using Ohlin's lemma and the Levin–Stečkin theorem.

In Sections 11.3 and 11.4, we study inequalities of the Hermite–Hadamard type involving numerical differentiation formulas of the first order and the second order, respectively.

In Section 11.5, we give simple proofs of Hermite–Hadamard type inequalities for higher-order convex functions, using the theorem of Denuit, Lefèvre, and Shaked, and a generalization of the Levin–Stečkin theorem to higher orders. These results are applied to derive some inequalities between quadrature operators.

11.2 Some Generalizations of the Hermite–Hadamard Inequality

Let $f\colon [a,b] \to \mathbb{R}$ be a convex function ($a, b \in \mathbb{R}$, $a < b$). The following double inequality

$$f\left(\frac{a+b}{2}\right) \le \frac{1}{b-a}\int_a^b f(x)\,dx \le \frac{f(a)+f(b)}{2} \tag{11.1}$$

is known as the Hermite–Hadamard inequality (see [16] for many generalizations and applications of (11.1)).

In many papers, the Hermite–Hadamard type inequalities are studied based on the convex stochastic ordering properties (see, for example, [19, 32, 35–37, 40, 41]). In the paper [36], the Ohlin lemma on sufficient conditions for convex stochastic ordering is used to get a simple proof of some known Hermite–Hadamard type inequalities as well as to obtain new Hermite–Hadamard type inequalities. Recently, the Ohlin lemma is also used to study the inequalities of the Hermite–Hadamard type for convex functions in [32, 35, 40, 41]. In [37], also the inequalities of the Hermite–Hadamard type for delta-convex functions are studied by using the Ohlin lemma. In the papers [32, 40, 41], furthermore, the Levin–Stečkin theorem [25] (see also [30]) is used to examine the Hermite–Hadamard type inequalities. This theorem gives necessary and sufficient conditions for the stochastic convex ordering.

Let us recall some basic notions and results on the stochastic convex order (see, for example, [13]). As usual, F_X denotes the distribution function of a random variable X and μ_X is the distribution corresponding to X. For real-valued random variables X, Y with a finite expectation, we say that X is dominated by Y in *convex ordering* sense, if

$$\mathbb{E}f(X) \leq \mathbb{E}f(Y)$$

for all convex functions $f: \mathbb{R} \to \mathbb{R}$ (for which the expectations exist). In that case, we write $X \leq_{cx} Y$, or $\mu_X \leq_{cx} \mu_Y$.

In the following Ohlin's lemma [31], are given sufficient conditions for convex stochastic ordering.

Lemma 11.1 (Ohlin [31]) *Let X, Y be two random variables such that $\mathbb{E}X = \mathbb{E}Y$. If the distribution functions F_X, F_Y cross exactly one time, i.e., for some x_0 holds*

$$F_X(x) \leq F_Y(x) \text{ if } x < x_0 \quad \text{and} \quad F_X(x) \geq F_Y(x) \text{ if } x > x_0,$$

then

$$\mathbb{E}f(X) \leq \mathbb{E}f(Y) \tag{11.2}$$

for all convex functions $f: \mathbb{R} \to \mathbb{R}$.

The inequality (11.1) may be easily proved with the use of the Ohlin lemma (see[36]). Indeed, let X, Y, Z be three random variables with the distributions $\mu_X = \delta_{(a+b)/2}$, μ_Y which is equally distributed in $[a, b]$ and $\mu_Z = \frac{1}{2}(\delta_a + \delta_b)$, respectively. Then, it is easy to see that the pairs (X, Y) and (Y, Z) satisfy the assumptions of the Ohlin lemma, and using (11.2), we obtain (11.1).

Let $a < c < d < b$. Let $f: I \to \mathbb{R}$ be a convex function, $a, b \in I$. Then (see [21]),

$$\frac{f(c)+f(d)}{2} - f\left(\frac{c+d}{2}\right) \leq \frac{f(a)+f(b)}{2} - f\left(\frac{a+b}{2}\right). \tag{11.3}$$

To prove (11.3) from the Ohlin lemma, it suffices to take random variables X, Y (see [27]) with

$$\mu_X = \frac{1}{4}(\delta_c + \delta_d) + \frac{1}{2}\delta_{(a+b)/2},$$

$$\mu_Y = \frac{1}{4}(\delta_a + \delta_b) + \frac{1}{2}\delta_{\delta_{(c+d)/2}}.$$

Then, by Lemma 11.1, we obtain

$$\frac{f(c)+f(d)}{2}+f\left(\frac{a+b}{2}\right)\leq\frac{f(a)+f(b)}{2}+f\left(\frac{c+d}{2}\right), \tag{11.4}$$

which implies (11.3).

Similarly, it can be proved the Popoviciu inequality

$$\frac{2}{3}\left[f\left(\frac{x+y}{2}\right)+f\left(\frac{y+z}{2}\right)+f\left(\frac{z+x}{2}\right)\right]\leq\frac{f(x)+f(y)+f(z)}{3}+f\left(\frac{x+y+z}{3}\right), \tag{11.5}$$

where $x, y, z \in I$ and $f: I \to \mathbb{R}$ is a convex function. To prove (11.5) from the Ohlin lemma, it suffices (assuming $x \leq y \leq z$) to take random variables X, Y (see [27]) with

$$\mu_X = \frac{1}{4}\left(\delta_{(x+y)/2} + \delta_{(y+z)/2} + \delta_{(z+x)/2}\right),$$

$$\mu_Y = \frac{1}{6}\left(\delta_x + \delta_y + \delta_z\right) + \frac{1}{2}\delta_{(x+y+z)/3}.$$

Convexity has a nice probabilistic characterization, known as Jensen's inequality (see [6]).

Proposition 11.1 ([6]) *A function $f: (a, b) \to \mathbb{R}$ is convex if, and only if,*

$$f(\mathbb{E}X) \leq \mathbb{E}f(X) \tag{11.6}$$

for all (a, b)-valued integrable random variables X.

To prove (11.6) from the Ohlin lemma, it suffices to take a random variable Y (see [35]) with

$$\mu_Y = \delta_{\mathbb{E}X},$$

then we have

$$\mathbb{E}f(Y) = f(\mathbb{E}X). \tag{11.7}$$

By the Ohlin lemma, we obtain $\mathbb{E}f(Y) \leq \mathbb{E}f(X)$, then taking into account (11.7), this implies (11.6).

Remark 11.1 Note that in [29], the Ohlin lemma was used to obtain a solution of the problem of Raşa concerning inequalities for Bernstein operators.

In [17], Fejér gave a generalization of the inequality (11.1).

Proposition 11.2 ([17]) *Let $f: I \to \mathbb{R}$ be a convex function defined on a real interval I, $a, b \in I$ with $a < b$ and let $g: [a, b] \to \mathbb{R}$ be nonnegative and symmetric with respect to the point $(a + b)/2$ (the existence of integrals is assumed in all formulas). Then,*

$$f\left(\frac{a+b}{2}\right)\cdot\int_a^b g(x)\,dx \le \int_a^b f(x)g(x)\,dx \le \frac{f(a)+f(b)}{2}\cdot\int_a^b g(x)\,dx. \qquad (11.8)$$

The double inequality (11.8) is known in the literature as the Fejér inequality or the Hermite–Hadamard-Fejér inequality (see [16, 28, 33] for the historical background).

Remark 11.2 ([36]) Using the Ohlin lemma (Lemma 11.1), we get a simple proof of (11.8). Let f and g satisfy the assumptions of Proposition 11.2. Let X, Y, Z be three random variables such that $\mu_X = \delta_{(a+b)/2}$, $\mu_Y(dx) = (\int_a^b g(x)dx)^{-1}g(x)dx$, $\mu_Z = \frac{1}{2}(\delta_a + \delta_b)$. Then, by Lemma 11.1, we obtain that $X \le_{cx} Y$ and $Y \le_{cx} Z$, which implies (11.8).

Remark 11.3 Note that for $g(x) = w(x)$ such that $\int_a^b w(x)dx = 1$, the inequality (11.8) can be rewritten in the form

$$f\left(\frac{a+b}{2}\right) \le \int_a^b f(x)w(x)dx \le \frac{f(a)+f(b)}{2}. \qquad (11.9)$$

Conversely, from the inequality (11.9), it follows (11.8). Indeed, if $\int_a^b g(x)dx > 0$, it suffices to take $w(x) = \left(\int_a^b g(x)dx\right)^{-1} g(x)$. If $\int_a^b g(x)dx = 0$, then (11.8) is obvious.

For various modifications of (11.1) and (11.8), see, e.g., [3–5, 10, 11, 16], and the references given there.

As Fink noted in [18], one wonders what the symmetry has to do with the inequality (11.8) and if such an inequality holds for other functions (cf. [16, p. 53]).

As an immediate consequence of Lemma 11.1, we obtain the following theorem, which is a generalization of the Fejér inequality.

Theorem 11.1 ([36]) *Let $0 < p < 1$. Let $f: I \to \mathbb{R}$ be a convex function, $a, b \in I$ with $a < b$. Let μ be a finite measure on $\mathscr{B}([a,b])$ such that: (i) $\mu([a, pa+qb]) \le pP_0$, (ii) $\mu((pa+qb, b]) \le qP_0$, and (iii) $\int_{[a,b]} x\mu(dx) = (pa+qb)P_0$, where $q = 1-p$, $P_0 = \mu([a,b])$. Then,*

$$f(pa+qb)P_0 \le \int_{[a,b]} f(x)\mu(dx) \le [pf(a)+qf(b)]P_0. \qquad (11.10)$$

Fink proved in [18] a general weighted version of the Hermite–Hadamard inequality. In particular, we have the following probabilistic version of this inequality.

Proposition 11.3 ([18]) *Let X be a random variable taking values in the interval $[a, b]$ such that m is the expectation of X and μ_X is the distribution corresponding to X. Then,*

$$f(m) \le \int_a^b f(x)\, \mu_X(dx) \le \frac{b-m}{b-a} f(a) + \frac{m-a}{b-a} f(b). \tag{11.11}$$

Moreover, in [19] it was proved that, starting from such a fixed random variable X, we can fill the whole space between the Hermite–Hadamard bounds by highlighting some parametric families of random variables. The authors propose two alternative constructions based on the convex ordering properties.

In [35], based on Lemma 11.1, a very simple proof of Proposition 11.3 is given. Let X be a random variable satisfying the assumptions of Proposition 11.3. Let Y, Z be two random variables such that $\mu_Y = \delta_m$, $\mu_Z = \frac{b-m}{b-a}\delta_a + \frac{m-a}{b-a}\delta_b)$. Then, by Lemma 11.1, we obtain that $Y \le_{cx} X$ and $X \le_{cx} Z$, which implies (11.11).

In [36], some results related to the Brenner–Alzer inequality are given. In the paper [23] by Klaričić Bakula, Pečarić, and Perić, some improvements of various forms of the Hermite–Hadamard inequality can be found; namely, that of Fejér, Lupas, Brenner–Alzer, and Beesack–Pečarić. These improvements imply the Hammer–Bullen inequality. In 1991, Brenner and Alzer [9] obtained the following result generalizing Fejér's result as well as the result of Vasić and Lacković [43] and Lupas [26] (see also [33]).

Proposition 11.4 ([9]) *Let p, q be the given positive numbers and $a_1 \le a < b \le b_1$. Then, the inequalities*

$$f\left(\frac{pa+qb}{p+q}\right) \le \frac{1}{2y}\int_{A-y}^{A+y} f(t)dt \le \frac{pf(a)+qf(b)}{p+q} \tag{11.12}$$

hold for $A = \frac{pa+qb}{p+q}$, $y > 0$, and all continuous convex functions $f\colon [a_1, b_1] \to \mathbb{R}$ if, and only if,

$$y \le \frac{b-a}{p+q} \min\{p, q\}.$$

Remark 11.4 It is known [33, p. 144] that under the same conditions Hermite–Hadamard's inequality holds, the following refinement of (11.12):

$$f\left(\frac{pa+qb}{p+q}\right) \le \frac{1}{2y}\int_{A-y}^{A+y} f(t)dt \le \frac{1}{2}\{f(A-y)+f(A+y)\} \le \frac{pf(a)+qf(b)}{p+q} \tag{11.13}$$

holds.

In the following theorem, we give some generalization of the Brenner and Alzer inequalities (11.13), which we prove using the Ohlin lemma.

Theorem 11.2 ([36]) *Let p, q be the given positive numbers, $a_1 \le a < b \le b_1$, $0 < y \le \frac{b-a}{p+q}\min\{p, q\}$ and let $f\colon [a_1, b_1] \to \mathbb{R}$ be a convex function. Then,*

$$f\left(\frac{pa+qb}{p+q}\right) \le$$

$$\frac{\alpha}{2}\{f(A-(1-\alpha)y)+f(A+(1-\alpha)y)\}+\frac{1}{2y}\int_{A-(1-\alpha)y}^{A+(1-\alpha)y} f(t)dt \le$$

$$\frac{\alpha}{2n}\sum_{k=1}^{n}\left\{f\left(A-y+k\frac{\alpha y}{n}\right)+f\left(A+y-k\frac{\alpha y}{n}\right)\right\}+\frac{1}{2y}\int_{A-(1-\alpha)y}^{A+(1-\alpha)y} f(t)dt \le$$

$$\frac{1}{2y}\int_{A-y}^{A+y} f(t)dt, \tag{11.14}$$

where $0 \le \alpha \le 1$, $n = 1, 2, \ldots$,

$$\begin{aligned}\frac{1}{2y}\int_{A-y}^{A+y} f(t)dt &\le \frac{\beta}{2}\{f(A-y)+f(A+y)\}+(1-\beta)\frac{1}{2y}\int_{A-y}^{A+y} f(t)dt \\ &\le \frac{1}{2}\{f(A-y)+f(A+y)\},\end{aligned} \tag{11.15}$$

where $0 \le \beta \le 1$,

$$\frac{1}{2}\{f(A-y)+f(A+y)\} \le$$

$$(\frac{1}{2}-\gamma)\{f(A-y-c)+f(A+y+c)\}+\gamma\{f(A-y)+f(A+y)\} \le$$

$$\frac{pf(a)+qf(b)}{p+q}, \tag{11.16}$$

where $c = \min\{b-(A+y), (A-y)-a\}$, $\gamma = \left|\frac{1}{2}-p\right|$.

To prove this theorem, it suffices to consider random variables X, Y, W, Z, ξ_n, η and λ such that:

$$\mu_X = \delta_{\frac{pa+qb}{p+q}},$$

$$\mu_Y(dx) = \frac{1}{2y}\chi_{[A-y,A+y]}(x)dx,$$

$$\mu_Z = \frac{p}{p+q}\delta_a + \frac{q}{p+q}\delta_b,\ \mu_W = \frac{1}{2}\delta_{A-y} + \frac{1}{2}\delta_{A+y},$$

$$\mu_{\xi_n}(dx) = \frac{\alpha}{2n}\sum_{k=1}^{n}\{\delta_{A-y+k\frac{\alpha y}{n}} + \delta_{A+y-k\frac{\alpha y}{n}}\} + \frac{1}{2y}\chi_{[A-(1-\alpha)y,A+(1-\alpha)y]}(x)dx,$$

$$\mu_\eta(dx) = \frac{\beta}{2}\{\delta_{A-y} + \delta_{A+y}\} + \frac{1-\beta}{2y}\chi_{[A-y,A+y]}(x)dx,$$

$$\mu_\lambda = (\frac{1}{2} - \gamma)\{\delta_{A-y-c} + \delta_{A+y+c}\} + \gamma\{\delta_{A-y} + \delta_{A+y}\}.$$

Then, using the Ohlin lemma, we obtain:

- $X \leq_{cx} Y$, $Y \leq_{cx} W$, and $W \leq_{cx} Z$, which implies the inequalities (11.13),
- $X \leq_{cx} \xi_1$, $\xi_1 \leq_{cx} \xi_n$, and $\xi_n \leq_{cx} Y$, which implies (11.14),
- $Y \leq_{cx} \eta$ and $\eta \leq_{cx} W$, which implies (11.15), and
- $W \leq_{cx} \lambda$ and $\lambda \leq_{cx} Z$, which implies (11.16).

Theorem 11.3 ([36]) *Let p, q be the given positive numbers, $0 < \alpha < 1$, $a_1 \leq a < b \leq b_1$, $0 < y \leq \frac{b-a}{p+q}\min\{p,q\}$ and $0 \leq \frac{\alpha}{1-\alpha}y \leq \frac{b-a}{p+q}\min\{p,q\}$. Let $f\colon [a_1,b_1] \to \mathbb{R}$ be a convex function. Then,*

$$\begin{aligned} f(A) &\leq \frac{\alpha}{y}\int_{A-y}^{A} f(t)dt + \frac{(1-\alpha)^2}{\alpha y}\int_{A}^{A+\frac{\alpha}{1-\alpha}y} f(t)dt \\ &\leq \alpha f(A-y) + (1-\alpha)f(A + \frac{\alpha}{1-\alpha}y) \\ &\leq \frac{p}{p+q}f(a) + \frac{q}{p+q}f(b), \end{aligned} \tag{11.17}$$

where $A = \frac{pa+qb}{p+q}$.

Let X, Y, Z, and W be random variables such that:

$$\begin{aligned} \mu_X &= \delta_A, \\ \mu_Y(dx) &= \frac{\alpha}{y}\chi_{[A-y,A]}(x)dx + \frac{(1-\alpha)^2}{\alpha y}\chi_{[A,A+\frac{\alpha}{1-\alpha}y]}(x)dx, \\ \mu_W &= \alpha\delta_{A-y} + (1-\alpha)\delta_{A+\frac{\alpha}{1-\alpha}y}, \\ \mu_Z &= \frac{p}{p+q}\delta_a + \frac{q}{p+q}\delta_b. \end{aligned}$$

Then, using the Ohlin lemma, we obtain $X \leq_{cx} Y$, $Y \leq_{cx} W$, $W \leq_{cx} Z$, which implies the inequalities (11.17).

Remark 11.5 If we choose $\alpha = \frac{1}{2}$ in Theorem 11.3, then the inequalities (11.17) reduce to the inequalities (11.15).

Remark 11.6 If we choose $\alpha = \frac{p}{p+q}$ and $y = (1-p)z$ in Theorem 11.3, then we have

$$f(A) \leq \frac{p}{qz}\int_{A-\frac{q}{p+q}z}^{A} f(t)dt + \frac{q}{pz}\int_{A}^{A+\frac{p}{p+q}z} f(t)dt$$

$$\leq \frac{p}{p+q}f(A - \frac{q}{p+q}z) + \frac{q}{p+q}f(A + \frac{p}{p+q}z)$$
$$\leq \frac{p}{p+q}f(a) + \frac{q}{p+q}f(b),$$

where $A = \frac{pa+qb}{p+q}$, $0 < z \leq b - a$.

In the paper [40], the author used Ohlin's lemma to prove some new inequalities of the Hermite–Hadamard type, which are a generalization of known Hermite–Hadamard type inequalities.

Theorem 11.4 ([40]) *The inequality*

$$af(\alpha x + (1-\alpha)y) + (1-a)f(\beta x + (1-\beta)y) \leq \frac{1}{y-x}\int_x^y f(t)dt, \tag{11.18}$$

with some $a, \alpha, \beta \in [0, 1]$, $\alpha > \beta$ *is satisfied for all* $x, y \in \mathbb{R}$ *and all continuous and convex functions* $f : [x, y] \to \mathbb{R}$ *if, and only if,*

$$a\alpha + (1-a)\beta = \frac{1}{2}, \tag{11.19}$$

and one of the following conditions holds true:

(i) $a + \alpha \leq 1$,
(ii) $a + \beta \geq 1$, *and*
(iii) $a + \alpha > 1$, $a + \beta < 1$, *and* $a + 2\alpha \leq 2$.

Theorem 11.5 ([40]) *Let* $a, b, c, \alpha \in (0, 1)$ *be numbers such that* $a + b + c = 1$. *Then, the inequality*

$$af(x) + bf(\alpha x + (1-\alpha)y) + cf(y) \geq \frac{1}{y-x}\int_x^y f(t)dt \tag{11.20}$$

is satisfied for all $x, y \in \mathbb{R}$ *and all continuous and convex functions* $f : [x, y] \to \mathbb{R}$ *if, and only if,*

$$b(1-\alpha) + c = \frac{1}{2} \tag{11.21}$$

and one of the following conditions holds true:

(i) $a + \alpha \geq 1$,
(ii) $a + b + \alpha \leq 1$, *and*
(iii) $a + \alpha < 1$, $a + b + \alpha > 1$, *and* $2a + \alpha \geq 1$.

Note that the original Hermite–Hadamard inequality consists of two parts. We treated these cases separately. However, it is possible to formulate a result containing both inequalities.

Corollary 11.1 ([40]) *If $a, \alpha, \beta \in (0, 1)$ satisfy (11.19) and one of the conditions (i)–(iii) of Theorem 11.4, then the inequality*

$$af(\alpha x + (1 - \alpha y) + (1 - a)f(\beta x + (1 - \beta)y) \leq \frac{1}{y - x}\int_x^y f(t)dt \leq$$
$$(1 - \alpha)f(x) + (\alpha - \beta)f(ax + (1 - a)y) + \beta f(y)$$

is satisfied for all $x, y \in \mathbb{R}$ and for all continuous and convex functions $f : \mathbb{R} \to \mathbb{R}$.

As we can see, the Ohlin lemma is very useful; however, it is worth noticing that in the case of some inequalities, the distribution functions cross more than once. Therefore, a simple application of the Ohlin lemma is impossible.

In the papers [32, 41], the authors used the Levin–Stečkin theorem [25] (see also [30, Theorem 4.2.7]), which gives necessary and sufficient conditions for convex ordering of functions with bounded variation, which are distribution functions of signed measures.

Theorem 11.6 (Levin, Stečkin [25]) *Let $a, b \in \mathbb{R}$, $a < b$ and let $F_1, F_2\colon [a, b] \to \mathbb{R}$ be functions with bounded variation such that $F_1(a) = F_2(a)$. Then, in order that*

$$\int_a^b f(x)dF_1(x) \leq \int_a^b f(x)dF_2(x) \tag{11.22}$$

for all continuous convex functions $f\colon [a, b] \to \mathbb{R}$, it is necessary and sufficient that F_1 and F_2 verify the following three conditions:

$$F_1(b) = F_2(b), \tag{11.23}$$

$$\int_a^b F_1(x)dx = \int_a^b F_2(x)dx, \tag{11.24}$$

$$\int_a^x F_1(t)dt \leq \int_a^x F_2(t)dt \quad \textit{for all} \quad x \in (a, b). \tag{11.25}$$

Define the number of sign changes of a function $\varphi\colon \mathbb{R} \to \mathbb{R}$ by

$$S^-(\varphi) = \sup\{S^-[\varphi(x_1), \varphi(x_2), \ldots, \varphi(x_k)]\colon x_1 < x_2 < \ldots x_k \in \mathbb{R},\ k \in \mathbb{N}\},$$

where $S^-[y_1, y_2, \ldots, y_k]$ denotes the number of sign changes in the sequence y_1, y_2,…, y_k (zero terms are being discarded). Two real functions φ_1, φ_2 are said to have n crossing points (or cross each other n-times) if $S^-(\varphi_1 - \varphi_2) = n$. Let $a = x_0 < x_1 < \ldots < x_n < x_{n+1} = b$. We say that the functions φ_1, φ_2 cross n-times at the points $x_1, x_2, \ldots, x_n$ (or that $x_1, x_2, \ldots, x_n$ are the points of sign changes of $\varphi_1 - \varphi_2$) if $S^-(\varphi_1 - \varphi_2) = n$ and there exist $a < \xi_1 < x_1 < \ldots < \xi_n < x_n < \xi_{n+1} < b$ such that $S^-[\xi_1, \xi_2, \ldots, \xi_{n+1}] = n$.

Szostok [41] used Theorem 11.6 to make an observation, which is more general than Ohlin's lemma and concerns the situation when the functions F_1 and F_2 have more crossing points than one. In [41] is given some useful modification of the Levin–Stečkin theorem [25], which can be rewritten in the following form.

Lemma 11.2 ([41]) *Let $a, b \in \mathbb{R}$, $a < b$ and let $F_1, F_2\colon (a, b) \to \mathbb{R}$ be functions with bounded variation such that $F(a) = F(b) = 0$, $\int_a^b F(x)dx = 0$, where $F = F_2 - F_1$. Let $a < x_1 < \ldots < x_m < b$ be the points of sign changes of the function F. Assume that $F(t) \geq 0$ for $t \in (a, x_1)$.*

- *If m is even, then the inequality*

$$\int_a^b f(x)dF_1(x) \leq \int_a^b f(x)dF_2(x) \tag{11.26}$$

is not satisfied by all continuous convex functions $f\colon [a, b] \to \mathbb{R}$.
- *If m is odd, define A_i ($i = 0, 1, \ldots, m$, $x_0 = a$, $x_{m+1} = b$)*

$$A_i = \int_{x_i}^{xi+1} |F(x)|dx.$$

Then, the inequality (11.26) is satisfied for all continuous convex functions $f\colon [a, b] \to \mathbb{R}$, if, and only if, the following inequalities hold true:

$$\begin{gathered} A_0 \geq A_1, \\ A_0 + A_2 \geq A_1 + A_3, \\ \vdots \\ A_0 + A_2 + \ldots + A_{m-3} \geq A_1 + A_3 + \ldots + A_{m-2}. \end{gathered} \tag{11.27}$$

Remark 11.7 ([38]) Let

$$H(x) = \int_a^x F(t)dt.$$

Then, the inequalities (11.27) are equivalent to the following inequalities

$$H(x_2) \geq 0,\ H(x_4) \geq 0,\ H(x_6) \geq 0,\ \ldots,\ H(x_{m-1}) \geq 0.$$

In [41], Lemma 11.2 is used to prove results, which extend the inequalities (11.18) and (11.20) and inequalities between quadrature operators.

Theorem 11.7 ([41]) *Let numbers $a_1, a_2, a_3, \alpha_1, \alpha_2, \alpha_3 \in (0, 1)$ satisfy $a_1 + a_2 + a_3 = 1$ and $\alpha_1 > \alpha_2 > \alpha_3$.*

Then, the inequality

$$\sum_{i=1}^{3} a_i f(\alpha_i x + (1-\alpha_i)y) \le \frac{1}{y-x}\int_x^y f(t) \tag{11.28}$$

is satisfied by all convex functions $f : [x, y] \to \mathbb{R}$ *if, and only if, we have*

$$\sum_{i=1}^{3} a_i(1-\alpha_i) = \frac{1}{2} \tag{11.29}$$

and one of the following conditions is satisfied

(i) $a_1 \le 1-\alpha_1$ *and* $a_1 + a_2 \ge 1 - \alpha_3$,
(ii) $a_1 \ge 1-\alpha_2$ *and* $a_1 + a_2 \ge 1 - \alpha_3$,
(iii) $a_1 \le 1-\alpha_1$ *and* $a_1 + a_2 \le 1 - \alpha_2$,
(iv) $a_1 \le 1-\alpha_1, a_1 + a_2 \in (1-\alpha_2, 1-\alpha_3)$, *and* $2\alpha_3 \ge a_3$,
(v) $a_1 \ge 1-\alpha_2, a_1 + a_2 < 1-\alpha_3$, *and* $2\alpha_3 \ge a_3$,
(vi) $a_1 > 1-\alpha_1, a_1 + a_2 \le 1-\alpha_2$, *and* $1-\alpha_1 \ge \frac{a_1}{2}$,
(vii) $a_1 \in (1-\alpha_1, 1-\alpha_2), a_1 + a_2 \ge 1-\alpha_3$, *and* $1-\alpha_1 \ge \frac{a_1}{2}$, *and*
(viii) $a_1 \in (1-\alpha_1, 1-\alpha_2)$, $a_1 + a_2 \in (1-\alpha_2, 1-\alpha_3)$, $1-\alpha_1 \ge \frac{a_1}{2}$, *and* $2a_1(1-\alpha_1) + 2a_2(1-\alpha_2) \ge (a_1+a_2)^2$.

To prove Theorem 11.7, we note that, if the inequality (11.28) is satisfied for every convex function f defined on the interval $[0, 1]$, then it is satisfied by every convex function f defined on a given interval $[x, y]$. Therefore, without loss of generality, it suffices to consider the interval $[0, 1]$ in place of $[x, y]$.

To prove Theorem 11.7, we consider the functions $F_1, F_2 : \mathbb{R} \to \mathbb{R}$ given by the following formulas

$$F_1(t) := \begin{cases} 0, & t < 1-\alpha_1, \\ a_1, & t \in [1-\alpha_1, 1-\alpha_2), \\ a_1 + a_2, & t \in [1-\alpha_2, 1-\alpha_3), \\ 1, & t \ge 1-\alpha_3, \end{cases} \tag{11.30}$$

and

$$F_2(t) := \begin{cases} 0, & t < 0, \\ t, & t \in [0, 1), \\ 1, & t \ge 1. \end{cases} \tag{11.31}$$

Observe that the equality (11.29) gives us

$$\int_0^1 t dF_1(t) = \int_0^1 t dF_2(t).$$

Further, it is easy to see that in the cases (*i*)–(*iii*) the pair (F_1, F_2) crosses exactly once and, consequently, the inequality (11.28) follows from the Ohlin lemma.

In the case (*iv*), the pair (F_1, F_2) crosses three times. Let $A_0, \ldots, A_3$ be defined as in Lemma 11.2. In order to prove the inequality (11.28), we note that $A_0 \geq A_1$. However, since $A_0 - A_1 + A_2 - A_3 = 0$, we shall show that $A_2 \leq A_3$. We have

$$A_2 = \int_{a_1+a_2}^{1-\alpha_3} (t - a_1 - a_2)dt = \frac{(1-\alpha_3 - a_1 - a_2)^2}{2} = \frac{a_3^2 - 2a_3\alpha_3 + \alpha_3^2}{2}$$

and

$$A_3 = \int_{1-\alpha_3}^{1} (1-t)dt = \frac{\alpha_3^2}{2}.$$

This means that $A_2 \leq A_3$ is equivalent to $2\alpha_3 \geq a_3$, as claimed.

We omit similar proofs in the cases (*v*)–(*vii*) and we pass to the case (*vii*). In this case, the pair (F_1, F_2) crosses five times. We have

$$A_0 = \int_0^{1-\alpha_1} t dt = \frac{(1-\alpha_1)^2}{2}$$

and

$$A_1 = \int_{1-\alpha_1}^{a_1} (a_1 - t)dt = a_1(a_1 - (1-\alpha_1)) - \frac{a_1^2 - (1-\alpha_1)^2}{2} = \frac{[a_1 - (1-\alpha_1)]^2}{2}.$$

This means that the inequality $A_0 \geq A_1$ is satisfied if, and only if, $1 - \alpha_1 \geq \frac{a_1}{2}$.

Further,

$$A_2 = \int_{a_1}^{1-\alpha_2} (t - a_1)dt = \frac{(1-\alpha_2)^2 - a_1^2}{2} - a_1(1 - \alpha_2 - a_1)$$

and

$$A_3 = \int_{1-\alpha_2}^{a_1+a_2} (a_1+a_2-t)dt = (a_1+a_2)(a_1+a_2-(1-\alpha_2)) - \frac{(a_1+a_2)^2 - (1-\alpha_2)^2}{2},$$

therefore, the inequality $A_0 + A_2 \geq A_3 + A_1$ is satisfied if, and only if,

$$(1-\alpha_1)^2 + (1-\alpha_2-a_1)^2 \geq (a_1 - 1 - \alpha_1)^2 + (a_1 + a_2 - 1 + \alpha_2)^2,$$

which, after some calculations, gives us the last inequality from (*vii*).

Using assertions (i) and (vii) of Theorem 11.7, it is easy to get the following example.

Example 11.1 ([41]) Let $x, y \in \mathbb{R}$, $\alpha \in (\frac{1}{2}, 1)$, and $a, b \in (0,1)$ be such that $2a + b = 1$. Then, the inequality

$$af(\alpha x+(1-\alpha)y)+bf\left(\frac{x+y}{2}\right)+af((1-\alpha)x+\alpha y) \le \frac{1}{y-x}\int_x^y f(t)dt \qquad (11.32)$$

is satisfied by all convex functions $f : [x, y] \to \mathbb{R}$ if, and only if, $a \le 2 - 2\alpha$.

In the next theorem, we obtain inequalities, which extend the second of the Hermite–Hadamard inequalities.

Theorem 11.8 ([41]) *Let numbers* $a_1, a_2, a_3, a_4 \in (0, 1), \alpha_1, \alpha_2, \alpha_3, \alpha_4 \in [0, 1]$ *satisfy* $a_1 + a_2 + a_3 + a_4 = 1$ *and* $1 = \alpha_1 > \alpha_2 > \alpha_3 > \alpha_4 = 0$.

Then, the inequality

$$\sum_{i=1}^{4} a_i f(\alpha_i x + (1 - \alpha_i)y) \ge \frac{1}{y-x}\int_x^y f(t) \qquad (11.33)$$

is satisfied by all convex functions $f : [x, y] \to \mathbb{R}$ *if, and only if, we have*

$$\sum_{i=1}^{4} a_i(1 - \alpha_i) = \frac{1}{2} \qquad (11.34)$$

and one of the following conditions is satisfied:

(i) $a_1 \ge 1 - \alpha_2$ *and* $a_1 + a_2 \ge 1 - \alpha_3$,
(ii) $a_1 + a_2 \le 1 - \alpha_2$ *and* $a_1 + a_2 + a_3 \le 1 - \alpha_3$,
(iii) $1 - \alpha_2 \le a_1$ *and* $1 - \alpha_3 \ge a_1 + a_2 + a_3$,
(iv) $1 - \alpha_2 \le a_1$, $1 - \alpha_3 \in (a_1 + a_2, a_1 + a_2 + a_3)$, *and* $\alpha_3 \le 2a_4$,
(v) $1 - \alpha_2 \ge a_1 + a_2$, $a_1 + a_2 + a_3 > 1 - \alpha_3$, *and* $\alpha_3 \le 2a_4$,
(vi) $a_1 < 1 - \alpha_2$, $a_1 + a_2 \ge 1 - \alpha_3$, *and* $2a_1 + \alpha_2 \ge 1$,
(vii) $a_1 < 1 - \alpha_2$, $a_1 + a_2 > 1 - \alpha_2$, $a_1 + a_2 + a_3 \le 1 - \alpha_3$, *and* $2a_1 + \alpha_2 \ge 1$,
(viii) $1 - \alpha_2 \in (a_1, a_1 + a_2)$, $1 - \alpha_3 \in (a_1 + a_2, a_1 + a_2 + a_3)$, $2a_1 + \alpha_2 \ge 1$, *and* $2a_1(1 - \alpha_3) + 2a_2(\alpha_2 - \alpha_3) \ge (1 - \alpha_3)^2$.

To prove Theorem 11.8, we assume that $F_1 : \mathbb{R} \to \mathbb{R}$ is the function given by the following formula

$$F_1(t) := \begin{cases} 0, & t < 0, \\ a_1, & t \in [0, 1 - \alpha_1), \\ a_1 + a_2, & t \in [1 - \alpha_1, 1 - \alpha_2), \\ a_1 + a_2 + a_3, & t \in [1 - \alpha_2, 1), \\ 1, & t \ge 1. \end{cases} \qquad (11.35)$$

and let F_2 be the function given by (11.31). In view of (11.34), we have

$$\int_0^1 F_1(t)dt = \int_0^1 F_2(t)dt.$$

In cases (i)–(iii), there is only one crossing point of (F_2, F_1) and our assertion is a consequence of the Ohlin lemma.

In the cases (iv)–(vii), the pair (F_2, F_1) crosses three times and, therefore, we have to use Lemma 11.2.

In the case (iv), the inequality (11.33) is satisfied by all convex functions f if, and only if, $A_0 \geq A_1$. Further, we know that

$$A_0 - A_1 + A_2 - A_3 = 0,$$

which implies that the inequality $A_0 \geq A_1$ is equivalent to $A_3 \geq A_2$. Clearly, we have

$$\begin{aligned} A_2 = & \int_{1-\alpha_3}^{1-a_4} (F_1(t) - F_2(t))dt = (\alpha_3 - a_4)(1 - a_4) - \frac{(1-a_4)^2 - (1-\alpha_3)^2}{2} \\ =& (\alpha_3 - a_4)\left(1 - a_4 + \frac{2 - (\alpha_3 + a_4)}{2}\right) \end{aligned} \tag{11.36}$$

and

$$A_3 = \int_{1-a_4}^{1} (t - (1 - a_4))dt = \frac{1 - (1-a_4)^2}{2} - (1 - a_4)a_4 \tag{11.37}$$

that is, $A_3 \geq A_2$ is equivalent to $\alpha_3 \leq 2a_4$.

We omit similar reasoning in the cases (v)–(vii) and we pass to the most interesting case $(viii)$. In this case, (F_2, F_1) has five crossing points and, therefore, we must check that the inequalities

$$A_0 \geq A_1 \text{ and } A_0 - A_1 + A_2 \geq A_3$$

are equivalent to the inequalities of the condition $(viii)$, respectively. To this end, we write

$$A_0 = \int_0^{a_1} (a_1 - t)dt = \frac{a_1^2}{2},$$

$$A_1 = \int_{a_1}^{1-\alpha_1} (t - (a_1 + a_2))dt = \frac{(a_1 + a_2 - 1 + \alpha_1)^2}{2},$$

which means that $A_0 \geq A_1$ if, and only if, $2a_1 + \alpha_2 \geq 1$. Further, A_2 and A_3 are given by formulas (11.36) and (11.37). Thus, $A_0 - A_1 + A_2 \geq A_3$ is equivalent to

$$a_1^2 + (a_1 + a_2 - (1 - \alpha_2))^2 \geq (1 - \alpha_2 - a_1)^2 + (1 - \alpha_3 - a_1 - a_2)^2,$$

which yields

$$2a_1(1-\alpha_3)+2a_2(\alpha_2-\alpha_3)\geq(1-\alpha_3)^2.$$

Using assertions (*ii*) and (*vii*) of Theorem 11.8, we get the following example.

Example 11.2 ([41]) Let $x, y \in \mathbb{R}$, let $\alpha \in (\frac{1}{2}, 1)$, and let $a, b \in (0, 1)$ be such that $2a + 2b = 1$. Then, the inequality

$$af(x)+bf(\alpha x+(1-\alpha)y)+bf\left((1-\alpha)x+\alpha y\right)+af(y)\geq\frac{1}{y-x}\int_x^y f(t)dt$$

is satisfied by all convex functions $f : [x, y] \to \mathbb{R}$ if, and only if, $a \geq \frac{1-\alpha}{2}$.

In the next theorem, we show that the same tools may be used to obtain some inequalities between quadrature operators, which do not involve the integral mean.

Theorem 11.9 ([41]) *Let $a, \alpha_1, \alpha_2, \beta \in (0, 1)$ and let $b_1, b_2, b_3 \in (0, 1)$ satisfy $b_1 + b_2 + b_3 = 1$.*

Then, the inequality

$$af(\alpha_1 x+(1-\alpha_1)y)+(1-a)f(\alpha_2 x+(1-\alpha_2)y)\leq$$

$$b_1f(x)+b_2f(\beta x+(1-\beta)y)+b_3f(y) \tag{11.38}$$

is satisfied by all convex functions $f : [x, y] \to \mathbb{R}$ if, and only if, we have

$$b_2(1-\beta)+b_3=a(1-\alpha_1)+(1-a)(1-\alpha_2) \tag{11.39}$$

and one of the following conditions is satisfied:

(i) $a \leq b_1$,
(ii) $a \geq b_1 + b_2$, *and*
(iii) $\alpha_2 \geq \beta$

or

(iv) $a \in (b_1, b_1 + b_2)$, $\alpha_2 < \beta$, *and* $(1-\alpha_1)b_1 \geq (\alpha_1-\beta)(a-b_1)$.

Now, using this theorem, we shall present positive and negative examples of inequalities of the type (11.38).

Example 11.3 ([41]) Let $\alpha \in \left(\frac{1}{2}, 1\right)$. The inequality

$$\frac{f(\alpha x+(1-\alpha)y)+f((1-\alpha)x+\alpha y)}{2}\leq\frac{f(x)+f\left(\frac{x+y}{2}\right)+f(y)}{3}$$

is satisfied by all convex functions $f : [x, y] \to \mathbb{R}$ if, and only if, $\alpha \leq \frac{5}{6}$.

Example 11.4 ([41]) Let $\alpha \in \left(\frac{1}{2}, 1\right)$. The inequality

$$\frac{f(\alpha x + (1-\alpha)y) + f((1-\alpha)x + \alpha y)}{2} \leq \frac{1}{6}f(x) + \frac{2}{3}f\left(\frac{x+y}{2}\right) + \frac{1}{6}f(y)$$

is satisfied by all convex functions $f : [x, y] \to \mathbb{R}$ if, and only if, $\alpha \leq \frac{2}{3}$.

11.3 Inequalities of the Hermite–Hadamard Type Involving Numerical Differentiation Formulas of the First Order

In the paper [32], expressions connected with numerical differentiation formulas of order 1 are studied. The authors used the Ohlin lemma and the Levin–Stečkin theorem to study inequalities of the Hermite–Hadamard type connected with these expressions.

First, we recall the classical Hermite–Hadamard inequality

$$f\left(\frac{x+y}{2}\right) \leq \frac{1}{y-x}\int_x^y f(t)dt \leq \frac{f(x)+f(y)}{2}. \tag{11.40}$$

Now, let us write (11.40) in the form

$$f\left(\frac{x+y}{2}\right) \leq \frac{F(y)-F(x)}{y-x} \leq \frac{f(x)+f(y)}{2}. \tag{11.41}$$

Clearly, this inequality is satisfied by every convex function f and its primitive function F. However, (11.41) may be viewed as an inequality involving two types of expressions used, in numerical integration and differentiation, respectively. Namely, $f\left(\frac{x+y}{2}\right)$ and $\frac{f(x)+f(y)}{2}$ are the simplest quadrature formulas used to approximate the definite integral, whereas $\frac{F(y)-F(x)}{y-x}$ is the simplest expression used to approximate the derivative of F. Moreover, as it is known from numerical analysis, if $F' = f$, then the following equality is satisfied

$$f(x) = \frac{F(x+h) - F(x-h)}{2h} - \frac{h^2}{6}f''(\xi) \tag{11.42}$$

for some $\xi \in (x-h, x+h)$. This means that (11.42) provides an alternate proof of (11.41) (for twice differentiable f).

This new formulation of the Hermite–Hadamard inequality was inspiration in [32] to replace the middle term of Hermite–Hadamard inequality by more complicated expressions than those used in (11.40). In [32], the authors study inequalities of the form

$$f\left(\frac{x+y}{2}\right) \le \frac{a_1F(x) + a_2F(\alpha x + (1-\alpha)y) + a_3F(\beta x + (1-\beta)y) + a_4F(y)}{y-x}$$

and

$$\frac{a_1F(x) + a_2F(\alpha x + (1-\alpha)y) + a_3F(\beta x + (1-\beta)y) + a_4F(y)}{y-x} \le \frac{f(x)+f(y)}{2},$$

where $f : [x, y] \to \mathbb{R}$ is a convex function, $F' = f$, $\alpha, \beta \in (0, 1)$, and $a_1 + a_2 + a_3 + a_4 = 0$.

Proposition 11.5 ([32]) *Let $n \in \mathbb{N}$, $\alpha_i \in (0, 1)$, $a_i \in \mathbb{R}$, $i = 1, \dots, n$ be such that $\alpha_1 > \alpha_2 > \dots > \alpha_n$ and $a_1 + a_2 + \dots + a_n = 0$, and let F be a differentiable function with $F' = f$. Then,*

$$\frac{\sum_{i=1}^{n} a_iF(\alpha_i x + (1-\alpha_i)y)}{y-x} = \int f d\mu,$$

with

$$\mu(A) = -\frac{1}{y-x}\sum_{i=1}^{n-1}(a_1 + \dots + a_i)l_1(A \cap [\alpha_i x + (1-\alpha_i)y, \alpha_{i+1}x + (1-\alpha_{i+1})y]),$$

where l_1 stands for the one-dimensional Lebesgue measure.

Remark 11.8 ([32]) Taking $F_1(t) := \mu((-\infty, t])$ with μ from Proposition 11.5, we can see that

$$\frac{\sum_{i=1}^{n} a_iF(\alpha_i x + (1-\alpha_i)y)}{y-x} = \int f dF_1. \tag{11.43}$$

Next proposition will show that, in order to get some inequalities of the Hermite–Hadamard type, we have to use sums containing more than three summands.

Proposition 11.6 ([32]) *There are no numbers $\alpha_i, a_i \in \mathbb{R}, i = 1, 2, 3$, satisfying $1 = \alpha_1 > \alpha_2 > \alpha_3 = 0$ such that any of the inequalities*

$$f\left(\frac{x+y}{2}\right) \le \frac{\sum_{i=1}^{3} a_iF(\alpha_i x + (1-\alpha_i)y)}{y-x}$$

or

$$\frac{\sum_{i=1}^{3} a_iF(\alpha_i x + (1-\alpha_i)y)}{y-x} \le \frac{f(x)+f(y)}{2}$$

is fulfilled by every continuous and convex function f and its antiderivative F.

To prove Proposition 11.6, we note that by Proposition 11.5, we can see that

$$\frac{\sum_{i=1}^{3} a_i F(\alpha_i x + (1-\alpha_i)y)}{y-x} = \int_x^y f d\mu,$$

with

$$\mu(A) = -\frac{1}{y-x}\big(a_1 l_1(A \cap [x, \alpha_2 x + (1-\alpha_2)y]) +$$
$$(a_2 + a_1) l_1(A \cap [\alpha_2 x + (1-\alpha_2)y, y])\big),$$

and

$$\frac{\sum_{i=1}^{3} a_i F(\alpha_i x + (1-\alpha_i)y)}{y-x} = \int_x^y f(t) dF_1(t),$$

where

$$F_1(t) = \mu\{(-\infty, t]\}. \tag{11.44}$$

Now, if

$$F_2(t) = \frac{1}{y-x} l_1\{(-\infty, t] \cap [x, y]\},$$

then F_1 lies strictly above or below F_2 (on $[x, y]$). This means that

$$\int_x^y F_2(t)dt \neq \int_x^y F_1(t)dt. \tag{11.45}$$

But, on the other hand, if

$$F_3(t) := \begin{cases} 0, & t < x, \\ \frac{1}{2}, & t \in [x, y), \\ 1, & t \geq y, \end{cases} \tag{11.46}$$

and

$$F_4(t) := \begin{cases} 0, & t < \frac{x+y}{2}, \\ 1, & t \geq \frac{x+y}{2}, \end{cases} \tag{11.47}$$

then

$$\int_x^y F_2(t)dt = \int_x^y F_3(t)dt = \int_x^y F_4(t)dt = \frac{y-x}{2}.$$

This, together with (11.45), shows that neither

$$\int_x^y fdF_2 \le \int_x^y fdF_3$$

nor

$$\int_x^y fdF_2 \ge \int_x^y fdF_4$$

is satisfied. To complete the proof, it suffices to observe that

$$\int_x^y fdF_3 = \frac{f(x)+f(y)}{2},$$
$$\int_x^y fdF_4 = f\left(\frac{x+y}{2}\right).$$

Remark 11.9 ([32]) Observe that the assumptions of Proposition 11.6, $\alpha_1 = 1$ and $\alpha_3 = 0$, are essential. For example, it follows from the Ohlin lemma that the inequality

$$f\left(\frac{x+y}{2}\right) \le \frac{-3F(\frac{3}{4}x+\frac{1}{4}y)+\frac{25}{11}F(\frac{11}{20}x+\frac{9}{20}y)+\frac{8}{11}F(y)}{y-x} \le \frac{1}{y-x}\int f(t)dt$$

is satisfied by all continuous and convex functions f (where $F' = f$). Clearly, there are many more examples of inequalities of this type.

Lemma 11.3 ([32]) *If any of the inequalities*

$$f\left(\frac{x+y}{2}\right) \le \frac{\sum_{i=1}^4 a_i F(\alpha_i x + (1-\alpha_i)y)}{y-x} \tag{11.48}$$

or

$$\frac{\sum_{i=1}^4 a_i F(\alpha_i x + (1-\alpha_i)y)}{y-x} \le \frac{f(x)+f(y)}{2} \tag{11.49}$$

is satisfied for all continuous and convex functions $f : [x,y] \to \mathbb{R}$ (where $F' = f$), then

$$a_1(\alpha_2-\alpha_1)+(a_2+a_1)(\alpha_3-\alpha_2)+(a_3+a_2+a_1)(\alpha_4-\alpha_3) = 1 \tag{11.50}$$

and

$$a_1(\alpha_2^2-\alpha_1^2)+(a_2+a_1)(\alpha_3^2-\alpha_2^2)+(a_3+a_2+a_1)(\alpha_4^2-\alpha_3^2) = 1. \tag{11.51}$$

To prove this lemma, we take $x = 0$, $y = 1$. Then, using Proposition 11.5, we can see that

$$\sum_{i=1}^{4} a_i F(1-\alpha_i) = \int_0^1 f d\mu = -a_1 \int_{1-\alpha_1}^{1-\alpha_2} f(x)dx +$$

$$-(a_1+a_2)\int_{1-\alpha_3}^{1-\alpha_2} f(x)dx - (a_1+a_2+a_3)\int_{1-\alpha_4}^{1-\alpha_3} f(x)dx.$$

Now, we consider the functions F_1, F_3, and F_4 given by the formulas (11.44), (11.46), and (11.47), respectively. Then, the inequalities (11.48) and (11.49) may be written in the form

$$\int fdF_4 \leq \int fdF_1$$

and

$$\int fdF_1 \leq \int fdF_3.$$

This means that, if, for example, the inequality (11.48) is satisfied, then we have $F_1(1) = F_4(1) = 1$, which yields (11.50). Further,

$$\int_0^1 F_1(t)dt = \int_0^1 F_4(t)dt = \frac{1}{2},$$

which gives us (11.51).

Proposition 11.7 ([32]) *Let $\alpha_i \in (0,1)$, $a_i \in \mathbb{R}$, $i = 1,\dots,4$, be such that $1 = \alpha_1 > \alpha_2 > \alpha_3 > \alpha_4 = 0$, $a_1 + a_2 + a_3 + a_4 = 0$, and the equalities* (11.50) *and* (11.51) *are satisfied. If F_1 is such that*

$$\frac{\sum_{i=1}^{4} a_i F(\alpha_i x + (1-\alpha_i)y)}{y-x} = \int_x^y fdF_1$$

and F_2 is the distribution function of a measure which is uniformly distributed in the interval $[x,y]$, then (F_1,F_2) crosses exactly once.

Indeed, from (11.50) we can see that $F_1(x) = F_2(x) = 0$ and $F_1(y) = F_2(y) = 1$. Note that, in view of Proposition 11.5, the graph of the restriction of F_1 to the interval $[x,y]$ consists of three segments. Therefore, F_1 and F_2 cannot have more than one crossing point. On the other hand, if graphs F_1 and F_2 do not cross, then

$$\int_x^y tdF_1(t) \neq \int_x^y tdF_1(t)$$

that is, (11.51) is not satisfied.

Theorem 11.10 *Let $\alpha_i \in (0,1)$, $a_i \in \mathbb{R}$, $i = 1, \ldots, 4$, be such that $1 = \alpha_1 > \alpha_2 > \alpha_3 > \alpha_4 = 0$, $a_1 + a_2 + a_3 + a_4 = 0$, and the equalities* (11.50) *and* (11.51) *are satisfied. Let $F, f : [x,y] \to \mathbb{R}$ be functions such that f is continuous and convex and $F' = f$. Then,*

(i) If $a_1 > -1$, then

$$\frac{\sum_{i=1}^{4} a_i F(\alpha_i x + (1-\alpha_i)y)}{y-x} \le \frac{1}{y-x}\int_x^y f(t)dt \le \frac{f(x)+f(y)}{2},$$

(ii) If $a_1 < -1$, then

$$f\left(\frac{x+y}{2}\right) \le \frac{1}{y-x}\int_x^y f(t)dt \le \frac{\sum_{i=1}^{4} a_i F(\alpha_i x + (1-\alpha_i)y)}{y-x},$$

(iii) If $a_1 \in (-1,0]$, then

$$f\left(\frac{x+y}{2}\right) \le \frac{\sum_{i=1}^{4} a_i F(\alpha_i x + (1-\alpha_i)y)}{y-x} \le \frac{1}{y-x}\int_x^y f(t)dt, \text{ and}$$

(iv) If $a_1 < -1$ and $a_2 + a_1 \le 0$, then

$$\frac{1}{y-x}\int_x^y f(t)dt \le \frac{\sum_{i=1}^{4} a_i F(\alpha_i x + (1-\alpha_i)y)}{y-x} \le \frac{f(x)+f(y)}{2}.$$

We shall prove the first assertion. Other proofs are similar and will be omitted. It is easy to see that if inequalities which we consider are satisfied by every continuous and convex function defined on the interval $[0,1]$, then they are true for every continuous and convex function on a given interval $[x,y]$. Therefore, we assume that $x = 0$ and $y = 1$. Let F_1 be such that (11.43) is satisfied and let F_2 be the distribution function of a measure, which is uniformly distributed in the interval $[0,1]$. From Proposition 11.5 and Remark 11.8, we can see that the graph of F_1 consists of three segments and, since $a_1 > -1$, the slope of the first segment is smaller than 1, i.e., F_1 lies below F_2 on some right-hand neighborhood of x. In view of the Proposition 11.7, this means that the assumptions of the Ohlin lemma are satisfied and we get our result from this lemma.

Now, we shall present examples of inequalities, which may be obtained from this theorem.

Example 11.5 ([32]) Using (i), we can see that the inequality

$$\frac{1}{3}F(x) - \frac{8}{3}F\left(\frac{3x+y}{4}\right) + \frac{8}{3}F\left(\frac{x+3y}{4}\right) - \frac{1}{3}F(y) \le \frac{\int_x^y f(t)dt}{y-x}$$

is satisfied for every continuous and convex f and its antiderivative F.

Example 11.6 ([32]) Using (ii), we can see that the inequality

$$-2F(x) + 3F\left(\frac{2x+y}{3}\right) - 3F\left(\frac{x+2y}{3}\right) + 2F(y) \geq \frac{\int_x^y f(t)dt}{y-x}$$

is satisfied by every continuous and convex function f and its antiderivative F.

Example 11.7 ([32]) Using (iii), we can see that the inequality

$$\frac{\int_x^y f(t)dt}{y-x} \geq \frac{-\frac{1}{2}F(x) - \frac{3}{2}F\left(\frac{2x+y}{3}\right) + \frac{3}{2}F\left(\frac{x+2y}{3}\right) + \frac{1}{2}F(y)}{y-x} \geq f\left(\frac{x+y}{2}\right)$$

is satisfied by every continuous and convex function f and its antiderivative F.

Example 11.8 ([32]) Using (iv), we can see that the inequality

$$\frac{\int_x^y f(t)dt}{y-x} \leq \frac{-\frac{3}{2}F(x) + 2F\left(\frac{3x+y}{4}\right) - 2F\left(\frac{x+3y}{4}\right) + \frac{3}{2}F(y)}{y-x} \leq \frac{f(x)+f(y)}{2}$$

is satisfied by every continuous and convex function f and its antiderivative F.

In all cases considered in the above theorem, we used only the Ohlin lemma. Using Lemma 11.2, it is possible to obtain more subtle inequalities. However (for the sake of simplicity), in the next result, we shall restrict our considerations to expressions of the simplified form. Note that the inequality between $f\left(\frac{x+y}{2}\right)$ and expressions which we consider is a bit unexpected.

Theorem 11.11 ([32]) *Let* $\alpha \in \left(0, \frac{1}{2}\right)$, $a, b \in \mathbb{R}$.

(i) If $a > 0$, *then the inequality*

$$f\left(\frac{x+y}{2}\right) \geq \frac{aF(x) + bF(\alpha x + (1-\alpha)y) - bF((1-\alpha)x + \alpha y) - aF(y)}{y-x}$$

is satisfied by every continuous and convex f *and its antiderivative* F *if, and only if,*

$$(1-\alpha)^2 \frac{ab}{a+b} > \frac{1}{2} - (1-\alpha)\frac{b}{a+b}, \text{ and} \tag{11.52}$$

(ii) If $a < -1$ *and* $a_1 + a_2 > 0$, *then the inequality*

$$\frac{aF(x) + bF(\alpha x + (1-\alpha)y) - bF((1-\alpha)x + \alpha y) - aF(y)}{y-x} \leq \frac{f(x)+f(y)}{2}$$

is satisfied by every continuous and convex f and its antiderivative F if, and only if,

$$-\frac{1}{4a} > \left(-a(1-\alpha) - \frac{1}{2}\right)\left(\frac{1}{2} + \frac{1}{2a}\right).$$

We shall prove the assertion (i) of Theorem 11.11. The proof of (ii) is similar and will be omitted. Similarly as before, we may assume without loss of generality that $x = 0, y = 1$. Let F_1 be such that

$$aF(0) + bF(1-\alpha) - bF(\alpha) + aF(1) = \int_0^1 f dF_1$$

and let F_4 be given by (11.47). Then, it is easy to see that (F_1, F_4) crosses three times: at $\frac{(1-\alpha)b}{a+b}$, $\frac{1}{2}$, and at $\frac{a+\alpha b}{a+b}$.

We are going to use Lemma 11.2. Since, from (11.51), we have that

$$A_0 + A_1 + A_2 + A_3 = 0,$$

it suffices to check that $A_0 \geq A_1$ if, and only if, the inequality (11.52) is satisfied. Since, $F_4(x) = 0$, for $x \in \left(0, \frac{1}{2}\right)$, we get

$$A_0 = -\int_0^{\frac{(1-\alpha)b}{a+b}} F_1(t)dt$$

and

$$A_1 = \int_{\frac{(1-\alpha)b}{a+b}}^{\frac{1}{2}} F_1(t)dt,$$

which yields our assertion.

Example 11.9 ([32]) Neither inequality

$$f\left(\frac{x+y}{2}\right) \leq \frac{\frac{1}{3}F(x) - \frac{8}{3}F\left(\frac{3x+y}{4}\right) + \frac{8}{3}F\left(\frac{x+3y}{4}\right) - \frac{1}{3}F(y)}{y-x} \tag{11.53}$$

nor

$$f\left(\frac{x+y}{2}\right) \geq \frac{\frac{1}{3}F(x) - \frac{8}{3}F\left(\frac{3x+y}{4}\right) + \frac{8}{3}F\left(\frac{x+3y}{4}\right) - \frac{1}{3}F(y)}{y-x} \tag{11.54}$$

is satisfied for all continuous and convex $f : [x, y] \to \mathbb{R}$. Indeed, if F_1 is such that

$$\int_x^y f(t)dF_1(t) = \frac{\frac{1}{3}F(x) - \frac{8}{3}F\left(\frac{3x+y}{4}\right) + \frac{8}{3}F\left(\frac{x+3y}{4}\right) - \frac{1}{3}F(y)}{y-x},$$

then

$$\int_x^{\frac{3x+y}{4}} F_1(t)dt < \int_x^{\frac{3x+y}{4}} F_4(t)dt,$$

thus inequality (11.53) cannot be satisfied. On the other hand, the coefficients and nodes of the expression considered do not satisfy (11.52). Therefore, (11.54) is also not satisfied for all continuous and convex $f : [x, y] \to \mathbb{R}$.

Example 11.10 ([32]) Using assertion (i) of Theorem 11.11, we can see that the inequality

$$\frac{2F(x) - 3F\left(\frac{3x+y}{4}\right) + 3F\left(\frac{x+3y}{4}\right) - 2F(y)}{y-x} \leq f\left(\frac{x+y}{2}\right)$$

is satisfied for every continuous and convex f and its antiderivative F.

Example 11.11 ([32]) Using assertion (ii) of Theorem 11.11, we can see that the inequality

$$\frac{-2F(x) + 3F\left(\frac{2x+y}{3}\right) - 3F\left(\frac{x+2y}{3}\right) + 2F(y)}{y-x} \leq \frac{f(x) + f(y)}{2}$$

is satisfied for every continuous and convex f and its antiderivative F.

11.4 Inequalities of the Hermite–Hadamard Type Involving Numerical Differentiation Formulas of Order Two

In the paper [42], expressions connected with numerical differentiation formulas of order 2 are studied. The author used the Ohlin lemma and the Levin–Stečkin theorem to study inequalities connected with these expressions. In particular, the author presents a new proof of the inequality

$$f\left(\frac{x+y}{2}\right) \leq \frac{1}{(y-x)^2}\int_x^y\int_x^y f\left(\frac{s+t}{2}\right) ds\, dt \leq \frac{1}{y-x}\int_x^y f(t)dt, \qquad (11.55)$$

satisfied by every convex function $f : \mathbb{R} \to \mathbb{R}$ and he obtains extensions of (11.55). In the previous section, inequalities involving expressions of the form

$$\frac{\sum_{i=1}^{n} a_i F(\alpha_i x + \beta_i y)}{y - x},$$

where $\sum_{i=1}^{n} a_i = 0$, $\alpha_i + \beta_i = 1$, and $F' = f$ were considered. In this section, we study inequalities for expressions of the form

$$\frac{\sum_{i=1}^{n} a_i F(\alpha_i x + \beta_i y)}{(y - x)^2},$$

which we use to approximate the second order derivative of F and, surprisingly, we discover a connection between our approach and the inequality (11.55) (see [42]).

First, we make the following simple observation.

Remark 11.10 ([42]) Let $f, F, \Phi : [x, y] \to \mathbb{R}$ be such that $\Phi' = F, F' = f$. Let $n_i, m_i \in \mathbb{N} \cup \{0\}$, $i = 1, 2, 3$; $a_{i,j} \in \mathbb{R}$, $\alpha_{i,j}, \beta_{i,j} \in [0, 1]$, $i = 1, 2, 3$; $j = 1, \ldots, n_i$, $b_{i,j} \in \mathbb{R}$, $\gamma_{i,j}, \delta_{i,j} \in [0, 1]$, $i = 1, 2, 3$; $j = 1, \ldots, m_i$. If the inequality

$$\begin{aligned} &\sum_{i=1}^{n_1} a_{1,i} f(\alpha_{1,i}x + \beta_{1,i}y) + \frac{\sum_{i=1}^{n_2} a_{2,i} F(\alpha_{2,i}x + \beta_{2,i}y)}{y - x} \\ &\quad + \frac{\sum_{i=1}^{n_3} a_{3,i} \Phi(\alpha_{3,i}x + \beta_{3,i}y)}{(y - x)^2} \le \sum_{i=1}^{m_1} b_{1,i} f(\gamma_{1,i}x + \delta_{1,i}y) \\ &\quad + \frac{\sum_{i=1}^{m_2} b_{2,i} F(\gamma_{2,i}x + \delta_{2,i}y)}{y - x} + \frac{\sum_{i=1}^{m_3} b_{3,i} \Phi(\gamma_{3,i}x + \delta_{3,i}y)}{(y - x)^2} \end{aligned} \tag{11.56}$$

is satisfied for $x = 0, y = 1$ and for all continuous and convex functions $f : [0, 1] \to \mathbb{R}$, then it is satisfied for all $x, y \in \mathbb{R}$, $x < y$ and for each continuous and convex function $f : [x, y] \to \mathbb{R}$. To see this, it is enough to observe that expressions from (11.56) remain unchanged if we replace $f : [x, y] \to \mathbb{R}$ by $\varphi : [0, 1] \to \mathbb{R}$ given by $\varphi(t) := f\left(x + \frac{t}{y-x}\right)$.

The simplest expression used to approximate the second order derivative of f is of the form

$$f''\left(\frac{x + y}{2}\right) \approx \frac{f(x) - 2f\left(\frac{x+y}{2}\right) + f(y)}{\left(\frac{y-x}{2}\right)^2}.$$

Remark 11.11 ([42]) From numerical analysis, it is known that

$$f''\left(\frac{x + y}{2}\right) = \frac{f(x) - 2f\left(\frac{x+y}{2}\right) + f(y)}{\left(\frac{y-x}{2}\right)^2} - \frac{\left(\frac{y-x}{2}\right)^2}{12} f^{(4)}(\xi).$$

This means that for a convex function g and for G such that $G'' = g$ we have

$$g\left(\frac{x+y}{2}\right) \leq \frac{G(x) - 2G\left(\frac{x+y}{2}\right) + G(y)}{\left(\frac{y-x}{2}\right)^2}.$$

In the paper [42], some inequalities for convex functions which do not follow from formulas used in numerical differentiation are obtained.

Let now $f : [x, y] \to \mathbb{R}$ be any function and let $F, \Phi : [x, y] \to \mathbb{R}$ be such that $F' = f$ and $\Phi'' = f$. We need to write the expression

$$\frac{\Phi(x) - 2\Phi\left(\frac{x+y}{2}\right) + \Phi(y)}{\left(\frac{y-x}{2}\right)^2} \tag{11.57}$$

in the form

$$\int_x^y f dF_1$$

for some F_1. In the next proposition, we show that it is possible—here for the sake of simplicity we shall work on the interval [0, 1].

Proposition 11.8 ([42]) *Let $f : [0, 1] \to \mathbb{R}$ be any function and let $\Phi : [0, 1] \to \mathbb{R}$ be such that $\Phi'' = f$. Then, we have*

$$4\left(\Phi(0) - 2\Phi\left(\frac{1}{2}\right) + \Phi(1)\right) = \int_x^y f dF_1,$$

where $F_1 : [0, 1] \to \mathbb{R}$ is given by

$$F_1(t) := \begin{cases} 2x^2, & x \leq \frac{1}{2}, \\ -2x^2 + 4x - 1, & x > \frac{1}{2}. \end{cases} \tag{11.58}$$

Now, we observe that the following equality is satisfied

$$\frac{\Phi(x) - 2\Phi\left(\frac{x+y}{2}\right) + \Phi(y)}{\left(\frac{y-x}{2}\right)^2} = \frac{1}{(y-x)^2} \int_x^y \int_x^y f\left(\frac{s+t}{2}\right) ds\, dt.$$

After this observation, it turns out that inequalities involving the expression (11.57) were considered in the paper of Dragomir [14], where (among others) the following inequalities were obtained

$$f\left(\frac{x+y}{2}\right) \leq \frac{1}{(y-x)^2} \int_x^y \int_x^y f\left(\frac{s+t}{2}\right) ds\, dt \leq \frac{1}{y-x} \int_x^y f(t) dt. \tag{11.59}$$

As we already know (Remark 11.11), the first one of the above inequalities may be obtained using the numerical analysis results.

Now, the inequalities from the Dragomir's paper easily follow from the Ohlin lemma but there are many possibilities of generalizations and modifications of inequalities (11.59). These generalizations will be discussed in this section.

First, we consider the symmetric case. We start with the following remark.

Remark 11.12 ([42]) Let $F_*(t) = at^2 + bt + c$ for some $a, b, c \in \mathbb{R}, a \neq 0$. It is impossible to obtain inequalities involving $\int_x^y f dF_*$ and any of the expressions:

$$\frac{1}{y-x}\int_x^y f(t)dt, \quad f\left(\frac{x+y}{2}\right), \quad \frac{f(x)+f(y)}{2},$$

which are satisfied for all convex functions $f : [x, y] \to \mathbb{R}$. Indeed, suppose that we have

$$\int_x^y f dF_* \leq \frac{1}{y-x}\int_x^y f(t)dt$$

for all convex $f : [x, y] \to \mathbb{R}$. Without loss of generality, we may assume that $F_*(x) = 0$, then from Theorem 11.6 we have $F_*(y) = 1$. Also from Theorem 11.6 we get

$$\int_x^y F_*(t)dt = \int_x^y F_0 dt,$$

where $F_0(t) = \frac{t-x}{y-x}$, $t \in [x, y]$, which is impossible, because F_* is either strictly convex or concave.

This remark means that in order to get some new inequalities of the Hermite–Hadamard type we have to integrate with respect to functions constructed with the use of (at least) two quadratic functions.

Now, we present the main result of this section.

Theorem 11.12 ([42]) *Let x, y be some real numbers such that $x < y$ and let $a \in \mathbb{R}$. Let $f, F, \Phi : [x, y] \to \mathbb{R}$ be any functions such that $F' = f$ and $\Phi' = F$ and let $T_a f(x, y)$ be the function defined by the following formula*

$$T_a f(x, y) = \left(1 - \frac{a}{2}\right)\frac{F(y) - F(x)}{y-x} + 2a\frac{\Phi(x) - 2\Phi\left(\frac{x+y}{2}\right) + \Phi(x)}{(y-x)^2}.$$

Then, the following inequalities hold for all convex functions $f : [x, y] \to \mathbb{R}$:

- *If $a \geq 0$, then*

$$T_a f(x, y) \leq \frac{1}{y-x}\int_x^y f(t)dt, \tag{11.60}$$

- *If $a \leq 0$, then*

$$T_a f(x,y) \geq \frac{1}{y-x}\int_x^y f(t)dt, \tag{11.61}$$

- *If $a \leq 2$, then*

$$f\left(\frac{x+y}{2}\right) \leq T_a f(x,y), \tag{11.62}$$

- *If $a \geq 6$, then*

$$T_a f(x,y) \leq f\left(\frac{x+y}{2}\right), \tag{11.63}$$

- *If $a \geq -6$, then*

$$T_a f(x,y) \leq \frac{f(x)+f(y)}{2}, \tag{11.64}$$

Furthermore,

- *If $a \in (2,6)$, then the expressions $T_a f(x,y), f\left(\frac{x+y}{2}\right)$ are not comparable in the class of convex functions, and*
- *If $a < -6$, then expressions $T_a f(x,y)$, $\frac{f(x)+f(y)}{2}$ are not comparable in the class of convex functions.*

To prove Theorem 11.12, we note that we may restrict ourselves to the case $x = 0, y = 1$. Take $a \in \mathbb{R}$, let $f : [0,1] :\rightarrow \mathbb{R}$ be any convex function, and let $F, \Phi : [0,1] \rightarrow \mathbb{R}$ be such that $F' = f, \Phi' = F$. Define $F_1 : [0,1] \rightarrow \mathbb{R}$ by the formula

$$F_1(t) := \begin{cases} at^2 + \left(1 - \frac{a}{2}\right)t, & t < \frac{1}{2}, \\ -at^2 + \left(1 + \frac{3a}{2}\right)t - \frac{a}{2}, & t \geq \frac{1}{2}. \end{cases} \tag{11.65}$$

First, we prove that $T_a f(0,1) = \int_0^1 f dF_1$. Now, let $F_2(t) = t, t \in [0,1]$. Then, the functions F_1, F_2 have exactly one crossing point (at $\frac{1}{2}$) and

$$\int_0^1 F_1(t)dt = \frac{1}{2} = \int_0^1 t dt.$$

Moreover, if $a > 0$, then the function F_1 is convex on the interval $(0, \frac{1}{2})$ and concave on $(\frac{1}{2}, 1)$. Therefore, it follows from the Ohlin lemma that for $a > 0$ we have

$$\int_0^1 f dF_1 \leq \int_0^1 f dF_2,$$

which, in view of Remark 11.10, yields (11.60) and for $a < 0$ the opposite inequality is satisfied, which gives (11.61). Take

$$F_3(t) := \begin{cases} 0, \ t \leq \frac{1}{2}, \\ 1, \ t > \frac{1}{2}. \end{cases}$$

It is easy to calculate that for $a \leq 2$ we have $F_1(t) \geq F_3(t)$ for $t \in \left[0, \frac{1}{2}\right]$, and $F_1(t) \leq F_3(t)$ for $t \in \left[\frac{1}{2}, 1\right]$, and this means that from the Ohlin lemma we get (11.62). Let now

$$F_4(t) := \begin{cases} 0, \ t = 0, \\ \frac{1}{2}, \ t \in (0, 1), \\ 1, \ t = 1. \end{cases}$$

Similarly as before, if $a \geq -2$, then we have $F_1(t) \geq F_4(t)$ for $t \in \left[0, \frac{1}{2}\right]$ and $F_1(t) \leq F_4(t)$ for $t \in \left[\frac{1}{2}, 1\right]$. Therefore, from the Ohlin lemma, we get (11.63).

Suppose that $a > 2$. Then there are three crossing points of the functions F_1 and $F_3 : x_0, \frac{1}{2}, x_1$, where $x_0 \in (0, \frac{1}{2}), x_1 \in (\frac{1}{2}, 1)$. The function

$$\varphi(s) := \int_0^s (F_3(t) - F_1(t))dt, \ s \in [0, 1]$$

is increasing on the intervals $[0, x_0], [\frac{1}{2}, x_1]$ and decreasing on $[x_0, \frac{1}{2}]$ and on $[x_1, 1]$. This means that φ takes its absolute minimum at $\frac{1}{2}$. It is easy to calculate that $\varphi\left(\frac{1}{2}\right) \geq 0$, if $a \geq 6$, which, in view of Theorem 11.6, gives us (11.63).

To see that, for $a \in (2, 6)$, the expressions $T_a f(x, y)$ and $f\left(\frac{x+y}{2}\right)$ are not comparable in the class of convex functions, it is enough to observe that in this case $\varphi(x_0) > 0$ and $\varphi\left(\frac{1}{2}\right) < 0$.

Analogously (using functions F_1 and F_4), we show that for $a \in (-2, -6]$ we have (11.64), and in the case $a < -6$ the expressions $T_a f(x, y)$ and $\frac{f(x)+f(y)}{2}$ are not comparable in the class of convex functions. This theorem provides us with a full description of inequalities, which may be obtained using Stieltjes integral with respect to a function of the form (11.65). Some of the obtained inequalities are already known. For example, from (11.60) and (11.61) we obtain the inequality

$$\frac{1}{(y-x)^2} \int_x^y \int_x^y f\left(\frac{s+t}{2}\right) ds \, dt \leq \frac{1}{y-x} \int_x^y f(t)dt,$$

whereas from (11.62) for $a = 2$ we get the inequality

$$f\left(\frac{x+y}{2}\right) \leq \frac{1}{(y-x)^2} \int_x^y \int_x^y f\left(\frac{s+t}{2}\right) ds \, dt.$$

However, inequalities obtained for "critical" values of a, i.e., $-6, 6.$ are here particularly interesting. In the following corollary, we explicitly write these inequalities.

Corollary 11.2 ([42]) *For every convex function $f : [x, y] \to \mathbb{R}$, the following inequalities are satisfied*

$$3\frac{1}{(y-x)^2}\int_x^y\int_x^y f\left(\frac{s+t}{2}\right)dsdt \leq \frac{2}{y-x}\int_x^y f(t)dt + f\left(\frac{x+y}{2}\right), \tag{11.66}$$

$$\frac{4}{y-x}\int_x^y f(t)dt \leq 3\frac{1}{(y-x)^2}\int_x^y\int_x^y f\left(\frac{s+t}{2}\right)dsdt + \frac{f(x)+f(y)}{2}. \tag{11.67}$$

Remark 11.13 ([42]) In the paper [15], Dragomir and Gomm obtained the following inequality

$$3\int_x^y f(t)dt \leq 2\frac{1}{(y-x)^2}\int_x^y\int_x^y f\left(\frac{s+t}{2}\right)dsdt + \frac{f(x)+f(y)}{2}. \tag{11.68}$$

Inequality (11.67) from Corollary 11.2 is stronger than (11.68). Moreover, as it was observed in Theorem 11.12, the inequalities (11.66) and (11.67) cannot be improved, i.e., the inequality

$$\frac{1}{y-x}\int_x^y f(t)dt \leq \lambda\frac{1}{(y-x)^2}\int_x^y\int_x^y f\left(\frac{s+t}{2}\right)dsdt + (1-\lambda)\frac{f(x)+f(y)}{2}$$

for $\lambda > \frac{3}{4}$ is not satisfied by every convex function $f : [x, y] \to \mathbb{R}$ and the inequality

$$\frac{1}{(y-x)^2}\int_x^y\int_x^y f\left(\frac{s+t}{2}\right)dsdt \leq \gamma\frac{1}{y-x}\int_x^y f(t)dt + (1-\gamma)f\left(\frac{x+y}{2}\right)$$

with $\gamma > \frac{2}{3}$ is not true for all convex functions $f : [x, y] \to \mathbb{R}$.

In Corollary 11.2, we obtained inequalities for the triples:

$$\frac{1}{(y-x)^2}\int_x^y\int_x^y f\left(\frac{s+t}{2}\right)dsdt, \quad \int_x^y f(t)dt, \quad \frac{f(x)+f(y)}{2}$$

and

$$\frac{1}{(y-x)^2}\int_x^y\int_x^y f\left(\frac{s+t}{2}\right)dsdt, \quad \int_x^y f(t)dt, \quad f\left(\frac{x+y}{2}\right).$$

In the next remark, we present an analogous result for expressions

$$\frac{1}{(y-x)^2}\int_x^y\int_x^y f\left(\frac{s+t}{2}\right)dsdt,\quad \frac{f(x)+f(y)}{2},\quad f\left(\frac{x+y}{2}\right).$$

Remark 11.14 ([42]) Using the functions: F_1 defined by (11.58) and F_5 given by

$$F_5(t) := \begin{cases} 0, & t = 0, \\ \frac{1}{6}, & t \in \left(0, \frac{1}{2}\right), \\ \frac{5}{6}, & t \in \left[\frac{1}{2}, 1\right), \\ 1, & t = 1, \end{cases}$$

we can see that

$$\frac{1}{6}f(x) + \frac{2}{3}f\left(\frac{x+y}{2}\right) + \frac{1}{6}f(y) \geq \frac{1}{(y-x)^2}\int_x^y\int_x^y f\left(\frac{s+t}{2}\right)dsdt$$

for all convex functions $f : [x, y] \to \mathbb{R}$.

Moreover, it is easy to see that the above inequality cannot be strengthened, which means that if $a, b \geq 0$, $2a + b = 1$ and $a < \frac{1}{6}$, then the inequality

$$af(x) + bf\left(\frac{x+y}{2}\right) + af(y) \geq \frac{1}{(y-x)^2}\int_x^y\int_x^y f\left(\frac{s+t}{2}\right)dsdt,$$

is not satisfied by all convex functions f.

In [42], inequalities for $f(\alpha x + (1-\alpha)y)$ and for $\alpha f(x) + (1-\alpha)f(y)$, where α is not necessarily equal to $\frac{1}{2}$ (the nonsymmetric case), are also obtained.

Theorem 11.13 ([42]) *Let x, y be some real numbers such that $x < y$ and let $\alpha \in [0, 1]$. Let $f : [x, y] \to \mathbb{R}$ be a convex function, let F be such that $F' = f$, and let Φ satisfy $\Phi' = F$. If $S^2_\alpha f(x, y)$ is defined by*

$$S^2_\alpha f(x,y) := \frac{(4-6\alpha)F(y) + (2-6\alpha)F(x)}{y-x} + \frac{(6-12\alpha)(\Phi(y)-\Phi(x))}{(y-x)^2},$$

then the following conditions hold true:

- $$S^2_\alpha f(x,y) \leq \alpha f(x) + (1-\alpha)f(y),$$
- *If $\alpha \in \left[\frac{1}{3}, \frac{2}{3}\right]$, then*

$$S^2_\alpha f(x,y) \geq f(\alpha x + (1-\alpha)y),$$

- *If* $\alpha \in [0,1] \setminus \left[\frac{1}{3}, \frac{2}{3}\right]$, *then the expressions* $S_{\alpha}^{2}f(x,y)$ *and* $f(\alpha x + (1-\alpha)y)$ *are incomparable in the class of convex functions,*
- *If* $\alpha \in \left(0, \frac{1}{3}\right] \cup \left[\frac{2}{3}, 1\right)$, *then*

$$S_{\alpha}^{2}f(x,y) \leq S_{\alpha}^{1}f(x,y), \text{ and}$$

- *If* $\alpha \in \left(\frac{1}{3}, \frac{1}{2}\right) \cup \left(\frac{1}{2}, \frac{2}{3}\right)$, *then* $S_{\alpha}^{1}f(x,y)$ *and* $S_{\alpha}^{2}f(x,y)$ *are incomparable in the class of convex functions.*

11.5 The Hermite–Hadamard Type Inequalities for *n*-th Order Convex Functions

Now, we are going to study Hermite–Hadamard type inequalities for higher-order convex functions. Many results on higher-order generalizations of the Hermite–Hadamard type inequality one can find, among others, in [1–5, 16, 20, 36, 37]. In recent papers [36, 37], the theorem of Denuit, Lefèvre, and Shaked [13] on sufficient conditions for s-convex ordering was used, to prove Hermite–Hadamard type inequalities for higher-order convex functions.

Let us review some notations. The convexity of n-th order (or n-convexity) was defined in terms of divided differences by Popoviciu [34]; however, we will not state it here. Instead, we list some properties of n-th order convexity which are equivalent to Popoviciu's definition (see [24]).

Proposition 11.9 *A function* $f\colon (a,b) \to \mathbb{R}$ *is* n*-convex on* (a,b) $(n \geq 1)$ *if, and only if, its derivative* $f^{(n-1)}$ *exists and is convex on* (a,b) *(with the convention* $f^{(0)}(x) = f(x)$*).*

Proposition 11.10 *Assume that* $f\colon [a,b] \to \mathbb{R}$ *is* $(n+1)$*-times differentiable on* (a,b) *and continuous on* $[a,b]$ $(n \geq 1)$*. Then,* f *is* n*-convex if, and only if,* $f^{(n+1)}(x) \geq 0$, $x \in (a,b)$.

For real-valued random variables X, Y and any integer $s \geq 2$, we say that X is dominated by Y in *s-convex ordering* sense if $\mathbb{E}f(X) \leq \mathbb{E}f(Y)$ for all $(s-1)$-convex functions $f\colon \mathbb{R} \to \mathbb{R}$, for which the expectations exist [13]. In that case, we write $X \leq_{s-cx} Y$, or $\mu_X \leq_{s-cx} \mu_Y$, or $F_X \leq_{s-cx} F_Y$. Then, the order $\leq_{2-cx}$ is just the usual convex order $\leq_{cx}$.

A very useful criterion for the verification of the s-convex order is given by Denuit, Lefèvre, and Shaked in [13].

Proposition 11.11 ([13]) *Let* X *and* Y *be two random variables such that* $\mathbb{E}(X^j - Y^j) = 0$, $j = 1, 2, \ldots, s-1$ $(s \geq 2)$*. If* $S^{-}(F_X - F_Y) = s-1$ *and the last sign of* $F_X - F_Y$ *is positive, then* $X \leq_{s-cx} Y$.

We now apply Proposition 11.11 to obtain the following results.

Theorem 11.14 ([36]) *Let $n \geq 1$, $a_1 \leq a < b \leq b_1$.*

Let $a(n) = \left[\frac{n}{2}\right] + 1$, $b(n) = \left[\frac{n+1}{2}\right] + 1$.

Let $\alpha_1, \ldots, \alpha_{a(n)}$, $x_1, \ldots, x_{a(n)}$, $\beta_1, \ldots, \beta_{b(n)}$, $y_1, \ldots, y_{b(n)}$ be real numbers such that

- *If n is even, then*

$$0 < \beta_1 < \alpha_1 < \beta_1 + \beta_2 < \alpha_1 + \alpha_2 < \ldots < \alpha_1 + \ldots + \alpha_{a(n)} = \beta_1 + \ldots + \beta_{b(n)} = 1,$$
$$a \leq y_1 < x_1 < y_2 < x_2 < \ldots < x_{a(n)} < y_{b(n)} \leq b,$$

- *If n is odd, then*

$$0 < \beta_1 < \alpha_1 < \beta_1 + \beta_2 < \alpha_1 + \alpha_2 < \ldots < \beta_1 + \ldots + \beta_{b(n)} < \alpha_1 + \ldots + \alpha_{a(n)} = 1$$
$$a \leq y_1 < x_1 < y_2 < x_2 < \ldots < y_{b(n)} < x_{a(n)} \leq b;$$

and

$$\sum_{k=1}^{a(n)} x_i^k \alpha_i = \sum_{j=1}^{b(n)} y_j^k \beta_j$$

for any $k = 1, 2, \ldots, n$.

Let $f\colon [a_1, b_1] \to \mathbb{R}$ be an n-convex function. Then, we have the following inequalities:

- *If n is even, then*

$$\sum_{i=1}^{a(n)} \alpha_i f(x_i) \leq \sum_{j=1}^{b(n)} \beta_j f(y_j),$$

- *If n is odd, then*

$$\sum_{j=1}^{b(n)} \beta_j f(y_j) \leq \sum_{i=1}^{a(n)} \alpha_i f(x_i).$$

Theorem 11.15 ([36]) *Let $n \geq 1$, $a_1 \leq a < b \leq b_1$. Let $a(n), b(n) \in \mathbb{N}$. Let $\alpha_1, \ldots, \alpha_{a(n)}$, $\beta_1, \ldots, \beta_{b(n)}$ be positive real numbers such that $\alpha_1 + \ldots + \alpha_{a(n)} = \beta_1 + \ldots + \beta_{b(n)} = 1$. Let $x_1, \ldots, x_{a(n)}$, $y_1, \ldots, y_{b(n)}$ be real numbers such that*

- *$a \leq x_1 \leq x_2 \leq \ldots \leq x_{a(n)} \leq b$ and $a \leq y_1 \leq y_2 \leq \ldots \leq y_{b(n)} \leq b$,*
- *$\sum_{k=1}^{a(n)} x_i^k \alpha_i = \sum_{j=1}^{b(n)} y_j^k \beta_j$, for any $k = 1, 2, \ldots, n$.*

Let $\alpha_0 = \beta_0 = 0$, $x_0 = y_0 = -\infty$. *Let* $F_1, F_2 \colon \mathbb{R} \to \mathbb{R}$ *be two functions given by the following formulas:* $F_1(x) = \alpha_0 + \alpha_1 + \ldots + \alpha_k$ *if* $x_k < x \leq x_{k+1}$ $(k = 0, 1, \ldots, a(n) - 1)$ *and* $F_1(x) = 1$ *if* $x > x_{a(n)}$; $F_2(x) = \beta_0 + \beta_1 + \ldots + \beta_k$ *if* $y_k < x \leq y_{k+1}$ $(k = 0, 1, \ldots, b(n) - 1)$ *and* $F_2(x) = 1$ *if* $x > y_{b(n)}$. *If the functions* F_1, F_2 *have* n *crossing points and the last sign of* $F_1 - F_2$ *is a*$+$, *then for any* n*-convex function* $f \colon [a_1, b_1] \to \mathbb{R}$ *we have the following inequality*

$$\sum_{i=1}^{a(n)} \alpha_i f(x_i) \leq \sum_{j=1}^{b(n)} \beta_j f(y_j).$$

Theorem 11.16 ([36]) *Let* $n \geq 1$, $a_1 \leq a < b \leq b_1$. *Let* $a(n) = \left[\frac{n}{2}\right] + 1$, $b(n) = \left[\frac{n+1}{2}\right] + 1$. *Let* $x_1, \ldots, x_{a(n)}, y_1, \ldots, y_{b(n)}$ *be real numbers, and* $\alpha_1, \ldots, \alpha_{a(n)}$, $\beta_1, \ldots, \beta_{b(n)}$ *be positive numbers, such that* $\alpha_1 + \ldots + \alpha_{a(n)} = 1$, $\beta_1 + \ldots + \beta_{b(n)} = 1$,

$$\frac{1}{b-a} \int_a^b x^k dx = \sum_{j=1}^{b(n)} y_j^k \beta_j = \sum_{i=1}^{a(n)} x_i^k \alpha_i \quad (k = 1, 2, \ldots, n),$$

$a \leq x_1 < x_2 < \ldots < x_{a(n)} \leq b$, $a \leq y_1 < y_2 < \ldots < y_{b(n)} < b$,

$$\begin{gathered}
\tfrac{x_1 - a}{b-a} < \alpha_1 < \tfrac{x_2 - a}{b-a}, \\
\tfrac{x_2 - a}{b-a} < \alpha_1 + \alpha_2 < \tfrac{x_3 - a}{b-a}, \\
\ldots \\
\tfrac{x_{a(n)-1} - a}{b-a} < \alpha_1 + \ldots + \alpha_{a(n)-1} < \tfrac{x_{a(n)} - a}{b-a},
\end{gathered}$$

$$\begin{gathered}
\tfrac{y_1 - a}{b-a} < \beta_1 < \tfrac{y_2 - a}{b-a}, \\
\tfrac{y_2 - a}{b-a} < \beta_1 + \beta_2 < \tfrac{y_2 - a}{b-a}, \\
\ldots \\
\tfrac{y_{b(n)-1} - a}{b-a} < \beta_1 + \ldots + \beta_{b(n)-1} < \tfrac{y_{b(n)} - a}{b-a};
\end{gathered}$$

if n *is even, then* $y_1 = a$, $y_{b(n)} = b$, $x_1 > a$, $x_{a(n)} < b$;
if n *is odd, then* $y_1 = a$, $y_{b(n)} < b$, $x_1 > a$, $x_{a(n)} = b$.

Let $f \colon [a_1, b_1] \to \mathbb{R}$ *be an* n*-convex function. Then, we have the following inequalities:*

- *If* n *is even, then*

$$\sum_{i=1}^{a(n)} \alpha_i f(x_i) \leq \frac{1}{b-a} \int_a^b f(x) dx \leq \sum_{j=1}^{b(n)} \beta_j f(y_j),$$

- *If n is odd, then*

$$\sum_{j=1}^{b(n)} \beta_j f(y_j) \le \frac{1}{b-a} \int_a^b f(x)dx \le \sum_{i=1}^{a(n)} \alpha_i f(x_i).$$

Note that Proposition 11.11 can be rewritten in the following form.

Proposition 11.12 ([13]) *Let X and Y be two random variables such that*

$$\mathbb{E}(X^j - Y^j) = 0, \quad j = 1, 2, \ldots, s \ (s \ge 1).$$

If the distribution functions F_X and F_Y cross exactly s-times at points $x_1 < x_2 < \ldots < x_s$ and

$$(-1)^{s+1} \left(F_Y(x) - F_X(x)\right) \ge 0 \quad \textit{for all } x \le x_1,$$

then

$$\mathbb{E}f(X) \le \mathbb{E}f(Y) \tag{11.69}$$

for all s-convex functions $f \colon \mathbb{R} \to \mathbb{R}$.

Proposition 11.11 is a counterpart of the Ohlin lemma concerning convex ordering. This proposition gives sufficient conditions for s-convex ordering and is very useful for the verification of higher-order convex orders. However, it is worth noticing that in the case of some inequalities, the distribution functions cross more than s-times. Therefore, a simple application of this proposition is impossible.

In the paper [38], a theorem on necessary and sufficient conditions for higher-order convex stochastic ordering is given. This theorem is a counterpart of the Levin–Stečkin theorem [25] concerning convex stochastic ordering. Based on this theorem, useful criteria for the verification of higher-order convex stochastic ordering are given. These results can be useful in the study of Hermite–Hadamard type inequalities for higher-order convex functions, and in particular inequalities between the quadrature operators. It is worth noticing that these criteria can be easier to checking of higher-order convex orders, than those given in [13, 22].

Let $F_1, F_2 \colon [a,b] \to \mathbb{R}$ be two functions with bounded variation and μ_1, μ_2 be the signed measures corresponding to F_1, F_2, respectively. We say that F_1 is dominated by F_2 in $(n+1)$-*convex ordering* sense $(n \ge 1)$ if

$$\int_{-\infty}^{\infty} f(x)dF_1(x) \le \int_{-\infty}^{\infty} f(x)dF_2(x)$$

for all n-convex functions $f \colon [a,b] \to \mathbb{R}$. In that case, we write $F_1 \le_{(n+1)-cx} F_2$, or $\mu_1 \le_{(n+1)-cx} \mu_2$. In the following theorem, we give necessary and sufficient conditions for $(n+1)$-convex ordering of two functions with bounded variation.

Theorem 11.17 ([38]) *Let $a, b \in \mathbb{R}$, $a < b$, $n \in \mathbb{N}$ and let $F_1, F_2 \colon [a, b] \to \mathbb{R}$ be two functions with bounded variation such that $F_1(a) = F_2(a)$. Then, in order that*

$$\int_a^b f(x)dF_1(x) \le \int_a^b f(x)dF_2(x)$$

for all continuous n-convex functions $f \colon [a, b] \to \mathbb{R}$, it is necessary and sufficient that F_1 and F_2 verify the following conditions:

$$F_1(b) = F_2(b),$$

$$\int_a^b F_1(x)dx = \int_a^b F_2(x)dx,$$

$$\int_a^b \int_a^{x_{k-1}} \cdots \int_a^{x_1} F_1(t)dtdx_1 \dots dx_{k-1} =$$
$$\int_a^b \int_a^{x_{k-1}} \cdots \int_a^{x_1} F_2(t)dtdx_1 \dots dx_{k-1} \quad \text{for } k = 2, \dots, n, \tag{11.70}$$

$$(-1)^{n+1} \int_a^x \int_a^{x_{n-1}} \cdots \int_a^{x_1} F_1(t)dtdx_1 \dots dx_{n-1} \le$$
$$(-1)^{n+1} \int_a^x \int_a^{x_{n-1}} \cdots \int_a^{x_1} F_2(t)dtdx_1 \dots dx_{n-1} \quad \text{for all } x \in (a, b). \tag{11.71}$$

Corollary 11.3 ([38]) *Let μ_1, μ_2 be two signed measures on $\mathscr{B}(\mathbb{R})$, which are concentrated on (a, b), and such that $\int_a^b |x|^n \mu_i(dx) < \infty$, $i = 1, 2$. Then, in order that*

$$\int_a^b f(x)d\mu_1(x) \le \int_a^b f(x)d\mu_2(x)$$

for continuous n-convex functions $f \colon [a, b] \to \mathbb{R}$, it is necessary and sufficient that μ_1, μ_2 verify the following conditions:

$$\mu_1((a, b)) = \mu_2((a, b)), \tag{11.72}$$

$$\int_a^b x^k \mu_1(dx) = \int_a^b x^k \mu_2(dx) \quad \text{for } k = 1, \dots, n, \tag{11.73}$$

$$\int_a^b (t - x)_+^n \mu_1(dt) = \int_a^b (t - x)_+^n \mu_2(dt) \quad \text{for all } x \in (a, b), \tag{11.74}$$

where $y_+^n = \left\{\max\{y, 0\}\right\}^n$, $y \in \mathbb{R}$.

In [13], it can be found the following necessary and sufficient conditions for the verification of the $(s+1)$-convex order.

Proposition 11.13 ([13]) *If X and Y are two real-valued random variables such that $\mathbb{E}|X|^s < \infty$ and $\mathbb{E}|Y|^s < \infty$, then*

$$\mathbb{E}f(X) \leq \mathbb{E}f(Y)$$

for all continuous s-convex functions $f: \mathbb{R} \to \mathbb{R}$ if, and only if,

$$\mathbb{E}X^k = \mathbb{E}Y^k \quad \text{for } k = 1, 2, \ldots, s, \tag{11.75}$$

$$\mathbb{E}(X-t)^s_+ \leq \mathbb{E}(Y-t)^s_+ \quad \text{for all } t \in \mathbb{R}. \tag{11.76}$$

Remark 11.15 ([38]) Note that if the measures μ_X, μ_Y, corresponding to the random variables X, Y, respectively, occurring in Proposition 11.13, are concentrated on some interval $[a, b]$, then this proposition is an easy consequence of Corollary 11.3.

Theorem 11.17 can be rewritten in the following form.

Theorem 11.18 ([38]) *Let $F_1, F_2: [a, b] \to \mathbb{R}$ be two functions with bounded variation such that $F_1(a) = F_2(a)$. Let*

$$H_0(t_0) = F_2(t_0) - F_1(t_0) \quad \text{for } t_0 \in [a, b],$$

$$H_k(t_k) = \int_a^{t_{k-1}} H_{k-1}(t_{k-1}) dt_{k-1} \quad \text{for } t_k \in [a, b],\ k = 1, 2, \ldots, n.$$

Then, in order that

$$\int_a^b f(x) dF_1(x) \leq \int_a^b f(x) dF_2(x)$$

for all continuous n-convex functions $f: [a, b] \to \mathbb{R}$, it is necessary and sufficient that the following conditions are satisfied:

$$H_k(b) = 0 \quad \text{for } k = 0, 1, 2, \ldots, n,$$

$$(-1)^{n+1} H_n(x) \geq 0 \quad \text{for all } x \in (a, b).$$

Remark 11.16 ([38]) The functions $H_1, \ldots, H_n$ that appear in Theorem 11.18 can be obtained from the following formulas

$$H_n(x) = (-1)^{n+1} \int_a^b \frac{(t-x)^n_+}{n!} d(F_2(t) - F_1(t)), \tag{11.77}$$

$$H_{k-1}(x) = H_k^{'}(x), \quad k = 2, 3, \ldots, n. \tag{11.78}$$

Note that the function $(-1)^{n+1}H_{n-1}$, that appears in Theorem 11.18, plays a role similar to the role of the function $F = F_2 - F_1$ in Lemma 11.2. Consequently, from Theorem 11.18, Lemma 11.2, and Remarks 11.7, 11.16, we obtain immediately the following criterion, which can be useful for the verification of higher-order convex ordering.

Corollary 11.4 ([38]) *Let $F_1, F_2 \colon [a,b] \to \mathbb{R}$ be functions with bounded variation such that $F_1(a) = F_2(a)$, $F_1(b) = F_2(b)$ and $H_k(b) = 0$ $(k = 1, 2, \ldots, n)$, where $H_k(x)$ $(k = 1, 2, \ldots, n)$ are given by* (11.77) *and* (11.78)*. Let $a < x_1 < \ldots < x_m < b$ be the points of sign changes of the function H_{n-1} and let $(-1)^{n+1}H_{n-1}(x) \geq 0$ for $x \in (a, x_1)$.*

- *If m is even, then the inequality*

$$\int_a^b f(x)dF_1(x) \leq \int_a^b f(x)dF_2(x), \tag{11.79}$$

  is not satisfied by all continuous n-convex functions $f \colon [a,b] \to \mathbb{R}$.
- *If m is odd, then the inequality* (11.79) *is satisfied for all continuous n-convex functions $f \colon [a,b] \to \mathbb{R}$ if, and only if,*

$$(-1)^{n+1}H_n(x_2) \geq 0, \quad (-1)^{n+1}H_n(x_4) \geq 0, \quad \ldots, \quad (-1)^{n+1}H_n(x_{m-1}) \geq 0. \tag{11.80}$$

In the numerical analysis, some inequalities, which are connected with quadrature operators, are studied. These inequalities, called extremalities, are a particular case of the Hermite–Hadamard type inequalities. Many extremalities are known in the numerical analysis (cf. [1, 7, 8] and the references therein). The numerical analysts prove them using the suitable differentiability assumptions. As proved by Wąsowicz in the papers [44, 45, 47], for convex functions of higher order, some extremalities can be obtained without assumptions of this kind, using only the higher-order convexity itself. The support-type properties play here the crucial role. As we show in [36, 37], some extremalities can be proved using a probabilistic characterization. The extremalities, which we study, are known; however, our method using the Ohlin lemma [31] and the Denuit–Lefèvre–Shaked theorem [13] on sufficient conditions for the convex stochastic ordering seems to be quite easy. It is worth noticing that these theorems concern only the sufficient conditions, and they cannot be used to the proof some extremalities (see [36, 37]). In these cases, results given in the paper [38] may be useful.

For a function $f : [-1,1] \to \mathbb{R}$, we consider six operators approximating the integral mean value

$$\mathscr{I}(f) := \tfrac{1}{2}\int_{-1}^{1} f(x)dx.$$

They are given by

$$\begin{aligned}
C(f) &:= \tfrac{1}{3}\Big(f\big(-\tfrac{\sqrt{2}}{2}\big)+f(0)+f\big(\tfrac{\sqrt{2}}{2}\big)\Big),\\
\mathcal{G}_2(f) &:= \tfrac{1}{2}\Big(f\big(-\tfrac{\sqrt{3}}{3}\big)+f\big(\tfrac{\sqrt{3}}{3}\big)\Big),\\
\mathcal{G}_3(f) &:= \tfrac{4}{9}f(0)+\tfrac{5}{18}\Big(f\big(-\tfrac{\sqrt{15}}{5}\big)+f\big(\tfrac{\sqrt{15}}{5}\big)\Big),\\
\mathcal{L}_4(f) &:= \tfrac{1}{12}\big(f(-1)+f(1)\big)+\tfrac{5}{12}\Big(f\big(-\tfrac{\sqrt{5}}{5}\big)+f\big(\tfrac{\sqrt{5}}{5}\big)\Big),\\
\mathcal{L}_5(f) &:= \tfrac{16}{45}f(0)+\tfrac{1}{20}\big(f(-1)+f(1)\big)+\tfrac{49}{180}\Big(f\big(-\tfrac{\sqrt{21}}{7}\big)+f\big(\tfrac{\sqrt{21}}{7}\big)\Big), \textit{and}\\
S(f) &:= \tfrac{1}{6}\big(f(-1)+f(1)\big)+\tfrac{2}{3}f(0).
\end{aligned}$$

The operators $\mathcal{G}_2$ and $\mathcal{G}_3$ are connected with Gauss–Legendre rules. The operators $\mathcal{L}_4$ and $\mathcal{L}_5$ are connected with Lobatto quadratures. The operators S and C concern Simpson and Chebyshev quadrature rules, respectively. The operator $\mathcal{I}$ stands for the integral mean value (see, e.g., [39, 48–51]).

We will establish all possible inequalities between these operators in the class of higher-order convex functions.

Remark 11.17 Let X_2, X_3, Y_4, Y_5, U, V, and Z be random variables such that

$$\begin{aligned}
\mu_{X_2} &= \frac{1}{2}\left(\delta_{-\frac{\sqrt{3}}{3}}+\delta_{\frac{\sqrt{3}}{3}}\right),\\
\mu_{X_3} &= \frac{4}{9}\,\delta_0+\frac{5}{18}\left(\delta_{-\frac{\sqrt{15}}{5}}+\delta_{\frac{\sqrt{15}}{5}}\right),\\
\mu_{Y_4} &= \frac{1}{12}\,(\delta_{-1}+\delta_1)+\frac{5}{12}\left(\delta_{-\frac{\sqrt{5}}{5}}+\delta_{\frac{\sqrt{5}}{5}}\right),\\
\mu_{Y_5} &= \frac{16}{45}\,\delta_0+\frac{1}{20}(\delta_{-1}+\delta_1)+\frac{49}{180}\left(\delta_{-\frac{\sqrt{21}}{7}}+\delta_{\frac{\sqrt{21}}{7}}\right),\\
\mu_U &= \frac{2}{3}\,\delta_0+\frac{1}{6}(\delta_{-1}+\delta_1),\\
\mu_V &= \frac{1}{3}\left(\delta_{-\frac{\sqrt{2}}{2}}+\delta_0+\delta_{\frac{\sqrt{2}}{2}}\right), \textit{and}\\
\mu_Z(dx) &= \frac{1}{2}\,\chi_{[-1,1]}(x)dx.
\end{aligned}$$

Then, we have

$$\mathcal{G}_2(f)=\mathbb{E}[f(X_2)],\quad \mathcal{G}_3(f)=\mathbb{E}[f(X_3)],$$

$$\mathcal{L}_4(f) = \mathbb{E}[f(Y_4)], \quad \mathcal{L}_5(f) = \mathbb{E}[f(Y_5)],$$
$$S(f) = \mathbb{E}[f(U)], \quad C(f) = \mathbb{E}[f(V)], \quad \mathcal{I}(f) = \mathbb{E}[f(Z)].$$

Theorem 11.19 *Let $f: [-1, 1] \to \mathbb{R}$ be 5-convex. Then,*

$$\mathcal{G}_3(f) \leq \mathcal{I}(f) \leq \mathcal{L}_4(f), \tag{11.81}$$

$$\mathcal{G}_3(f) \leq \mathcal{L}_5(f) \leq \mathcal{L}_4(f). \tag{11.82}$$

Note that the inequalities (11.81) and (11.82) can be simply derived from Theorems 11.16 and 11.15 (see [38]).

Remark 11.18 The inequalities (11.82) can be found in [45, 47]. Wąsowicz [45] proved that in the class of 5-convex functions the operators $\mathcal{G}_2, C, S$ are not comparable both with each other and with $\mathcal{G}_3, \mathcal{L}_4, \mathcal{L}_5$.

Theorem 11.20 *Let $f: [-1, 1] \to \mathbb{R}$ be 3-convex. Then,*

$$\mathcal{G}_2(f) \leq \mathcal{I}(f) \leq S(f), \tag{11.83}$$

$$\mathcal{G}_2(f) \leq C(f) \leq T(f) \leq S(f), \tag{11.84}$$

where $T \in \{\mathcal{G}_3, \mathcal{L}_5\}$.

In [38] is given a new simple proof of Theorem 11.20. Note that from Theorem 11.16, we obtain $\mathcal{G}_3(f) \leq \mathcal{I}(f)$ and $\mathcal{I}(f) \leq S(f)$, which implies (11.83). From Theorem 11.14, we obtain $\mathcal{G}_2(f) \leq C(f)$. By Theorem 11.15, we get $C(f) \leq \mathcal{G}_3(f)$, $C(f) \leq \mathcal{L}_5(f)$, $\mathcal{G}_3(f) \leq S(f)$, $\mathcal{L}_5(f) \leq S(f)$.

Remark 11.19 The inequalities (11.84) can be found in [44]. Wąsowicz [44] proved that the quadratures $\mathcal{L}_4$, $\mathcal{L}_5$, and $\mathcal{G}_3$ are not comparable in the class of 3-convex functions.

Remark 11.20 Moreover, Wąsowicz [44, 46] proved that

$$C(f) \leq \mathcal{L}_4(f), \tag{11.85}$$

if f is 3-convex.

The proof given in [44] is rather complicated. This was done using computer software. In [46], can be found a new proof of (11.85), without the use of any computer software, based on the spline approximation of convex functions of higher order. It is worth noticing that Proposition 11.11 does not apply to proving (11.85), because the distribution functions F_V and F_{Y_4} cross exactly five-times.

In [38], the following new proof of (11.85) is given. In this proof of (11.85), we use Corollary 11.4. Note that we have $F_1 = F_V$, $F_2 = F_{Y_4}$, and $H_0 = F = F_{Y_4} - F_V$. By (11.77) and (11.78), we obtain

$$H_3(x) = \frac{1}{72}\left\{(-1-x)_+^3 + (1-x)_+^3 + 5\left[\left(-\frac{\sqrt{5}}{5}-x\right)_+^3 + \left(\frac{\sqrt{5}}{5}-x\right)_+^3\right]\right.$$

$$\left. -4\left[(-1-x)_+^3 + \left(-\frac{\sqrt{2}}{2}-x\right)_+^3 + (-x)_+^3 + \left(\frac{\sqrt{2}}{2}-x\right)_+^3\right]\right\},$$

$$H_2(x) = \frac{1}{24}\left\{-(-1-x)_+^2 - (1-x)_+^2 - 5\left[\left(-\frac{\sqrt{5}}{5}-x\right)_+^2 + \left(\frac{\sqrt{5}}{5}-x\right)_+^2\right]\right.$$

$$\left. +4\left[(-1-x)_+^2 + \left(-\frac{\sqrt{2}}{2}-x\right)_+^2 + (-x)_+^2 + \left(\frac{\sqrt{2}}{2}-x\right)_+^2\right]\right\}.$$

Similarly, $H_1(x)$ can be obtained from the equality $H_1(x) = H_2'(x)$. We compute that $x_1 = -1 - \sqrt{5} + 2\sqrt{2}$, $x_2 = 0$, and $x_3 = 1 + \sqrt{5} - 2\sqrt{2}$ are the points of sign changes of the function $H_2(x)$. It is not difficult to check that the assumptions of Corollary 11.4 are satisfied. Since

$$(-1)^{3+1}H_3(x_2) = (-1)^{3+1}H_3(0) = \frac{1}{72} + \frac{\sqrt{5}}{360} - \frac{\sqrt{2}}{72} > 0,$$

it follows that the inequalities (11.80) are satisfied. From Corollary 11.4, we conclude that the relation (11.85) holds.

References

1. Bessenyei, M.: Hermite–Hadamard-type inequalities for generalized convex functions. J. Inequal. Pure Appl. Math. **9**, 1–51 (2008)
2. Bessenyei, M., Páles, Z.: Higher-order generalizations of Hadamard's inequality. Publ. Math. Debr. **61**, 623–643 (2002)
3. Bessenyei, M., Páles, Z.: Hadamard-type inequalities for generalized convex functions. Math. Inequal. Appl. **6**, 379–392 (2003)
4. Bessenyei, M., Páles, Z.: On generalized higher-order convexity and Hermite–Hadamard-type inequalities. Acta Sci. Math. (Szeged) **70**, 13–24 (2004)
5. Bessenyei, M., Páles, Z.: Characterization of higher-order monotonicity via integral inequalities. Proc. R. Soc. Edinb. Sect. A **140**, 723–736 (2010)
6. Billingsley, P.: Probability and Measure. Wiley, New York (1995)
7. Brass, H., Petras, K.: Quadrature Theory. The Theory of Numerical Integration on a Compact Interval. American Mathematical Society, Providence, RI (2011)
8. Brass, H., Schmeisser, G.: Error estimates for interpolatory quadrature formulae. Numer. Math. **37**, 371–386 (1981)
9. Brenner, J.L., Alzer, H.: Integral inequalities for concave functions with applications to special functions. Proc. R. Soc. Edinb. Sect. A **118**, 173–192 (1991)
10. Czinder, P.: A weighted Hermite-Hadamard-type inequality for convex-concave symmetric functions. Publ. Math. Debr. **68**, 215–224 (2006)

11. Czinder, P., Páles, Zs.: An extension of the Hermite-Hadamard inequality and an application for Gini and Stolarsky means. J. Inequal. Pure Appl. Math. **5**, Article 42, 8 pp. (2004)
12. de la Cal, J., Cárcamo, J.: Multidimensional Hermite-Hadamard inequalities and the convex order. J. Math. Anal. Appl. **324**, 248–261 (2006)
13. Denuit, M., Lefevre, C., Shaked, M.: The s-convex orders among real random variables, with applications. Math. Inequal. Appl. **1**, 585–613 (1998)
14. Dragomir, S.S.: Two mappings in connection to Hadamard's inequalities. J. Math. Anal. Appl. **167**, 49–56 (1992)
15. Dragomir, S.S., Gomm, I.: Some new bounds for two mappings related to the Hermite-Hadamard inequality for convex functions. Numer. Algebra Control Optim. **2**, 271–278 (2012)
16. Dragomir, S.S., Pearce, C.E.M.: Selected Topics on Hermite-Hadamard Inequalities and Applications. Victoria University (2000). Online: http://rgmia.vu.edu.au/monographs/
17. Fejér, L.: Über die Fourierreihen, II. Math. Naturwiss. Anz. Ungar. Akad. Wiss. **24**, 369–390 (1906)
18. Fink, A.M.: A best possible Hadamard inequality. Math. Inequal. Appl. **2**, 223–230 (1998)
19. Florea, A., Păltănea, E., Bălă, D.: Convex ordering properties and applications. J. Math. Inequal. **9**, 1245–1257 (2015)
20. Gilányi, A., Páles, Zs.: On convex functions of higher order. Math. Inequal. Appl. **11**, 271–282 (2008)
21. Hardy, G.H., Littlewood, J.E., Pólya, G.: Inequalities. Cambridge University Press, Cambridge (1952)
22. Karlin, S., Novikoff, A.: Generalized convex inequalities. Pacific J. Math. **13**, 1251–1279 (1963)
23. Klaričić Bakula, M., Pečarić, J., Perić, J.: Extensions of Hermite–Hadamard inequality with applications. Math. Inequal. Appl. **15**, 899–921 (2012)
24. Kuczma, M.: An Introduction to the Theory of Functional Equations and Inequalities. Polish Scientific Publishers & Silesian University, Warszawa-Kraków-Katowice (1985)
25. Levin, V.I., Stečkin, S.B.: Inequalities. Am. Math. Soc. Transl. **14**, 1–22 (1960)
26. Lupas, A.: A generalisation of Hadamard's inequality for convex functions. Univ. Beograd. Publ. Elek. Fak. Ser. Mat. Fiz. **544–576**, 115–121 (1976)
27. Mihai, M.V.: Rezultate de tip Hermite Hadamard pentru funcştiile convexe generalizate. PhD thesis, Craiova University (2015)
28. Mitrinović, D.S., Lacković, I.B.: Hermite and convexity. Aequationes Math. **28**, 229–232 (1985)
29. Mrowiec, J., Rajba, T., Wąsowicz, Sz.: A solution to the problem of Raşa connected with Bernstein polynomials. J. Math. Anal. Appl. **446**, 864–878 (2017) .
30. Niculescu, C.P., Persson, L.E.: Convex Functions and their Applications. A Contemporary Approach. Springer, New York (2006)
31. Ohlin, J.: On a class of measures of dispersion with application to optimal reinsurance. ASTIN Bulletin **5**, 249–266 (1969)
32. Olbryś, A., Szostok, T.: Inequalities of the Hermite-Hadamard type involving numerical differentiation formulas. Results. Math. **67**, 403–416 (2015)
33. Pečarić, J.E., Proschan, F., Tong, Y.L.: Convex Functions. Academic, Boston (1992)
34. Popoviciu, T.: Sur quelques proprietes des fonctions d'une ou de deux variables reelles. Mathematica **8**, 1–85 (1934)
35. Rajba, T.: On probabilistic characterizations of convexity and delta-convexity. Conference on Inequalities and Applications '14, September 7–13, 2014, Hajdúszoboszló (Hungary)
36. Rajba, T.: On the Ohlin lemma for Hermite-Hadamard-Fejér type inequalities. Math. Inequal. Appl. **17**, 557–571 (2014)
37. Rajba, T.: On strong delta-convexity and Hermite-Hadamard type inequalities for delta-convex functions of higher order. Math. Inequal. Appl. **18**, 267–293 (2015)
38. Rajba, T.: On a generalization of a theorem of Levin and Stečkin and inequalities of the Hermite-Hadamard type. arXiv:1509.01716 [math.CA]

39. Ralston, A.: A First Course in Numerical Analysis. McGraw-Hill Book Company, New York, St. Louis, San Francisco, Toronto, London, Sydney (1965)
40. Szostok, T.: Ohlin's lemma and some inequalities of the Hermite-Hadamard type. Aequationes Math. **89**, 915–926 (2015)
41. Szostok, T.: Levin-Stečkin theorem and inequalities of the Hermite-Hadamard type. arXiv:1411.7708v1 [math.CA]
42. Szostok, T.: Functional inequalities involving numerical differentiation formulas of order two. Bull. Malays. Math. Sci. Soc., 1–4 (2017). doi:10.1007/s40840-017-0462-3
43. Vasić, P.M., Lacković, I.B.: Some complements to the paper: on an inequality for convex functions. Univ. Beograd Publ. Elek. Fak., Ser. Mat. Fiz. **544–576**, 59–62 (1976)
44. Wąsowicz, S.: Inequalities between the quadrature operators and error bounds of quadrature rules. J. Inequal. Pure Appl. Math. **8**, Article 42, 8 pp. (2007)
45. Wąsowicz, S.: On quadrature rules, inequalities and error bounds. J. Inequal. Pure Appl. Math. **9**, Article 36, 4 pp. (2008)
46. Wąsowicz, S.: A new proof of some inequality connected with quadratures. J. Inequal. Pure Appl. Math. **9**, Article 7, 3 pp. (2008)
47. Wąsowicz, S.: On some extremalities in the approximate integration. Math. Inequal. Appl. **13**, 165–174 (2010)
48. Weisstein, E.W.: Chebyshev Quadrature. Online: http://mathworld.wolfram.com/ChebyshevQuadrature.html
49. Weisstein, E.W.: Legendre-Gauss Quadrature. Online: http://mathworld.wolfram.com/Legendre-GaussQuadrature.html
50. Weisstein, E.W.: Lobatto Quadrature. Online: http://mathworld.wolfram.com/LobattoQuadrature.html
51. Weisstein, E.W.: Simpson's Rule. Online: http://mathworld.wolfram.com/SimpsonsRule.html

Chapter 12
On the Construction of the Field of Reals by Means of Functional Equations and Their Stability and Related Topics

Jens Schwaiger

Abstract There are certain approaches to the construction of the field of real numbers which do not refer to the field of rationals. Two of these ideas are closely related to stability investigations for the Cauchy equation and for some homogeneity equation. The a priory different subgroups of $\mathbb{Z}^{\mathbb{Z}}$ used are shown to be more or less identical. Extension of these investigations shows that given a commutative semigroup G and a normed space X with completion X_c the group $\mathrm{Hom}(G, X_c)$ is isomorphic to $\mathscr{A}(G,X)/\mathscr{B}(G,X)$ where $\mathscr{B}(G,X)$ is the subgroup of X^G of all bounded functions and $\mathscr{A}(G,X)$ the subgroup of those $f: G \to X$ for which the Cauchy difference $(x,y) \mapsto f(x+y) - f(x) - f(y)$ is bounded.

The space $\mathrm{Hom}(\mathbb{N}, X_c)$ may be identified with X_c itself. With this in mind, we are able to show directly that $\mathscr{A}(\mathbb{N},X)/\mathscr{B}(\mathbb{N},X)$ is a completion of the normed space X.

Keywords Stability of the Cauchy equation • Completion of normed spaces

Mathematics Subject Classification (2010) 39B82, 46B99, 54D35

12.1 Introduction

Stability of functional equations is a very active and topical field of research. The example par excellence is the famous result found in Hyers [6].

Theorem 12.1 *Let G be an abelian semigroup and X a complete normed space. Given $f: G \to X$, assume that $\gamma_f: G \times G \to X$, $\gamma_f(x,y) := f(x+y) - f(x) - f(y)$, is bounded. Then, there is a unique $g \in Hom(G,X)$ such that $f - g$ is bounded.*

J. Schwaiger (✉)
Institute for Mathematics and Scientific Computing, University of Graz, Heinrichstraße 36, 8010 Graz, Austria
e-mail: jens.schwaiger@uni-graz.at

J. Brzdęk et al. (eds.), *Developments in Functional Equations and Related Topics*, Springer Optimization and Its Applications 124,
DOI 10.1007/978-3-319-61732-9_12

Moreover, $\|\gamma_f\|_\infty := \sup\{\|\gamma_f(x,y)\| \mid x,y \in G\} \leq \varepsilon$ *implies that* $\|f(x) - g(x)\| \leq \varepsilon$ *for all* $x \in G$.

Similar results hold true if the abelian group $(G,\cdot)$ operates on some set M when considering the inequality

$$\|f(g\cdot x) - \varphi(g)f(x)\| \leq c\psi(g), \quad x \in M,\ g \in G, \tag{12.1}$$

where $\varphi, \psi \in \mathrm{Hom}(G, \mathbb{R}\setminus\{0\})$ and $c \geq 0$. See Jabłoński and Schwaiger [7] for even more general results.

There are many construction methods for the field $\mathbb{R}$ of real numbers. They use many different principles and ideas. Contrasting most of these ideas some of them do not require the field of rationals as a tool. The starting point there is the ring of integers. In Faltin et al. [4], a subring of the ring of formal Laurent series over $\mathbb{Z}$ factored by some maximal ideal, namely the principal ideal generated by a certain *carry string*, is used.

Two other ones are closely related to functional equations and their stability. Schönhage [11] uses the subgroup $\mathscr{S} := \bigcup_{c\in\mathbb{N}} \mathscr{S}_c$ of $\mathbb{Z}^{\mathbb{N}}$, where

$$\mathscr{S}_c := \{g \in \mathbb{Z}^{\mathbb{N}} \mid |g(kn) - kg(n)| \leq ck \text{ for all } n,k \in \mathbb{N}\}. \tag{12.2}$$

In A'Campo [1], the starting point for the construction is the subgroup $\mathscr{A} := \bigcup_{c\in\mathbb{N}} \mathscr{A}_c$ of $\mathbb{Z}^{\mathbb{Z}}$, where

$$\mathscr{A}_c := \{f \in \mathbb{Z}^{\mathbb{Z}} \mid |f(n+m) - f(n) - f(m)| \leq c \text{ for all } n,m \in \mathbb{Z}\}. \tag{12.3}$$

Some basic tools are the following ones.

Theorem 12.2 *Let G be an abelian semigroup and X a normed vector space over the field $\mathbb{Q}$ of rationals. Given $c \geq 0$, let $f \in \mathscr{A}_c(G,X) := \{h \in X^G \mid \|\gamma_h\|_\infty \leq c\}$, where $\gamma_h(x,y) := h(x+y) - h(x) - h(y)$ is the* Cauchy difference *of h and*

$$\|\gamma_h\|_\infty := \sup_{x,y\in G}\|\gamma_h(x,y)\|.$$

Then

$$\|f(nx) - nf(x)\| \leq (n-1)c, \qquad x \in G, n \in \mathbb{N}, \tag{12.4}$$

and the sequence $(f(nx)/n)_{n\in\mathbb{N}}$ is a Cauchy sequence, since it satisfies

$$\left\|\frac{f(nx)}{n} - \frac{f(mx)}{m}\right\| \leq \left(\frac{1}{n} + \frac{1}{m}\right)c, \qquad x \in G, n,m \in \mathbb{N}.$$

Moreover, if this sequence converges for all $x \in G$, the limit function a,

$$a(x) := \lim_{n\to\infty} \frac{f(nx)}{n},$$

is an element of Hom(G, X), the set of homomorphisms from G to X. This a satisfies $\|f - a\|_\infty \leq c$ and it is the only homomorphism b, such that f − b is bounded.

Proof The first assertion is clear for $n = 1$. If it is true for n, we get

$$\begin{aligned}\|f((n+1)x) - (n+1)f(x)\| &= \|f(nx+x) - f(nx) - f(x) + f(nx) - nf(x)\| \\ &\leq \|f(nx+x) - f(nx) - f(x)\| + \|f(nx) - nf(x)\| \\ &\leq c + (n-1)c = nc.\end{aligned}$$

By the first part, $\|f(nmx) - nf(mx)\| \leq (n-1)c$ and $\|f(mnx) - mf(nx)\| \leq (m-1)c$. Thus,

$$\begin{aligned}\|nf(mx) - mf(nx)\| &\leq \|nf(mx) - f(nmx)\| + \|f(nmx) - mf(nx)\| \\ &\leq (n-1+m-1)c.\end{aligned}$$

Dividing by nm gives the desired result. So, the second assertion also is proved.

Finally, let

$$a_n(x) := \frac{f(nx)}{n}.$$

The properties of f imply

$$\|a_n(x+y) - a_n(x) - a_n(y)\| \leq \frac{c}{n}.$$

Thus a,

$$a(x) := \lim_{n\to\infty} a_n(x),$$

lies in $\mathrm{Hom}(G, X)$. Moreover, the second part with $n = 1$ gives

$$\|f(x) - a_m(x)\| \leq (1 + \frac{1}{m})c.$$

Taking the limit for m to ∞ shows that $\|f - a\|_\infty \leq c$.

Finally, assume that for $b \in \mathrm{Hom}(G, X)$ the difference $f - b$ is bounded. Then the homomorphism $b - a$ is bounded as well. This implies $b - a = 0$. □

Let now G be merely a set on which $(\mathbb{N}, \cdot)$ operates via $(n, x) \mapsto nx$ such that $n(mx) = (nm)x$ and $1x = 1$. Furthermore, let

$$\mathscr{S}_c(G, X) := \{f \in X^G \mid \|f(nx) - nf(x)\| \leq nc,\ x \in G, n \in \mathbb{N}\}.$$

Then we have the following result.

Theorem 12.3 *For any $f \in \mathscr{S}_c(G,X)$ and any $x \in G$, the sequence of the values $a_n(x) := f(nx)/n$ satisfies*

$$\|a_n(x) - a_m(x)\| \leq \left(\frac{1}{n} + \frac{1}{m}\right) c.$$

Therefore, it is a Cauchy sequence. If it converges for all x, the limit function a,

$$a(x) := \lim_{n\to\infty} a_n(x),$$

satisfies $a(nx) = na(x)$ for all n and all x. Moreover,

$$\|f - a\|_\infty \leq c$$

and a is the only homogeneous function b (i. e., $b(nx) = nb(x)$ for all $x \in G$ and all $n \in \mathbb{N}$), such that $f - b$ is bounded.

Proof $f \in \mathscr{S}_c(X)$ implies $\|f(nmx) - nf(mx)\| \leq nc$ and $\|f(mnx) - mf(nx)\| \leq mc$. As in the proof above, this means

$$\|a_n(x) - a_m(x)\| \leq \left(\frac{1}{n} + \frac{1}{m}\right) c.$$

This with $n = 1$ implies $\|f(x) - a(x)\| \leq c$. Using $\|f(mnx) - mf(nx)\| \leq mc$ or

$$\left\|\frac{f(nmx)}{n} - m\frac{f(nx)}{n}\right\| \leq \frac{m}{n}c$$

in the limit case $n \to \infty$ shows that $a(mx) - ma(x) = 0$ for all m and x. If $f - b$ is bounded and b homogeneous, then $a - b$ is also bounded and homogeneous. Thus, $a - b = 0$. □

Remark 12.1 Rational normed vector spaces are considered by Bourbaki, where in [3, TVS I.6] it is shown that the completion of such a space exists and that it is a real Banach space. Normed vector spaces over the rationals are special cases of normed abelian groups introduced in [13]: The homogeneity condition in normed abelian groups X reads as $\|nx\| = |n|\,\|x\|$ for all $n \in \mathbb{Z}$ and all $x \in X$.

Remark 12.2 For abelian semigroups G and normed abelian groups X, the set

$$\mathscr{A}(G,X) := \bigcup_{c\geq 0} \mathscr{A}_c(G,X)$$

is a subgroup of the abelian group X^G containing $\mathscr{B}(G,X)$ the subgroup of bounded functions in X^G.

If G is a set on which $\mathbb{N}$ operates in the above sense, the set

$$\mathscr{S}(G,X) := \bigcup_{c \geq 0} \mathscr{S}_c(G,X)$$

is also a subgroup of X^G containing $\mathscr{B}(G,X)$.

If additionally X is a module over some subring R of $\mathbb{C}$, the above subgroups are also modules over that ring. In particular, this applies to rational vector spaces X.

Proof $\mathscr{A}_c(G,X) \subseteq \mathscr{A}_d(G,X)$ for all $0 \leq c \leq d$ implies $0 \in \mathscr{A}_0(G,X) \subseteq \mathscr{A}(G,X)$. Moreover, by the triangle inequality

$$\mathscr{A}_c(G,X) + \mathscr{A}_d(G,X) \subseteq \mathscr{A}_{c+d}(G,X).$$

Finally,

$$r\mathscr{A}_c(G,X) \subseteq \mathscr{A}_{|r|c(G,X)} \text{ for all } r \in R.$$

Any $f \in \mathscr{B}(G,X)$ with $\|f(x)\| \leq c$ for all x is contained in $\mathscr{A}_{3c}(G,X)$.

The arguments are similar for $\mathscr{S}(G,X)$. In this case, $\|f(x)\| \leq c$ implies that $f \in \mathscr{S}_{2c}(G,X)$. □

Remark 12.3 (Least Absolute and Least Nonnegative Remainder) For further use, we also note that given $m \in \mathbb{N}$ any integer n may be written uniquely as $n = \alpha m + \rho$ with $\alpha, \rho \in \mathbb{Z}$ provided that ρ satisfies $-m \leq 2\rho < m$. The uniquely determined α will be denoted by $\langle n \colon m\rangle$. Thus,

$$-m \leq 2(n - \langle n \colon m\rangle m) < m$$

(and therefore a fortiori $|n - \langle n \colon m\rangle m| < m$).

n may also be written uniquely in the form $n = \beta m + \sigma$ with integers β, σ such that $0 \leq \sigma < m$. β will be denoted by $[n : m]$ and satisfies

$$0 \leq n - [n : m]m < m.$$

12.2 Two Constructions of the Reals and the Interplay Between Them

12.2.1 Schönhage

The base of construction in [11] is the set

$$\mathscr{S} := \bigcup_{c \in \mathbb{N}} \mathscr{S}_c$$

with

$$\mathscr{S}_c := \mathscr{S}_c(\mathbb{N}, \mathbb{Z}).$$

It is shown that $\mathscr{S}/\mathscr{B}(\mathbb{N}, \mathbb{Z})$ is an ordered field in which every subset bounded from above admits a supremum. Thus, this quotient group is a model of the set $\mathbb{R}$ of reals. All constructions are done completely in $\mathbb{Z}$, no noninteger rational numbers have to be used.

Addition is the one inherited from $\mathscr{S}$. A quasiorder on $\mathscr{S}$ is defined by

$$f \leq g :\Longleftrightarrow \text{ there is some integer } c \text{ such that } f(n) \leq g(n) + c \text{ for all } n \in \mathbb{N}.$$

This quasiorder is compatible with the addition. Both $f \leq g$ and $g \leq f$ are satisfied if and only if $f - g \in \mathscr{B}(\mathbb{N}, \mathbb{Z})$. This quasiorder is a total order, too. Accordingly, the relation

$$f + \mathscr{B}(\mathbb{N}, \mathbb{Z}) \leq g + \mathscr{B}(\mathbb{N}, \mathbb{Z}) :\Longleftrightarrow f \leq g$$

on $\mathscr{S}/\mathscr{B}(\mathbb{N}, \mathbb{Z})$ is well defined and, by the properties of the quasiorder on $\mathscr{S}$, it is a total order compatible with the addition of equivalence classes. A convenient fact,

$$\text{for any } f \in \mathscr{S}_c \text{ there is some } f' \in \mathscr{S}_2 \text{ such that } f - f' \in \mathscr{B}(\mathbb{N}, \mathbb{Z}),$$

is used several times, for instance, in the proof that any non-empty subset A of $\mathscr{S}/\mathscr{B}(\mathbb{N}, \mathbb{Z})$, which is bounded from above, admits a supremum. To this aim, A is written as

$$A = \{f + \mathscr{B}(\mathbb{N}, \mathbb{Z}) \mid f \in A'\}$$

with $A' \subseteq \mathscr{S}_2$. In the same manner, the set B of upper bounds of A is written as

$$B = \{f + \mathscr{B}(\mathbb{N}, \mathbb{Z}) \mid f \in B'\}$$

with $B' \subseteq \mathscr{S}_2$. Then, $f(n) \leq g(n) + 8$ for all $n \in \mathbb{N}, f \in A', g \in B'$. Accordingly, we may define $h \in \mathbb{Z}^{\mathbb{N}}$ by

$$h(n) := \max_{f \in A'}\{f(n)\}.$$

Then, $h(n) \leq g(n) + 8$ for all $n \in \mathbb{N}$, $g \in B'$. Moreover, it is shown that $h \in \mathscr{S}$. This and the definition of h implies that

$$h + \mathscr{B}(\mathbb{N}, \mathbb{Z})$$

is a supremum of A.

Remark 12.4 Given $f \in \mathcal{S}_c$, Schönhage uses the function

$$n \mapsto [f(kn) : k] =: g(n)$$

and shows that $g \in f + \mathcal{B}(\mathbb{N}, \mathbb{Z})$ and that $g \in \mathcal{S}_2$ for sufficiently large k. Using h, $h(n) := \langle f(kn) \colon k \rangle$, instead of g, it turns out that also $f - h$ is bounded and that even $h \in \mathcal{S}_1$ is true provided that k is suitably large.

In fact:

$$kh(n) = f(kn) + u(kn) \text{ with } 2\,|u(kn)| \leq k \text{ and}$$
$$2k\,|h(n) - f(n)| = |2(f(kn) - kf(n)) - 2u(kn)| \leq 2kc + k < 2(c+1)k.$$

Thus, $h - f$ is bounded. Moreover,

$$2k\,(h(mn) - mh(n)) = 2f(kmn) + 2u(kmn) - 2mf(kn) - 2mu(kn)$$

and

$$\begin{aligned} 2k\,|h(mn) - mh(n)| &\leq 2\,|f(kmn) - mf(kn)| + |2u(kmn) - 2mu(kn)| \\ &\leq 2cm + (m+1)k. \end{aligned}$$

For $k \geq 2c$, this implies $2k\,|h(mn) - mh(n)| \leq (2m+1)k$ or $2\,|h(mn) - mh(n)| \leq 2m + 1$. Since only integers are involved, this finally shows that

$$|h(mn) - mh(n)| \leq 1 \cdot m.$$

Multiplication for $f, g \in \mathcal{S}$ can be defined by

$$n \mapsto \langle f(n)g(n) \colon n \rangle n.$$

(Schönhage used $n \mapsto [f(n)g(n) : n]$ instead.) Denoting this by $f * g$, it is verified that $f * g \in \mathcal{S}$ in the following way:

Without loss of generality, we may assume that there is some c common to f and g such that $f, g \in \mathcal{S}_c$. Thus, $|f(n) - nf(1)|\,,\ |g(n) - ng(1)| \leq cn$ imply

$$|f(n)|\,,\ |g(n)| \leq c'n$$

for $c' := c + \max\{|f(1|\,,\ |g(1)|)\}$. According to

$$\begin{aligned} nk(\langle f(nk)g(nk) \colon nk \rangle &- k\langle f(n)g(n) \colon n \rangle) \\ &= (nk\langle f(nk)g(nk) \colon nk \rangle - f(nk)g(nk)) \\ &\quad + f(nk)\,(g(nk) - kg(n)) + kg(n)\,(f(nk) - kf(n)) \\ &\quad + k^2\,(f(n)g(n) - n\langle f(n)g(n) \colon n \rangle) \end{aligned}$$

we get for all n, k that

$$|nk\,(\langle f(nk)g(nk)\colon nk\rangle - k\langle f(n)g(n)\colon n\rangle)| \le nk + c'nkck + kc'nkc + k^2n,$$

implying that

$$|(\langle f(nk)g(nk)\colon nk\rangle - k\langle f(n)g(n)\colon n\rangle)| \le 1 + 2cc'k + k \le 2(1 + cc')k.$$

So,

$$f * g \in \mathscr{S}_{2(1+cc')}.$$

For bounded h_1, h_2, it is seen easily that

$$(f + h_1) * (g + h_2) = f * g + h_3$$

for some bounded h_3. Thus, a product on

$$\mathscr{S}/\mathscr{B}(\mathbb{N}, \mathbb{Z})$$

may be defined by

$$(f + \mathscr{B}(\mathbb{N}, \mathbb{Z})) \cdot (g + \mathscr{B}(\mathbb{N}, \mathbb{Z})) := f * g + \mathscr{B}(\mathbb{N}, \mathbb{Z}).$$

This product makes $\mathscr{S}/\mathscr{B}(\mathbb{N}, \mathbb{Z})$ to a commutative ring with unit element

$$\mathrm{id}_{\mathbb{N}} + \mathscr{B}(\mathbb{N}, \mathbb{Z}) =: 1.$$

Since $f * g \ge 0$ for $f, g \ge 0$, this is also an ordered ring. Finally, one verifies that this ring is even a field. If $f \in \mathscr{S}$ and $f + \mathscr{B}(\mathbb{N}, \mathbb{Z}) > 0$, we may additionally assume that $mf(n) \ge n, f(n) \ge 1$, and $f(n) \le dn$ for some m, d, and all n. Then, $f'\colon \mathbb{N} \to \mathbb{Z}$,

$$f'(n) := \langle n^2 \colon f(n)\rangle,$$

is contained in $\mathscr{S}$:

$$\begin{aligned} kn^2 &\left|\langle k^2n^2 \colon f(kn)\rangle - k\langle n^2 \colon f(n)\rangle\right| \\ &\le m^2 f(kn)f(n) \left|\langle k^2n^2 \colon f(kn)\rangle - k\langle n^2 \colon f(n)\rangle\right| \\ &= m^2 \left|f(n)\left(k^2n^2 - r_1\right) - kf(nk)\left(n^2 - r_2\right)\right| \end{aligned}$$

with $|r_1| < f(kn)$, $|r_2| < f(n)$. Therefore,

$$\begin{aligned} kn^2 &\left|\langle k^2n^2 \colon f(kn)\rangle - k\langle n^2 \colon f(n)\rangle\right| \\ &\le m^2 \left|f(n)r_1 + kf(nk)r_2\right| + kn^2 \left|kf(n) - f(nk)\right| \\ &\le m^2(dnnk + kdnkn + kn^2kc) = m^2n^2k(d + dk + kc) \end{aligned}$$

when we assume that $f \in \mathcal{S}_c$. This implies

$$\left|\langle k^2n^2 {:} f(kn)\rangle - k\langle n^2 {:} f(n)\rangle\right| \le m^2n^2k(2d+c)k$$

and

$$\left|\langle k^2n^2 {:} f(kn)\rangle - k\langle n^2 {:} f(n)\rangle\right| \le m^2(2d+c)k.$$

An (easier) calculation shows that

$$f * f' - \mathrm{id}_{\mathbb{N}} \in \mathcal{B}(\mathbb{N},\mathbb{Z}).$$

This implies

$$(f + \mathcal{B}(\mathbb{N},\mathbb{Z})) \cdot (f' + \mathcal{B}(\mathbb{N},\mathbb{Z})) = 1.$$

For $g < 0$, $g \notin \mathcal{B}(\mathbb{N},\mathbb{Z})$, choose $f \in g + \mathcal{B}(\mathbb{N},\mathbb{Z})$ such that

$$(-f) * (-f)' - \mathrm{id}_{\mathbb{N}} \in \mathcal{B}(\mathbb{N},\mathbb{Z}).$$

Then, $-(-f)' + \mathcal{B}(\mathbb{N},\mathbb{Z})$ is the multiplicative inverse of $g + \mathcal{B}(\mathbb{N},\mathbb{Z})$.

12.2.2 A'Campo et al.

Street [14] gave some hints on how to construct the reals using $\mathcal{A} := \mathcal{A}(\mathbb{Z},\mathbb{Z})$. Street [15] contains a report of what happened since then. Ross Street refers to several papers, in particular to A'Campo [1]. Nadine Manschek, one of my students, gave full worked out proofs of all important steps in [10].

From Remark 12.2, it immediately follows that $\mathcal{A}$ is an abelian group with subgroup $\mathcal{B}(\mathbb{Z},\mathbb{Z})$. Multiplication is defined by $(f,g) \mapsto f \circ g$. The proof that $f \circ g \in \mathcal{A}$ strongly depends on the fact that for $f \in \mathbb{Z}^{\mathbb{Z}}$ the boundedness of γ_f as defined in Theorem 12.1 implies that $\gamma_f(\mathbb{Z} \times \mathbb{Z})$ is finite.

Remark 12.5 This is not true for $\mathcal{A}(X,X)$ in general. In particular, there is some $f \in \mathcal{A}(\mathbb{R},\mathbb{R})$ such that $f \circ f \notin \mathcal{A}(\mathbb{R},\mathbb{R})$.

An example is given by $f = a + r$ with a additive and r bounded such that $a(\pi^{-n}) = 2^n$ for all n and $r(n) = \pi^{-n}$. (There is some additive a with this property, since $\{\pi^{-n} \mid n \in \mathbb{N}\}$ is linearly independent in the $\mathbb{Q}$-vector space $\mathbb{R}$.) Note that

$$f(f(x)) = a(a(x)) + a(r(x)) + r(f(x)).$$

Assuming $f \circ f \in \mathcal{A}(\mathbb{R},\mathbb{R})$ would imply the existence of some additive b such that $f \circ f - b$ were bounded. By Theorem 12.2,

$$b(x) = \lim_{n\to\infty} \frac{(f \circ f)(nx)}{n}.$$

But for $x = 1$

$$f(f(n \cdot 1))/n = a(a(1)) + 2^n/n + r(f(n))/n$$

produces a divergent sequence.

Then, it is shown that $f \circ g - g \circ f$ is bounded and that $f \circ g - f' \circ g$ is bounded provided that $f - f'$ is. Thus, $\mathscr{A}(\mathbb{Z}, \mathbb{Z})/\mathscr{B}(\mathbb{Z}, \mathbb{Z})$ becomes a commutative ring with unit $\mathrm{id}_{\mathbb{Z}} + \mathscr{B}(\mathbb{Z}, \mathbb{Z})$ and

$$(f + \mathscr{B}(\mathbb{Z}, \mathbb{Z})) \cdot (g + \mathscr{B}(\mathbb{Z}, \mathbb{Z})) := (f \circ g + \mathscr{B}(\mathbb{Z}, \mathbb{Z})).$$

$f \in \mathscr{A}_c$ implies $|f(nm) - nf(m)| \leq nc$ for all $m \in \mathbb{Z}$ and all $n \in \mathbb{N}$. In particular, the restriction of f to $\mathbb{N}$, $f|_{\mathbb{N}}$, is contained in $\mathscr{S}_c$ for $f \in \mathscr{A}_c$. Accordingly, the quasiorder defined in $\mathscr{S}$ is meaningful in $\mathscr{A}$ and defines a total order in $\mathscr{A}(\mathbb{Z}, \mathbb{Z})/\mathscr{B}(\mathbb{Z}, \mathbb{Z})$.

To make this ring a field, several additional steps have to be considered:

1. If $f > 0$ and $f \notin \mathscr{B}(\mathbb{Z}, \mathbb{Z})$, there is some $f' \in f + \mathscr{B}(\mathbb{Z}, \mathbb{Z})$ which is also > 0 and additionally odd.
2. For this f' and all $m \in \mathbb{Z}$, the set $M_m := f'^{-1}(\{k \in \mathbb{Z} \mid k \leq m\})$ is non-empty and bounded from above.
3. $g \in \mathbb{Z}^{\mathbb{Z}}$, $g(m) := \max M_m$, is contained in $\mathscr{A}(\mathbb{Z}, \mathbb{Z})$.
4. $f' \circ g - \mathrm{id}_{\mathbb{Z}} \in \mathscr{B}(\mathbb{Z}, \mathbb{Z})$.

Thus, any unbounded $f + \mathscr{B}(\mathbb{Z}, \mathbb{Z}) > 0$ is invertible. Inverses for $f + \mathscr{B}(\mathbb{Z}, \mathbb{Z}) < 0$ are constructed as the additive inverse of the multiplicative inverse of $(-f) + \mathscr{B}(\mathbb{Z}, \mathbb{Z})$.

Finally, it is shown that any non-empty subset A of $\mathscr{A}(\mathbb{Z}, \mathbb{Z})/\mathscr{B}(\mathbb{Z}, \mathbb{Z})$ has a supremum. Writing

$$A = \{f + \mathscr{B}(\mathbb{Z}, \mathbb{Z}) \mid f \in A'\}$$

with $A' \subseteq \mathscr{A}(\mathbb{Z}, \mathbb{Z})$, one may assume that all $f' \in A'$ are odd and elements of $\mathscr{A}_1(\mathbb{Z}, \mathbb{Z})$. For any odd $g \in \mathscr{A}(\mathbb{Z}, \mathbb{Z})$ such that $g + \mathscr{B}(\mathbb{Z}, \mathbb{Z})$ is an upper bound of A, it is seen that $f(n) \leq g(n) + 2$ for all $n \in \mathbb{N}_0$. Thus, $h \in \mathbb{Z}^{\mathbb{Z}}$ with

$$h(n) := \max\{f(n) \mid f \in A'\}$$

for $n \in \mathbb{N}_0$ and $h(n) := -h(-n)$ for $n \in \mathbb{Z}, n < 0$ is well defined. Then, some tedious calculation shows that $h \in \mathscr{A}_5$. Finally, it is shown that $h + \mathscr{B}(\mathbb{Z}, \mathbb{Z})$ is a least upper bound of A. (The definition of h is similar to the corresponding definition in Schönhage's approach.)

Remark 12.6 The abovementioned fact that f' may be chosen in $\mathscr{A}_1$ is shown by defining for given $f \in \mathscr{A}_c$ the function f' by

$$f'(n) := \langle f(nk) : k \rangle.$$

Then,

$$f - f' \in \mathscr{B}(\mathbb{Z}, \mathbb{Z})$$

and $f' \in \mathscr{A}_1$ for sufficiently large k. The proof for this is similar to that contained in Remark 12.4.

12.2.3 Synthesis

In Blatter [2], you may find the opinion that Schönhage's setting is a kind of predecessor of A'Campo's setting. In fact, in some sense, the settings are identical.

Theorem 12.4

$$\mathscr{S} = \mathscr{A}|_{\mathbb{N}} := \{f|_{\mathbb{N}} \mid f \in \mathscr{A}\} = \mathscr{A}(\mathbb{N}, \mathbb{Z})$$

and

$$\mathscr{A}/\mathscr{B}(\mathbb{Z}, \mathbb{Z}) \cong \mathscr{S}/\mathscr{B}(\mathbb{N}, \mathbb{Z}).$$

Proof Certainly, $\mathscr{A}|_{\mathbb{N}} \subseteq \mathscr{S}$ by the inequality (12.4) of Theorem 12.2, which also applies when X is an abelian normed group only. Now, take any $f \in \mathscr{S}_c$. Then,

$$|f(k) - \langle kf(m) \colon m\rangle| \leq 2c + 1$$

provided that $k \leq m$:

$$2m\,|f(k) - \langle kf(m) \colon m\rangle| \leq |2mf(k) - 2kf(m) + 2r|$$

for some r such that

$$2\,|r| \leq m.$$

Thus,

$$2m\,|f(k) - \langle kf(m) \colon m\rangle| \leq 2c(m + k) + m \leq (4c + 1)m < 2(2c + 1)m$$

since $|f(km) - kf(m)| \leq kc$ and $|f(km) - mf(k)| \leq mc$ imply

$$|mf(k) - kf(m)| \leq (k + m)c.$$

Using this and the easy to verify inequality

$$2m\,|\langle a + b \colon m\rangle - \langle a \colon m\rangle - \langle b \colon m\rangle| \leq 3m$$

we may estimate $f(k+l) - f(k) - f(l)$ as follows:

$$\begin{aligned}|f(k+l) - f(k) - f(l)| &\le |f(k+l) - \langle (k+l)f(m)\colon m\rangle|\\ &\quad + |f(k) - \langle kf(m)\colon m\rangle| + |f(l) - \langle lf(m)\colon m\rangle|\\ &\quad + |\langle (kf(m) + lf(m))\colon m\rangle - \langle (kf(m)\colon m\rangle - \langle (lf(m)\colon m\rangle|\\ &\le 3(2c+1) + 3 = 6(c+1) \text{ if } k, l \in \mathbb{N},\ k+l \le m.\end{aligned}$$

(Observe that the inequality $2m\,|\langle a+b\colon m\rangle - \langle a\colon m\rangle - \langle b\colon m\rangle| \le 3m$ implies

$$|\langle a+b\colon m\rangle - \langle a\colon m\rangle - \langle b\colon m\rangle| \le 1.)$$

Therefore,

$$f \in \mathscr{A}_{6(c+1)}(\mathbb{N}, \mathbb{Z}) \subseteq \mathscr{A}(\mathbb{N}, \mathbb{Z}).$$

For $g \in \mathscr{A}(\mathbb{N}, \mathbb{Z})$, we define (the odd extension) $g^* \in \mathbb{Z}^{\mathbb{Z}}$ by

$$g^*|_{\mathbb{N}} := g,$$

$g^*(0) := 0$, and $g^*(n) := -g(-n)$ for $n \in \mathbb{Z}, n < 0$. By considering the cases

(*a*) $n, m > 0$,
(*b*) $m = 0$ or $n = 0$,
(*c*) $n, m < 0$,
(*d*) $n < 0,\ m > 0,\ n + m = 0$,
(*e*) $n < 0,\ m > 0,\ n + m > 0$, and
(*f*) $n < 0,\ m > 0,\ n + m < 0$

and by observing that $g^*(n+m) - g^*(n) - g^*(m)$ is symmetric with respect to n and m, it can be verified that $f^* \in \mathscr{A}_{6(c+1)}(\mathbb{Z}, \mathbb{Z})$.

Altogether this shows that $\mathscr{S} = \mathscr{A}|_{\mathbb{N}}$ and $\mathscr{A}|_{\mathbb{N}} = \mathscr{A}(\mathbb{N}, \mathbb{Z})$.

Now, we prove the second assertion. Let $\varphi\colon \mathscr{A}(\mathbb{N}, \mathbb{Z}) \to \mathscr{A}(\mathbb{Z}, \mathbb{Z})/\mathscr{B}(\mathbb{Z}, \mathbb{Z})$ be defined by

$$\varphi(f) := f^* + \mathscr{B}(\mathbb{Z}, \mathbb{Z})$$

with f^* as above. Then, φ is a homomorphism of abelian groups. Since $f^* \in \mathscr{B}(\mathbb{Z}, \mathbb{Z})$ is equivalent to

$$f \in \mathscr{B}(\mathbb{N}, \mathbb{Z})$$

the kernel of φ equals $\mathscr{B}(\mathbb{N}, \mathbb{Z})$.

It remains to show that φ is surjective. For given $g + \mathscr{B}(\mathbb{Z}, \mathbb{Z})$ with $g \in \mathscr{A}(\mathbb{Z}, \mathbb{Z})$, assume that $g \in \mathscr{A}_c(\mathbb{Z}, \mathbb{Z})$. Then,

$$|g(0) - g(n) - g(-n)| \le c.$$

Thus, $|g(0)| \le c$ and

$$|g(n) - (-g(-n))| \le c + |g(0)| \le 2c$$

for all n which with

$$f := g|_{\mathbb{N}} \in \mathscr{A}(\mathbb{N}, \mathbb{Z})$$

implies $g - f^* \in \mathscr{B}(\mathbb{Z}, \mathbb{Z})$. Accordingly,

$$\varphi(f) = f^* + \mathscr{B}(\mathbb{Z}, \mathbb{Z}) = g + \mathscr{B}(\mathbb{Z}, \mathbb{Z}).$$

□

Remark 12.7 From the proof, it follows that an isomorphism $\widehat{\varphi}$ between $\mathscr{S}/\mathscr{B}(\mathbb{N}, \mathbb{Z})$ and $\mathscr{A}/\mathscr{B}(\mathbb{Z}, \mathbb{Z})$ is given by

$$f + \mathscr{B}(\mathbb{N}, \mathbb{Z}) \mapsto f^* + \mathscr{B}(\mathbb{Z}, \mathbb{Z}).$$

Till now, the usage of the set of real or even of the rational numbers has been avoided. Since at present we already have (two) models for the field $\mathbb{R}$ and since

$$\mathbb{Z} \subseteq \mathbb{Q} \subseteq \mathbb{R}$$

we may conclude from Theorem 12.2 that for any $f \in \mathscr{A}(\mathbb{N}, \mathbb{Z})$ and any $g \in \mathscr{A}(\mathbb{Z}, \mathbb{Z})$ there are real numbers α, β such that the set of $|f(n) - \alpha n|$, $n \in \mathbb{N}$, is bounded and that the same holds true for the set of $|g(n) - n\beta|$, $n \in \mathbb{Z}$. (The Cauchy sequences appearing in that theorem converge, since order completeness implies sequentially completeness; see, for example, Lang [9, Chapter 2, Theorem 1.5].) Moreover, any homomorphism a defined on $\mathbb{N}$ or $\mathbb{Z}$ is of the form $n \mapsto \gamma n$ with $\gamma = a(1)$. This will be used to show the following result, where, given any real x also the Gaussian bracket

$$[x] := \max\{m \in \mathbb{Z} \mid m \le x\}$$

is involved.

Theorem 12.5 *For any $f, g \in \mathscr{S} = \mathscr{A}(\mathbb{N}, \mathbb{Z})$, we have*

$$\widehat{\varphi}(f * g + \mathscr{B}(\mathbb{N}, \mathbb{Z})) = (f^* \circ g^*) + \mathscr{B}(\mathbb{Z}, \mathbb{Z}).$$

Thus, the multiplication in the sense of Schönhage and A'Campo coincides.

Proof Let $\alpha, \beta \in \mathbb{R}$ be such that

$$f(n) = \alpha n + u(n)$$

and

$$g(n) = \beta n + v(n)$$

for all $n \in \mathbb{N}$ with certain bounded functions u, v. Then,

$$f(n)g(n) = \alpha\beta n^2 + n(\alpha v(n) + \beta u(n)) + u(n)v(n).$$

Let

$$f(n)g(n) = n\langle f(n)g(n){:}\, n\rangle + r(n)$$

with some function r satisfying $2\,|r(n)| \leq n$. Then,

$$n\,|\langle f(n)g(n){:}\, n\rangle - \alpha\beta n| \leq cn$$

for all n with suitable c. Note that

$$\alpha\beta n = [\alpha\beta n] + s(n),$$

where $0 \leq s(n) < 1$. Thus,

$$n \mapsto ((f * g)(n) - [\alpha\beta n])$$

is bounded. Obviously,

$$\mathbb{Z} \ni n \mapsto [\alpha\beta n] \in \mathscr{A}(\mathbb{Z}, \mathbb{Z}).$$

Thus,

$$f * g - \mathrm{int}_{\alpha\beta}|_{\mathbb{N}} \in \mathscr{B}(\mathbb{N}, \mathbb{Z}),$$

where

$$\mathrm{int}_{\gamma}(n) := [\gamma n].$$

So,

$$\widehat{\varphi}(f * g + \mathscr{B}(\mathbb{N}, \mathbb{Z})) = \mathrm{int}_{\alpha\beta}|_{\mathbb{N}}^{*} + \mathscr{B}(\mathbb{N}, \mathbb{Z}).$$

If $f^*, g^* \in \mathscr{A}(\mathbb{Z}, \mathbb{Z})$ are the odd extensions of f, g, we may write

$$f^* = \alpha \mathrm{id}_{\mathbb{Z}} + u^*, g^* = \beta \mathrm{id}_{\mathbb{Z}} + v^*.$$

Then,

$$f^* \circ g^* = \alpha\beta \mathrm{id}_{\mathbb{Z}} + \alpha v^* + u^* \circ g^*$$

is implying that

$$f^* \circ g^* - \alpha\beta \mathrm{id}_{\mathbb{Z}}$$

is bounded. But, therefore also

$$f^* \circ g^* - \mathrm{int}_{\alpha\beta}$$

and

$$f^* \circ g^* - \mathrm{int}_{\alpha\beta}|_{\mathbb{N}}^*$$

are bounded. This finally implies the assertion. □

12.3 Stability and Completeness

Now, the interplay between the stability of the Cauchy equation and the completeness of the involved normed space will be investigated. In the following, Remark 12.2 should be taken into account.

Theorem 12.6 *Let G be an abelian semigroup, suppose X to be a normed vector space (over $\mathbb{Q}$) with completion X_c. Then, $\mathscr{A}(G,X)/\mathscr{B}(G,X) \cong Hom(G,X_c)$, the group of homomorphisms defined on G with values in X_c.*

Proof Since

$$\mathscr{A}(G,X) \subseteq \mathscr{A}(G,X_c)$$

Theorem 12.2 may be applied. Thus, given $f \in \mathscr{A}(G,X)$ the mapping a_f with

$$a_f(x) := \lim_{n\to\infty} \frac{f(nx)}{n}$$

is contained in $\mathrm{Hom}(G,X_c)$. Moreover, $f - a_f$ is bounded. Let

$$\varphi: \mathscr{A}(G,X) \to \mathrm{Hom}(G,X_c)$$

be defined by $\varphi(f) := a_f$. Then, obviously φ is a homomorphism. Since f is bounded iff a_f is bounded, the kernel of φ equals $\mathscr{B}(G,X)$. Since

$$\mathscr{A}(G,X)/\ker(\varphi) \cong \varphi(\mathscr{A}(G,X))$$

it remains to show that φ is surjective.

To this aim, let $a \in \mathrm{Hom}(G, X_c)$ be arbitrary. For any $x \in G$, we may choose some $f(x) \in X$ such that $\|a(x) - f(x)\| < 1$. Then, f is an element of $\mathscr{A}_3(G, X)$ because of

$$\|f(x+y) - f(x) - f(y)\| \le \|f(x+y) - a(x+y)\| \\ + \|f(x) - a(x)\| + \|f(y) - a(y)\| < 3.$$

The definition of f implies $\varphi(f) = a$. □

Corollary 12.1 *The groups $\mathscr{A}(\mathbb{N}, X)/\mathscr{B}(\mathbb{N}, X)$, $\mathscr{A}(\mathbb{Z}, X)/\mathscr{B}(\mathbb{Z}, X)$ both are isomorphic to X_c. In particular, for $X = \mathbb{Q}$ these groups are isomorphic to $\mathbb{R}$.*

Proof In both cases for G, the mapping $\mathrm{Hom}(G, X_c) \ni a \mapsto a(1) \in X_c$ is an isomorphism. □

It was proved for $G = \mathbb{Z}$ in Schwaiger [12] and for arbitrary abelian groups G containing at least one element of infinite order in Forti and Schwaiger [5] that the following theorem holds true.

Theorem 12.7 (Hyers' Theorem and Completeness) *If G is an abelian group as above and X a normed space such that for any $f \in \mathscr{A}(G, X)$ there is some $a \in \mathrm{Hom}(G, X)$ such that $f - a$ is bounded, then X necessarily must be complete.*

Proof (Alternative Method) In Forti and Schwaiger [5], it was shown that it is enough to prove the result for $G = \mathbb{Z}$. There the latter task was managed by constructing to any Cauchy sequence $(x_n)_{n\in\mathbb{N}}$ a suitable $f \in \mathscr{A}(\mathbb{Z}, X)$ and to use the hypotheses of the theorem for this f. Here, it is done in the following way: Choose any $\alpha \in X_c$ and $f\colon \mathbb{Z} \to X$ such that

$$\|f(n) - \alpha n\| < 1$$

for all $n \in \mathbb{Z}$. Then, $f \in \mathscr{A}(\mathbb{Z}, X)$ and by assumption there is some $a \in \mathrm{Hom}(\mathbb{Z}, X)$ such that $f - a$ is bonded. Since a is a homomorphism defined on $\mathbb{Z}$, there is some $\beta \in X$ such that

$$a(n) = \beta n$$

for all $n \in \mathbb{Z}$. Thus, also $\alpha \mathrm{id}_{\mathbb{Z}} - a$ is bounded implying $\alpha = \beta \in X$. Therefore, $X_c = X$. □

12.4 A Construction Method for the Completion of a Normed Space

There are well-known methods to construct the completion of metric and normed spaces. The most common ones use the set of Cauchy sequences on the underlying space. A different one, probably first mentioned by Kunugui [8], is contained in the following remark.

Remark 12.8 Let (X,d) be any (non-empty) metric space. Then, there is an isometry j from X into the Banach space $\mathscr{B}(X,\mathbb{R})$ of real-valued bounded functions defined on X. Thus, the closure of $j(X)$ is a completion of X.

The embedding is constructed by fixing some x_0 in X and by defining $j(x)$ pointwise as

$$j(x)(y) := d(y,x) - d(y,x_0).$$

If X has some additional structure, say of a normed space, this can be carried over easily to the completion $\overline{j(X)}$.

In the context of the present considerations, it has already been shown that for any normed space the factor group $\mathscr{A}(\mathbb{N},X)/\mathscr{B}(\mathbb{N},X)$ is isomorphic to the completion X_c of X. But, it seems to be desirable and interesting to give a proof that $\mathscr{A}(\mathbb{N},X)/(\mathbb{N},X)$ is a completion of X not using the existence of a completion of X a priori.

Theorem 12.8 *Let X be a real normed space, let $\mathscr{A} := \mathscr{A}(\mathbb{N},X)$ and $\mathscr{B} := \mathscr{B}(\mathbb{N},X)$. Then, $\mathscr{A}/\mathscr{B}$ is a completion of X, if addition and multiplication by a real number are defined as usual and if $\|f + \mathscr{B}\|$ is given by the well-defined limit*

$$\lim_{n\to\infty} \left\| \frac{f(n)}{n} \right\|.$$

Proof If X is a normed space, the abelian groups $\mathscr{A}$ and $\mathscr{B}$ are not only abelian groups but vector spaces by Remark 12.2. Thus, $\mathscr{A}/\mathscr{B}$ is a real vector space. Given $f \in \mathscr{A}$, we know by Theorem 12.2 that the sequence $(f(n)/n)_{n\in\mathbb{N}}$ is Cauchy in X. In detail

$$\left\| \frac{f(n)}{n} - \frac{f(m)}{m} \right\| \le c\left(\frac{1}{n} + \frac{1}{m}\right)$$

for $f \in \mathscr{A}_c := \mathscr{A}_c(\mathbb{N},X)$. Thus, by the reversed triangle inequality

$$|\|a\| - \|b\|| \le \|a - b\|$$

the sequence $(\|f(n)/n\|)_{n\in\mathbb{N}}$ is Cauchy in the complete normed space $\mathbb{R}$. Thus, we may define

$$\|f\| := \lim_{n\to\infty} \left\| \frac{f(n)}{n} \right\|.$$

Obviously, this is a seminorm on $\mathscr{A}$. Now, it is shown that

$$\{f \in \mathscr{A} \mid \|f\| = 0\} = \mathscr{B}.$$

If $f \in \mathscr{B}$, the sequence of $f(n)$ is bounded. Thus,

$$\|f\| = \lim_{n\to\infty} \left\| \frac{f(n)}{n} \right\| = 0.$$

On the other hand, let $f \in \mathscr{A}$ satisfy $\|f\| = 0$. Then, $\|f(n)/n\|$ tends to 0 for n tending to ∞. Therefore, the sequence of $f(n)/n$ converges to 0 in X. This implies that

$$\lim_{n\to\infty} \frac{f(nm)}{n} = m \lim_{n\to\infty} \frac{f(nm)}{nm} = 0$$

for all $m \in \mathbb{N}$. Theorem 12.2 thus implies that $f - 0 = f$ is bounded.

Accordingly, we may define a norm on $\mathscr{A}/\mathscr{B}$ by

$$\|f + \mathscr{B}\| := \|f\|.$$

To show that $\mathscr{A}/\mathscr{B}$ equipped with this norm is complete, we need the following result.

Claim For any $f \in \mathscr{A}$ and any $\varepsilon > 0$, there is some $g \in (f + \mathscr{B}) \cap \mathscr{A}_\varepsilon$.

For the proof, let $f \in \mathscr{A}_c$, i. e.,

$$\|f(n+m) - f(n) - f(m)\| \le c$$

for all $n, m \in \mathbb{N}$. Taking any $m \in \mathbb{N}$ such that $c/m \le \varepsilon$, the function g is defined by

$$g(n) := \frac{f(mn)}{m},$$

a construction similar to some used earlier. Then,

$$\|g(k+l) - g(k) - g(l)\| = \frac{1}{m}\|f(mk+ml) - f(ml) - f(mk)\| \le \frac{c}{m} \le \varepsilon.$$

Accordingly, $g \in \mathscr{A}_\varepsilon$. The estimation

$$\|f(mn) - mf(n)\| \le cm$$

from Theorem 12.2 implies that $g \in f + \mathscr{B}$.

To show the completeness of the normed space $\mathscr{A}/\mathscr{B}$, it is enough to show that any Cauchy sequence admits a convergent subsequence. So, let the Cauchy sequence $(f_n + \mathscr{B})$ with $f_n \in \mathscr{A}$ be given. By eventually passing to a subsequence, we may assume that

$$\|f_{n+1} + \mathscr{B} - (f_n + \mathscr{B})\| \le \frac{1}{2^{n+1}}, \qquad n \in \mathbb{N}.$$

Additionally, by the claim above, we may also assume that

$$f_n \in \mathscr{A}_{\frac{1}{2^n}}, \qquad n \in \mathbb{N}.$$

Using

$$\|f_{n+1} + \mathscr{B} - (f_n + \mathscr{B})\| \leq \frac{1}{2^{n+1}}$$

the triangle inequality shows that

$$\|f_{n+m} + \mathscr{B} - (f_n + \mathscr{B})\| \leq \frac{1}{2^{n+1}} + \ldots + \frac{1}{2^{n+m}} < \frac{1}{2^n}, \qquad n, m \in \mathbb{N}.$$

Moreover,

$$f_{n+m} - f_n \in \mathscr{A}_{\frac{1}{2^{n+m}}} + \mathscr{A}_{\frac{1}{2^n}} \subseteq \mathscr{A}_{\frac{1}{2^{n+1}} + \frac{1}{2^n}} = \mathscr{A}_{\frac{3}{2^{n+1}}}.$$

Thus, by Theorem 12.2

$$\|(f_{n+m} - f_n)(k) - k(f_{n+m} - f_n)(1)\| \leq k\frac{3}{2^{n+1}}$$

and

$$\|(f_{n+m} - f_n)(1)\| \leq \frac{3}{2^{n+1}} + \left\| \frac{(f_{n+m} - f_n)(k)}{k} \right\|,$$

which for $k \to \infty$ results in

$$\|f_{n+m}(1) - f_n(1)\| \leq \frac{5}{2^{n+1}}.$$

Now, let $f \in X^{\mathbb{N}}$ be defined by $f(n) := nf_n(1)$ for all $n \in \mathbb{N}$. Then,

$$\begin{aligned}(n+m)&f_{n+m}(1) - nf_n(1) - mf_m(1)\\ &= n(f_{n+m}(1) - f_n(1)) + m(f_{n+m}(1) - f_m(1)\end{aligned}$$

implies

$$\begin{aligned}\|f(n+m) - f(n) - f(m)\| &\leq \frac{5n}{2^{n+1}} + \frac{5m}{2^{m+1}} = \frac{5}{2}\left(\frac{n}{2^n} + \frac{m}{2^m}\right)\\ &\leq \frac{5}{2}2\frac{1}{2} = \frac{5}{2}\end{aligned}$$

and thus $f \in \mathscr{A}_{\frac{5}{2}}$.

Next, we consider $f - f_n$. The equality

$$\begin{aligned}(f - f_n)(n+m) &= (n+m)f_{n+m}(1) - (n+m)f_n(1)\\ &\quad + ((n+m)f_n(1) - f_n(n+m))\end{aligned}$$

implies

$$\|(f - f_n)(n+m)\| \leq (n+m)\,\|f_{n+m}(1) - f_n(1)\| + \frac{1}{2^{n+1}}(n+m)$$

and

$$\left\| \frac{(f - f_n)(n+m)}{n+m} \right\| \leq \|f_{n+m}(1) - f_n(1)\| + \frac{1}{2^{n+1}}$$
$$\leq \frac{5}{2^{n+1}} + \frac{1}{2^{n+1}} = \frac{3}{2^n}.$$

Thus,

$$\|(f - f_n) + \mathscr{B}\| = \lim_{m \to \infty} \left\| \frac{(f - f_n)(n+m)}{n+m} \right\| \leq \frac{3}{2^n}$$

implying that $f_n + \mathscr{B}$ tends to $f + \mathscr{B}$ in $\mathscr{X}$.

Finally, we find an isometry $j\colon X \to \mathscr{A}/\mathscr{B}$ such that $j(X)$ is dense in $\mathscr{A}/\mathscr{B}$. Given $x \in X$, the function α_x,

$$\alpha_x(n) := nx,$$

is contained in $\mathscr{A}_0$. Let

$$j(x) := \alpha_x + \mathscr{B}.$$

Then, j is a vector space homomorphism from X onto $j(X)$. This is also an isometry since $\|\alpha_x\| = \|x\|$. Let $f + \mathscr{B} \in \mathscr{A}/\mathscr{B}$ and $\varepsilon > 0$. We may assume that $f \in \mathscr{A}_\varepsilon$. Then, with $x := f(1)$ we have

$$\|f(n) - nx\| \leq n\varepsilon,$$

which implies that

$$\|f + \mathscr{B} - \alpha_x\| \leq \varepsilon.$$

□

References

1. A'Campo, N.: A natural construction for the real numbers. http://www.math.ethz.ch/~salamon/PREPRINTS/acampo-real.pdf (2003)
2. Blatter, Ch.: A concrete model of the real numbers. Elem. Math. **65**(2), 49–61 (2010). doi:10.4171/EM/140; Ein konkretes Modell der reellen Zahlen (German)
3. Bourbaki, N.: Elements of Mathematics. Topological Vector Spaces, Chaps. 1–5. Springer, Berlin (1987)
4. Faltin, F., Metropolis, N., Ross, B., Rota, G.-C.: The real numbers as a wreath product. Adv. Math. **16**, 278–304 (1975)
5. Forti, G.L., Schwaiger, J.: Stability of homomorphisms and completeness. C. R. Math. Acad. Sci. Soc. R. Can. **11**, 215–220 (1989)
6. Hyers, D.H.: On the stability of the linear functional equation. Proc. Natl. Acad. Sci. USA **27**, 222–224 (1941)
7. Jabłoński, W., Schwaiger, J.: Stability of the homogeneity and completeness. Sitzungsber. Abt. II Österr. Akad. Wiss. Math.-Naturwiss. Kl. **214**, 111–132 (2005)
8. Kunugui, K.: Applications des espaces à une infinité de dimensions à la théorie des ensembles. Proc. Imp. Acad. Jpn. **11**, 351–353 (1935)
9. Lang, S.: Undergraduate Analysis. Springer, New York (1983)
10. Manschek, N.: Verschiedene Zugänge zur Konstruktion der reellen Zahlen. AV Akademikerverlag (2015)
11. Schönhage, A.: Von den ganzen zu den reellen Zahlen. Math. Phys. Semesterber. N. F. **17**, 88–96 (1970)
12. Schwaiger, J.: Remark in Report of Meeting. Aequationes Math. **35**, 120–121 (1988)
13. Steprāns, J.: A characterization of free Abelian groups. Proc. Am. Math. Soc. **93**, 347–349 (1985)
14. Street, R.: An efficient construction of the real numbers. Gazette Austral. Math. Soc. **12**, 57–58 (1985)
15. Street, R.: Update on the efficient reals. http://www.maths.mq.edu.au/~street/reals.pdf (2003)

Chapter 13
Generalized Dhombres Functional Equation

Jaroslav Smítal and Marta Štefánková

Abstract We consider the equation $f(xf(x)) = \varphi(f(x))$, $x > 0$, where φ is given, and f is an unknown continuous function $(0, \infty) \to (0, \infty)$. This equation was for the first time studied in 1975 by Dhombres (with $\varphi(y) = y^2$), later it was considered for other particular choices of φ, and since 2001 for arbitrary continuous function φ. The main problem, a classification of possible solutions and a description of the structure of periodic points contained in the range of the solutions (which appeared to be important way of the classification of solutions), was basically solved. This process involved not only methods from one-dimensional dynamics but also some new methods which could be useful in other problems. In this paper we provide a brief survey.

Keywords Iterative functional equations • Invariant curves • Real solutions • Topological entropy • Periodic orbits

Mathematics Subject Classification (2010) Primary 39B12; Secondary 26A18

13.1 Introduction

We consider the equation $f(xf(x)) = \varphi(f(x))$, $x > 0$, where φ is given, and f is an unknown continuous function $(0, \infty) \to (0, \infty)$. This equation was for the first time studied in 1975 by Dhombres [1] (with $\varphi(y) = y^2$), later it was considered in many papers for other particular choices of φ, see, e.g., [2] or [3], and since 2001 in about ten papers for arbitrary continuous function φ. The main problem, a classification of possible solutions and a description of the structure of periodic points contained in the range of the solutions (which appeared to be important way of the classification of solutions), was basically solved. This process involved

J. Smítal • M. Štefánková (✉)
Mathematical Institute, Silesian University, CZ-746 01 Opava, Czech Republic
e-mail: jaroslav.smital@math.slu.cz; marta.stefankova@math.slu.cz

J. Brzdęk et al. (eds.), *Developments in Functional Equations and Related Topics*, Springer Optimization and Its Applications 124,
DOI 10.1007/978-3-319-61732-9_13

methods from one-dimensional dynamics but also some new methods which could be useful in other problems. In this paper we provide a brief survey. Note that the equation is considered also in the complex domain; in this case complex analysis and theory of formal power series are useful tools rather than dynamical systems theory. There is no space to go into details, the reader is referred, e.g., to survey papers [7] and [8].

It is a difficult task to give an explicit form of solutions of the equation, the main problem in the theory of functional equations, if we allow arbitrary given continuous φ. But important basic information like classification of solutions is possible. Notice that the equation has a continuous solution (possibly only trivial, i.e., constant) if φ has a fixed point. On the other hand, it is an easy exercise to find a φ without fixed points such that the equation has no solution. Therefore results of "classical type" concerning existence of continuous solutions are not interesting. Classification and basic characterizations of solutions are given in the next section; to prove such results, e.g., Sharkovsky's theorem on coexistence of periodic orbits or structure of continuous maps of the interval with zero topological entropy play an essential role. Section 13.3 is devoted to the special case when φ is an increasing homeomorphism; then a characterization of monotone solutions is possible. These results can be proved using standard tools from the theory of iterative functional equations like functions defined by infinite products; basic information on such equations can be found, e.g., in [6]. Section 13.4 contains survey for increasing homeomorphisms φ and non-monotone continuous solutions; it appears that they can be, e.g., strongly non-differentiable, see Theorem 13.6. Finally, Section 13.5 contains results concerning distribution of periodic points of φ in the range of regular solutions; they can have periods 1 and 2, only. Note that for singular solutions, which can have periodic points of arbitrary periods in the range, similar description would be complicated.

13.2 Equation with Arbitrary Continuous φ

Generalized Dhombres functional equation is the equation of the form

$$f(xf(x)) = \varphi(f(x)), \quad x \in \mathbb{R}_+ := (0, \infty), \tag{13.1}$$

where $\varphi : \mathbb{R}_+ \to \mathbb{R}_+$ is a given continuous map, and $f : \mathbb{R}_+ \to \mathbb{R}_+$ is an unknown continuous map. We denote by R_f the range of f, and by $\mathcal{S}(\varphi)$ the set of all solutions of (13.1). Notice that the point 1 has an important role for solutions of (13.1) since $1 \in R_f$ implies $\varphi(1) = 1$. This makes possible to introduce the notion of *conjugate equation* (see [7]) of the form

$$\widetilde{f}(x\widetilde{f}(x)) = \widetilde{\varphi}(\widetilde{f}(x)), \quad \text{where } \widetilde{f}(x) := 1/f(1/x),\ \widetilde{\varphi}(y) := 1/\varphi(1/y). \tag{13.2}$$

It is easy to verify that $f \in \mathcal{S}(\varphi)$ if and only if $\widetilde{f} \in \mathcal{S}(\widetilde{\varphi})$, that the transformation $f \mapsto \widetilde{f}$ is a bijection between the solutions of (13.1) and (13.2), and $R_f \subseteq (0, 1]$ if and only if $R_{\widetilde{f}} \subseteq [1, \infty)$. Finally, let

$$\Phi : \mathbb{R}_+ \to \mathbb{R}_+, \quad \Phi(x, y) = (xy, \varphi(y)). \tag{13.3}$$

This map is closely related to (13.1). Obviously, for $f \in \mathcal{S}(\varphi)$, $\Phi(f) \subseteq f$, if f is identified with its graph.

Definition 13.1 (See [7]) A map $f \in \mathcal{S}(\varphi)$ is

- *singular* if for some $0 < a \leq b < \infty$, $f(x) > 1$ if $x < a$, $f(x) < 1$ for $x > b$, and $f(x) = 1$ otherwise,
- *regular* if it is not singular.

It is easy to see that the conjugate equation preserves the classes of regular and singular solutions, respectively. Notice that the definition of regular solutions is implicit. Only recently the following explicit characterization was proved.

Theorem 13.1 (See [14]) *An $f \in \mathcal{S}(\varphi)$ is a regular solution of (13.1) if and only if one of the following conditions is satisfied:*

1. $R_f \subseteq (0, 1]$;
2. $R_f \subseteq [1, \infty)$;
3. *there are* $0 < a \leq b < \infty$ *such that* $f(x) < 1$ *for* $x < a$, $f(x) > 1$ *for* $x > b$, *and* $f(x) = 1$ *otherwise.*

It is difficult, if not even impossible, to describe the class $\mathcal{S}(\varphi)$ for arbitrary continuous φ. But some results are available, thanks to the Sharkovsky's result on coexistence of periodic orbits.

Theorem 13.2 (See [12]) *There is a singular $f \in \mathcal{S}(\varphi)$ such that $\varphi|_{R_f}$ has periodic orbits of all periods.*

In particular, $\varphi|_{R_f}$ can have positive topological entropy. Methods of constructing singular solutions possessing prescribed sets of periodic points $\varphi|_{R_f}$, compatible with the Sharkovsky's ordering, are indicated in [12]. On the other hand, properties of Equation (13.1) with regular solutions are quite different.

Theorem 13.3 (See [13]; cf also [14]) *Let $f \in \mathcal{S}(\varphi)$ be a regular solution of (13.1). Then every periodic point of $\varphi|_{R_f}$ has period* 1 *or* 2*.*

Theorem 13.4 (See [13]) *There is a φ, and an infinitely smooth function $f \in \mathcal{S}(\varphi)$ such that all points in $R_f \subset (0, 1)$ are periodic points of φ of period* 2*, except for one fixed point.*

Notice that the first example with similar properties, which, however, had not differentiable f, was given in [10].

A natural question arises whether $\varphi|_{R_f}$ can contain exactly one periodic point of period 2. We suspect that it is impossible.

Conjecture Let $f \in \mathscr{S}(\varphi)$ be a regular solution with $R_f \subseteq (0,1)$, and let P be the set of periodic points of $\varphi|_{R_f}$. Then P is a closed connected set, and contains exactly one fixed point p. Consequently, by Theorem 13.4, $P \setminus \{p\}$ is the set of periodic points, possibly empty, of $\varphi|_{R_f}$ of period 2.

Let us finish this section with the following result which is implicitly contained in [11].

Theorem 13.5 *For every regular $f \in \mathscr{S}(\varphi)$ with $(p,1] \subseteq R_f \subseteq [p,1]$ there is a continuous ψ and a regular $g \in \mathscr{S}(\psi)$ with $(p,1/p) \subseteq R_g = R_f \cup R_{\widetilde{f}} \subseteq [p,1/p]$, where $1/0$ means ∞, and $\widetilde{f}$ is the solution of the conjugate equation (13.2). If $f(x) = 1$ for some $x < 1$, then $\psi = \varphi$.*

Proof We may assume $p < 1$. Let $f(a) = 1$. If $a < 1/a =: b$ put $g(x) = f(x)$ for $0 < x \leq a$, $g(x) = \widetilde{f}(x)$ for $x \geq b$, and $g(x) = 1$ otherwise. Then $g \in \mathscr{S}(\varphi)$. If $b < a$ let $b' \leq b$ be a point where $f|_{(0,b]}$ attains maximum β; such a point exists since $\beta \geq \limsup_{x\to 0} f(x)$ (see, e.g., Proposition 2.2 in [7]), and by Theorem 13.1, $\beta := f(b') < 1$. Similarly find $a' \geq a$ with $\alpha := \widetilde{f}(a') > 1$. Using techniques described in [11], it is possible to connect the pieces $g_0 = f|(0,b']$ and $g_\infty = \widetilde{f}|[a',\infty)$ by a nondecreasing continuous function $g_m : [b',a'] \to [\beta,\alpha]$ and modify φ on the interval (α,β) to get ψ such that $g = g_0 \cup g_m \cup g_\infty \in \mathscr{S}(\psi)$. □

13.3 Monotone Solutions with φ an Increasing Homeomorphism

In this section we survey results in the case when φ is an increasing homeomorphism.

Theorem 13.6 (See [4]) *Let φ be an increasing homeomorphism of an interval $J \subseteq (0,\infty)$, and $R_f \subseteq J$. Then*

1. *any $f \in \mathscr{S}(\varphi)$ is regular;*
2. *Φ given by (13.3) is a homeomorphism;*
3. *for any $f \in \mathscr{S}(\varphi)$, $\Phi(f) = f$;*
4. *if $f \in \mathscr{S}(\varphi)$ and $p \in R_f$, $p \neq 1$, is a fixed point of φ, then $R_f = \{p\}$ and $f \equiv p$.*

Now we are able to characterize monotone solutions of (13.1), see [5]. They are assigned by a family of monotone continuous functions defined on a compact interval. Their iterations by Φ are then "pieces" composing a solution. Assume $\varphi : J \to J$ is an increasing homeomorphism,

$$0 \leq p < q \leq 1,\ J = (p,q),\ p,q \text{ are fixed points of } \varphi,\ \text{and } \varphi(y) \neq y \text{ otherwise.} \tag{13.4}$$

Note that the assumption $J \subseteq (0,1]$ is not too restrictive since using (13.2) we can get analogous results concerning solutions f with $R_f \subset (1,\infty)$. Moreover, for solutions with $1 \in R_f$ (like in Theorems 13.8 and 13.10), this follows by

Theorem 13.5. Actually, if φ is a homeomorphism with $1 \in R_\varphi$, then it is easy to modify the proof of Theorem 13.5 to get $\psi = \varphi$.

For $(x_0, y_0) \in \mathbb{R}_+ \times J$ define the *orbit* $\{x_n, y_n\}_{n=-\infty}^{\infty}$ of (x_0, y_0) by $(x_n, y_n) = \Phi^n(x_0, y_0)$, $n \in \mathbb{Z}$ (where Φ^n denotes the nth iterate of Φ). By Theorem 13.6 this definition is correct since Φ is a homeomorphism. Let

$$P(y, z, \varphi) = \prod_{k=1}^{\infty} \frac{\varphi^k(y)}{\varphi^k(z)}.$$

If $P(v, u, \varphi)$ is finite for some $u < v$, then $P(y, u, \varphi)$ is a continuous strictly increasing function of y in $[u, v]$. Finally, if $\varphi(y) < y$ on J, then $y_1 < y_0$. Let $\mathscr{M}(x_0, y_0)$ be the class of nondecreasing continuous functions g from $[x_1, x_0]$ onto $[y_1, y_0]$ such that

$$v/u \geq P(g(v), g(u), \varphi^{-1}) \quad \text{for } u < v \text{ in } [x_1, x_0].$$

In the dual case, if $\varphi(y) > y$ on J, then $y_1 > y_0$ and we can define $\mathscr{M}(x_0, y_0)$ in a similar way.

Theorem 13.7 (See [4]) *The class $\mathscr{M}(x_0, y_0)$ is nonempty and closed with respect to the uniform convergence. If $q < 1$, then it contains a continuum of functions.*

Theorem 13.8 (See [4]) *For $g \in \mathscr{M}(x_0, y_0)$ put $f := \bigcup_{n=-\infty}^{\infty} \Phi^n(g)$. Then f is a continuous monotone solution of (13.1) on its domain D_f. If $q < 1$, then $D_f = (0, \infty)$. If $q = 1$, then D_f is one of the intervals $(0, \infty)$, $(0, \alpha)$, or (α, ∞), with $\alpha > 0$ a real number. In the last two cases f can be extended, possibly not in a unique way, to a monotone solution defined on $(0, \infty)$.*

On the other hand, the class $\mathscr{M}(x_0, y_0)$ is complete in the following sense:

Theorem 13.9 (See [4]) *If $f \in \mathscr{S}(\varphi)$ is monotone with $f(x_0) = y_0$, then $g = f|_{[x_1, x_0]} \in \mathscr{M}(x_0, y_0)$.*

Even with φ an increasing homeomorphism, $\mathscr{S}(\varphi)$ can contain non-monotone solutions.

Theorem 13.10 (See [5]) *Assume (13.4) with $q = 1$. Then any $f \in \mathscr{S}(\varphi)$ is monotone. In particular if $\varphi(y) < y$ for any $y \in J$, then f is nondecreasing, and if $\varphi(y) > y$ for any $y \in J$, then f is nonincreasing.*

Theorem 13.11 (See [5]) *Assume (13.4) with $q < 1$. Then every $f \in \mathscr{S}(\varphi)$ is nondecreasing if and only if*

$$P(y, z, \varphi) = \infty \quad \textit{for any } y > z, \ y, z \in J,$$

and every $f \in \mathscr{S}(\varphi)$ is nonincreasing if and only if

$$P(y, z, \varphi^{-1}) = \infty \quad \textit{for any } y > z, \ y, z \in J.$$

13.4 Non-monotone Solutions with φ an Increasing Homeomorphism

When assuming (13.4) with $q = 1$ then, by Theorem 13.10, there are no non-monotone solutions. So throughout this section we assume $q < 1$. Then, by Theorem 13.11, there can exist non-monotone solutions (and such a solution can be non-monotone in a very strong sense), and every solution can be approximated almost uniformly (i.e., uniformly on every compact interval) by piecewise monotone solutions. (Recall that by a piecewise monotone function we mean a function with finite number of pieces of monotonicity.)

Theorem 13.12 (See [9]) *Assume (13.4) with $q < 1$. Then for every $f \in \mathscr{S}(\varphi)$ there is a sequence $f_n \in \mathscr{S}(\varphi)$ of functions piecewise monotone on every compact interval, which uniformly converges to f on every compact interval.*

Theorem 13.13 (See [9]) *Assume (13.4) and let $\mathscr{S}(\varphi)$ contain a non-monotone solution (hence $q < 1$). Then there is an $f \in \mathscr{S}(\varphi)$ and a compact interval $I \subset \mathbb{R}_+$ such that f is monotone on no subinterval of I.*

There are also results saying for which monotone functions $f : \mathbb{R}_+ \to \mathbb{R}_+$ there is a φ satisfying (13.4) with $q < 1$ such that $f \in \mathscr{S}(\varphi)$, the so-called converse problem (see [10]). Analogous result is available for continuous functions f which need not be monotone. The results are rather technical so we omit the details. In [10], there is also shown that for decreasing homeomorphisms φ it is difficult to obtain results of similar type as above; in particular, for decreasing homeomorphisms there can exist no non-constant solutions at all.

Problem Give a characterization of solutions of (13.1) in the case when φ is a decreasing homeomorphism of an open subinterval of (0, 1) hence with $R_f \subset (0, 1)$. The results could be extended to any open subinterval of $\mathbb{R}_+$ using the methods indicated in Theorem 13.5, see also [11].

13.5 Periodic Points in the Range of Regular Solutions

By Theorem 13.3, for regular solutions only periods 1 and 2 are possible. The results concerning fixed points can be summarized as follows:

Theorem 13.14 (See [3, 7], and [11], Respectively) *Let $f \in \mathscr{S}(\varphi)$, and $R_f \subseteq (0, 1]$. Denote by F the set of fixed points of φ contained in R_f. Then*

1. *if φ is an increasing homeomorphism, then $F \subseteq \{1\}$,*
2. *F contains at most one point $\neq 1$,*
3. *if every periodic point of φ is a fixed point, then*

$$F = \emptyset, \text{ or } F = \{p\},\ p \leq 1, \text{ or } F = \{p, 1\},\ p < 1,$$

and all types are possible.

By Theorem 13.4, Conjecture from Section 13.2, Theorem 13.5, and techniques described in its proof we can get a characterization of sets of periodic points in the range of a regular solutions: If R_f is a neighborhood of 1, then there can be at most three fixed points, and two families of periodic orbits, one contained in $(0, 1)$ and the other in $(1, \infty)$. It is easy to verify that there can be no periodic orbit $\{p, q\}$ of φ in R_f with $p < 1 < q$.

References

1. Dhombres, J.: Applications associatives ou commutatives. C.R. Acad. Sci. Paris Sér. A-B **281**, A809–A812 (1975)
2. Kahlig, P., Smítal, J.: On the solutions of a functional equation of Dhombres. Results Math. **27**, 362–367 (1995)
3. Kahlig, P., Smítal, J.: On a parametric functional equation of Dhombres type. Aequationes Math. **56**, 63–68 (1998)
4. Kahlig, P., Smítal, J.: On a generalized Dhombres functional equation. Aequationes Math. **62**, 18–29 (2001)
5. Kahlig, P., Smítal, J.: On a generalized Dhombres functional equation II. Math. Bohem. **127**, 547–555 (2002)
6. Kuczma, M., Choczewski, B., Ger, R.: Iterative Functional Equations. Cambridge University Press, Cambridge (1990)
7. Reich, L., Smítal, J.: Functional equation of Dhombres type – a simple equation with many open problems. J. Differ. Equ. Appl. **15**, 1179–1191 (2009)
8. Reich, L., Tomaschek, J.: Local solutions of the generalized Dhombres functional equation in a neighbourhood of infinity. In: Jarczyk, W., Fournier-Prunaret, D., Cabral, J.M.G. (eds.) ECIT 2012, 19th European Conference on Iteration Theory, pp. 233–246. ESAIM Proc. Surveys, vol. 46. EDP Sci., Les Ulis (2014)
9. Reich, L., Smítal, J., Štefánková, M.: The continuous solutions of a generalized Dhombres functional equation. Math. Bohem. **129**, 399–410 (2004)
10. Reich, L., Smítal, J., Štefánková, M.: The converse problem of the generalized Dhombres functional equation. Math. Bohem. **130**, 301–308 (2005)
11. Reich, L., Smítal, J., Štefánková, M.: Functional equation of Dhombres type in the real case. Publ. Math. Debr. **78**, 659–673 (2011)
12. Reich, L., Smítal, J., Štefánková, M.: Singular solutions of the generalized Dhombres functional equation. Results Math. **65**, 251–261 (2014)
13. Reich, L., Smítal, J., Štefánková, M.: On regular solutions of the generalized Dhombres equation II. Results Math. **67**, 521–528 (2015)
14. Smítal, J., Štefánková, M.: On regular solutions of the generalized Dhombres equation. Aequationes Math. **89**, 57–61 (2015)

Chapter 14
Functional Equations and Stability Problems on Hypergroups

László Székelyhidi

Abstract This is a survey paper about functional equations on hypergroups. We show how some fundamental functional equations can be treated on some types of hypergroups. We also present stability and superstability results using invariant means and other tools.

Keywords Hypergroup • Stability

Mathematics Subject Classification (2010) Primary 39B82; Secondary 39B52, 20N20

14.1 Basics on Hypergroups

The concept of DJS-hypergroup (according to the initials of Dunkl, Jewett and Spector) depends on a set of axioms which can be formulated in several different ways. The way of formulating these axioms we follow here is due to Lasser (see, e.g., [3, 20]). One begins with a locally compact Hausdorff space K and with the space $\mathscr{C}_c(K)$ of all compactly supported complex valued functions on the space K. The space $\mathscr{C}_c(K)$ will be topologized as the *inductive limit* of the spaces

$$\mathscr{C}_E(K) = \{f \in \mathscr{C}_c(K) \; : \; \operatorname{supp}(F) \subseteq E\},$$

where E is a compact subset of K carrying the uniform topology. A (complex) *Radon measure* μ is a *continuous linear functional* on $\mathscr{C}_c(K)$. Thus, for every compact subset E in K there exists a constant α_E such that $|\mu(f)| \le \alpha_E \|f\|_\infty$ for all f

---

L. Székelyhidi (✉)
Department of Mathematics, Institute of Mathematics, University of Debrecen,
H-4010 Debrecen, Hungary
e-mail: lszekelyhidi@gmail.com

J. Brzdęk et al. (eds.), *Developments in Functional Equations and Related Topics*, Springer Optimization and Its Applications 124,
DOI 10.1007/978-3-319-61732-9_14

in $\mathscr{C}_E(K)$. The set of Radon measures on K will be denoted by $\mathscr{M}(K)$. In the sequel by a *measure* we always mean a Radon measure. For each measure μ we write

$$\|\mu\| = \sup\{|\mu(f)| : f \in \mathscr{C}_c(K), \ \|f\|_\infty \leq 1\}.$$

A measure μ is said to be *bounded*, if $\|\mu\| < +\infty$. In addition, μ is called a *probability measure*, if μ is nonnegative and $\|\mu\| = 1$. The set of all bounded measures, the set of all compactly supported measures, the set of all probability measures, and the set of all probability measures with compact support in $\mathscr{M}(K)$ will be denoted by $\mathscr{M}_b(K)$, $\mathscr{M}_c(K)$, $\mathscr{M}_1(K)$, and $\mathscr{M}_{1,c}(K)$, respectively. The point mass concentrated at x is denoted by δ_x. Via integration theory we are able to consider measures as functions on the σ-algebra $\mathscr{B}(K)$ of *Borel subsets* of K and we use the notation $\int_K f\,d\mu$ rather than $\mu(f)$. We use the notation $\mathscr{M}_+(K)$ for the set of positive measures on the σ-algebra $\mathscr{B}(K)$ that means, for measures which take values in $[0, +\infty]$.

Now we formulate the first part of the axioms. Suppose that we have the following:

1. (H^*) There is a continuous mapping $(x, y) \mapsto \delta_x * \delta_y$ from $K \times K$ into $\mathscr{M}_{1,c}(K)$. This mapping is called *convolution.*
2. ($H^\vee$) There is an involutive homeomorphism $x \mapsto \check{x}$ from K to K. This mapping is called *involution.*
3. (*He*) There is a fixed element e in K. This element is called *identity*.

Identifying x by δ_x the mapping in (H^*) has a unique extension to a continuous bilinear mapping from $\mathscr{M}_b(K) \times \mathscr{M}_b(K)$ to $\mathscr{M}_b(K)$. The involution on K extends to a continuous involution on $\mathscr{M}_b(K)$. Convolution maps $\mathscr{M}_1(K) \times \mathscr{M}_1(K)$ into $\mathscr{M}_1(K)$ and involution maps $\mathscr{M}_1(K)$ onto $\mathscr{M}_1(K)$. Then a *DJS-hypergroup*, or simply a *hypergroup* is a quadruple $(K, *, \vee, e)$ satisfying the following axioms: for each x, y, z in K we have

1. (H1) $\delta_x * (\delta_y * \delta_z) = (\delta_x * \delta_y) * \delta_z$,
2. (H2) $(\delta_x * \delta_y)\check{} = \delta_{\check{y}} * \delta_{\check{x}}$,
3. (H3) $\delta_x * \delta_e = \delta_e * \delta_x = \delta_x$,
4. (H4) e is in the support of $\delta_x * \delta_{\check{y}}$ if and only if $x = y$,
5. (H5) the mapping $(x, y) \mapsto \operatorname{supp}(\delta_x * \delta_y)$ from $K \times K$ into the space of nonvoid compact subsets of K is continuous, the latter being endowed with the Michael topology (also called *finite topology*, see [3, 14]).

For arbitrary measures μ, ν in $\mathscr{M}_b(K)$ the symbol $\mu * \nu$ denotes their convolution, and $\check{\mu}$ denotes the involution of μ. With these operations $\mathscr{M}_b(K)$ is an *algebra with involution*. If the topology of K is discrete, then we call the hypergroup *discrete*. In case of discrete hypergroups the above axioms have a simpler form. Here we present a set of axioms for these types of hypergroups. Clearly, in the discrete case we can simply forget about the topological requirements in the previous axioms to get a purely algebraic system.

Let K be a set and suppose that the following properties are satisfied:

1. (D^*) There is a mapping $(x, y) \mapsto \delta_x * \delta_y$ from $K \times K$ into $\mathscr{M}_{1,c}(K)$, the space of all finitely supported probability measures on K. This mapping is called *convolution*.
2. ($D^\vee$) There is an involutive bijection $x \mapsto \check{x}$ from K to K. This mapping is called *involution*.
3. (De) There is a fixed element e in K. This element is called *identity*.

Identifying x by δ_x as above, and extending convolution and involution, a *discrete DJS-hypergroup* is a quadruple $(K, *, \check{}, e)$ satisfying the following axioms: for each x, y, z in K we have

1. (D1) $\delta_x * (\delta_y * \delta_z) = (\delta_x * \delta_y) * \delta_z$,
2. (D2) $(\delta_x * \delta_y)^\vee = \delta_{\check{y}} * \delta_{\check{x}}$,
3. (D3) $\delta_x * \delta_e = \delta_e * \delta_x = \delta_x$,
4. (D4) e is in the support of $\delta_x * \delta_{\check{y}}$ if and only if $x = y$.

If $\delta_x * \delta_y = \delta_y * \delta_x$ holds for all x, y in K, then we call the hypergroup *commutative*. If $\check{x} = x$ holds for all x in K, then we call the hypergroup *Hermitian*. By (H2), every Hermitian hypergroup is commutative. In any case we have $\check{e} = e$. For instance, if $K = G$ is a locally compact Hausdorff group, $\delta_x * \delta_y = \delta_{xy}$ for all x, y in K, $\check{x}$ is the inverse of x and e is the identity of G, then we obviously have a hypergroup $(K, *, ^\vee, e)$, which is commutative if and only if the group G is commutative. However, not every hypergroup originates in this way.

The simplest hypergroup is obviously the trivial one, consisting of a singleton. The next simplest hypergroup structure can be introduced on a set consisting of two elements. As an example we describe all hypergroups of this type.

Let $K = \{0, 1\}$. Clearly, the only Hausdorff topology on K is the discrete one. We specify $e = 0$ as the identity element. In this case the only involution satisfying the above axioms is the identity, that is, $\check{0} = 0$ and $\check{1} = 1$. Consequently, we have a Hermitian hypergroup, which is necessarily commutative. Now we have to define the four possible products $\delta_0 * \delta_0$, $\delta_0 * \delta_1$, $\delta_1 * \delta_0$, and $\delta_1 * \delta_1$. As δ_0 is the identity, the first three products are uniquely determined and the fourth one must have the form

$$\delta_1 * \delta_1 = \theta \cdot \delta_0 + (1 - \theta) \cdot \delta_1$$

with some number θ satisfying $0 \leq \theta \leq 1$. It turns out that $\theta \neq 0$, as a consequence of (D4). We shall denote this hypergroup by $D(\theta)$. It is clear that in this way we have a complete description of all possible hypergroup structures on a set consisting of two elements. Observe that in the case $\theta = 1$ we have a group isomorphic to $\mathbb{Z}_2$, the integers modulo 2, in any other case the resulting structure is not a group.

If K is any hypergroup and H is an arbitrary set, then for the function $f : K \to H$ we define $\check{f}$ by the formula

$$\check{f}(x) = f(\check{x})$$

for each x in K. Obviously, $\check{\check{f}} = f$. Each measure μ in $\mathcal{M}_b(K)$ satisfies

$$\check{\mu}(f) = \mu(\check{f})$$

whenever $f : K \to \mathbb{C}$ is a bounded Borel function.

Let K be an arbitrary hypergroup. Then, for each x, y in K the measure $\delta_x * \delta_y$ is a compactly supported probability measure on K which makes the measurable space $(K, \mathcal{B}(K), \delta_x * \delta_y)$ a *probability space*. An arbitrary function $f : K \mapsto \mathbb{C}$, which is $\delta_x * \delta_y$-measurable, can be considered as a *random variable* on this probability space. In particular, each continuous complex valued function on K is a random variable with respect to every measure of the form $\delta_x * \delta_y$. Clearly, each f is integrable with respect to every δ_x, and its *expectation* is

$$E_x(f) = \int f\, d\delta_x = f(x)\,,$$

hence it seems to be reasonable to define the "value" of f at δ_x as $f(x)$. This can be extended to an arbitrary probability measure μ on K by defining

$$f(\mu) = E_\mu(f) = \int f\, d\mu\,,$$

whenever f is integrable with respect to μ. In particular,

$$f(\delta_x * \delta_y) = \int f\, d(\delta_x * \delta_y)\,,$$

whenever f is integrable with respect to $\delta_x * \delta_y$. In this case we shall use the suggestive notation $f(x * y)$ for $f(\delta_x * \delta_y)$. In fact, in every hypergroup K we identify x by δ_x.

Here we call the attention to the fact that $f(x * y)$ has no meaning on its own, because $x * y$ is in general not an element of K, hence f is not defined at $x * y$. The expression $x * y$ denotes a kind of "blurred" product. If B is a Borel subset of K, then $\delta_x * \delta_y(B)$ expresses the probability of the event that this "blurred" product of x and y belongs to the set B. In the special case of groups this probability is 1, if B contains xy and is 0 otherwise, that is, exactly $\delta_{xy}(B)$.

We define the *right translation operator* τ_y by the element y in K according to the formula

$$\tau_y f(x) = \int_K f\, d(\delta_x * \delta_y)$$

for each f integrable with respect to $\delta_x * \delta_y$. In particular, τ_y is defined for every continuous complex valued function on K. Similarly, we can define *left translation operators*, denoted by ${}_y\tau$. In general one uses the above notation

$$f(x*y) = \int_K f\,d(\delta_x * \delta_y)\,,$$

for each x, y in K. Obviously, in case of commutative hypergroups the simple term *translation operator* is used. The function $\tau_y f$ is *the translate of f by y*.

As an example, we consider a function $f : D(\theta) \to \mathbb{C}$ on the $D(\theta)$ hypergroup. Then we have

$$f(1*1) = \int f(t)\,d(\delta_1 * \delta_1) =$$

$$\int f(t)\,d(\theta\cdot\delta_0 + (1-\theta)\cdot\delta_1) = \theta f(0) + (1-\theta)f(1)\,.$$

Convolution of functions and measures is defined in the following obvious way: for each measure μ in $\mathscr{M}_b(K)$ and for every continuous bounded function $f : K \to \mathbb{C}$ we let

$$f*\mu(x) = \int_K f(x*\check{y})\,d\mu(y)$$

whenever x is in K. Then $f*\mu$ is a continuous bounded function on K. For more details see [3].

In what follows we simply refer to the hypergroup $(K, *, \check{}, e)$ as a *hypergroup K*.

14.2 Functional Equations

The presence of translation operators makes it possible to introduce and to study some basic functional equations on hypergroups.

Let K be a hypergroup. The non-identically zero continuous function $m : K \to \mathbb{C}$ is called an *exponential*, if it satisfies

$$m(x*y) = m(x)m(y) \tag{14.1}$$

for each x, y in K. Exponentials play a basic role in harmonic analysis, spectral synthesis, and functional equations. More explicitly, the above functional equation can be written in the form of the integral equation

$$\int_K m(t)\,d(\delta_x * \delta_y)(t) = m(x)m(y)\,.$$

For each exponential m we have $m(e) = 1$. An exponential m with $m(\check{x}) = \overline{m(x)}$ for all x in K is called a *semi-character*, and bounded semi-characters are called *characters*. We note the inconvenient facts that, in contrast with the case of groups, exponentials can take the value zero, and the product of two exponentials is not necessary an exponential.

Here we give a simple illustration by describing all exponential functions on the hypergroup $D(\theta)$ with $0 < \theta \leq 1$.

Suppose that $m : D(\theta) \to \mathbb{C}$ is an exponential that is

$$\int_{D(\theta)} m(t)\, d(\delta_x * \delta_y)(t) = m(x)m(y)$$

holds for each x, y in $D(\theta)$. According to the definition of convolution in $D(\theta)$ the only nontrivial consequence of this equation we obtain in the case $x = y = 1$, by our above remark:

$$\theta m(0) + (1 - \theta)m(1) = m(1)m(1)\,.$$

Using $m(0) = 1$ and solving the quadratic equation for $m(1)$ we have the two possibilities: $m(1) = 1$ or $m(1) = -\theta$. The first case gives the trivial exponential which is identically 1, and the second case is the nontrivial one: $m(0) = 1$ and $m(1) = -\theta$. Obviously, both are characters.

Another important function class is the following. Given the commutative hypergroup K the continuous function $a : K \to \mathbb{C}$ is called *additive*, if it satisfies

$$a(x * y) = a(x) + a(y) \tag{14.2}$$

for each x, y in K. This equation can be written in the integral form

$$\int_K a(t)\, d(\delta_x * \delta_y)(t) = a(x) + a(y)\,.$$

Every additive function a satisfies $a(e) = 0$, and all additive functions on K form a complex linear space.

Considering the hypergroup $D(\theta)$ again, let $a : D(\theta) \to \mathbb{C}$ be an additive function. Then we have

$$\theta a(0) + (1 - \theta)a(1) = a(0) + a(1)\,,$$

and $a(0) = 0$ implies $a(1) = 0$, as $\theta \neq 0$. Hence every additive function is zero on $D(\theta)$. We note that, more generally, every additive function is zero on compact hypergroups.

Exponential and additive functions are fundamental—they are the solutions of the basic Cauchy functional equations on hypergroups. We remark that, obviously, these functions can be defined on non-commutative hypergroups, too. We can also consider their pexiderized versions.

Moment functions and moment function sequences play an important role in probability theory. These functions can be defined in the following manner. Let K

be a hypergroup. The sequence of continuous complex valued functions $(\varphi_n)_{n\in\mathbb{N}}$ on K is called a *generalized moment function sequence*, if φ_0 is not identically zero and for each natural number n we have

$$\varphi_n(x * y) = \sum_{k=0}^{n} \binom{n}{k} \varphi_k(x)\varphi_{n-k}(y) \tag{14.3}$$

for each x, y in K and for every natural number n. Obviously, φ_0 is an exponential. If $\varphi_0 = 1$, then this sequence is called a *moment function sequence*, and, in general, we say that the generalized moment function sequence $(\varphi_n)_{n\in\mathbb{N}}$ is *associated with the exponential* φ_0. Given a natural number N we say that the functions

$$\{\varphi_k : k = 0, 1, \ldots, N\}$$

form a *generalized moment function sequence of order N* if the above equations hold for $n = 0, 1, \ldots, N$.

The second equation of the above system is

$$\varphi_1(x * y) = \varphi_0(x)\varphi_1(y) + \varphi_1(x)\varphi_0(y)\,, \tag{14.4}$$

which is called *sine equation*, for obvious reasons.

An important functional equation related to involution is the *square norm functional equation* which has the form

$$f(x * y) + f(x * \check{y}) = 2f(x) + 2f(y) \tag{14.5}$$

for each x, y in K, where $f : K \to \mathbb{C}$ is a continuous function on the hypergroup K. Clearly, on Hermitian hypergroups this equation is identical with the additive Cauchy equation.

On Abelian groups an important function class is formed by the so-called *polynomial functions*. There are several different ways to introduce these functions, and in some cases they result in different function classes. In the subsequent sections we follow the approach used in [28, 30].

14.3 The Measure Algebra

Exponential monomials are the basic building blocks of spectral analysis and spectral synthesis on Abelian groups. Recently there have been some attempts to extend the most important results in spectral analysis and spectral synthesis from groups to hypergroups. For this purpose it is necessary to introduce a reasonable concept of exponential monomials. In the group case this concept arises from additive and exponential functions. Roughly speaking, in that case exponential monomials are the functions which can be represented as the product of an exponential and an ordinary polynomial in additive functions. It turns out that this definition will not work in the case of hypergroups. In fact, even in the simplest case, when we consider the square of an exponential, it will not be an exponential. In [30] we reconsidered this problem, and using a ring-theoretical approach we proved characterization theorems for particular function classes, which can be considered as "exponential monomials" on commutative hypergroups. Some of those ideas can be extended to the non-commutative case too. Nevertheless, here we consider the commutative setting only.

The basic structure is the measure algebra. If $K = K(*, \check{}, e)$ is a commutative hypergroup, then $\mathscr{C}(K)$ denotes the locally convex topological vector space of all continuous complex valued functions defined on K, equipped with the pointwise linear operations and the topology of compact convergence.

It is well known (see, e.g., [10, p. 551]) that the dual of $\mathscr{C}(K)$ can be identified with $\mathscr{M}_c(K)$, the space of all compactly supported complex measures on K. If K is discrete, then this space is also identified with the set of all finitely supported complex valued functions on K. The *pairing* between $\mathscr{C}(K)$ and $\mathscr{M}_c(K)$ is given by the formula

$$\langle \mu, f \rangle = \int f \, d\mu \,.$$

Convolution on $\mathscr{M}_c(K)$ is defined by

$$\mu * \nu(x) = \int \mu(x * \check{y}) \, d\nu(y),$$

for each μ, ν in $\mathscr{M}_c(K)$ and x in K. Convolution converts the space $\mathscr{M}_c(K)$ into a commutative algebra with unit δ_e. We call this algebra the *measure algebra* of K. If K is discrete, then we call it the *hypergroup algebra* of K, since in the case when K is a group it is identical with the *group algebra* of this group.

We also define convolution of measures in $\mathscr{M}_c(K)$ with arbitrary functions in $\mathscr{C}(K)$ by the same formula

$$\mu * f(x) = \int f(x * \check{y}) \, d\mu(y)$$

for each μ in $\mathscr{M}_c(K), f$ in $\mathscr{C}(K)$, and x in K. It is easy to see that equipped with this action $\mathscr{C}(K)$ turns into a module over the measure algebra.

Translation operators are closely related to convolution. In fact, τ_y is a *convolution operator*, namely it is the convolution with the measure $\delta_{\check{y}}$. A subset of $\mathscr{C}(K)$ is called *translation invariant*, if it contains all translates of its elements. A closed linear subspace of $\mathscr{C}(K)$ is called a *variety* on K, if it is translation invariant. For each function f the smallest variety containing f is called the *variety generated by* f, or simply the *variety of* f and is denoted by $\tau(f)$. It is the intersection of all varieties containing f.

We recall the concept of the annihilator. Given a subset H in $\mathscr{C}(K)$ its *annihilator* in $\mathscr{M}_c(K)$ is the set

$$\operatorname{Ann} H = \{\mu : \mu \in \mathscr{M}_c(K), \mu * f = 0 \text{ for each } f \in H\}.$$

It is easy to see that this is an ideal in $\mathscr{M}_c(K)$. Analogously, for each subset L in $\mathscr{M}_c(K)$ its *annihilator* in $\mathscr{C}(K)$ is defined by

$$\operatorname{Ann} L = \{f : f \in \mathscr{C}(K), \mu * f = 0 \text{ for each } \mu \in L\}.$$

It follows that $\operatorname{Ann} L$ is a variety in $\mathscr{C}(K)$.

The concept of the annihilator is closely related to the notion of the orthogonal complement. Given a subset H in $\mathscr{C}(K)$ its *orthogonal complement* in $\mathscr{M}_c(K)$ is the set

$$H^\perp = \{\mu : \mu \in \mathscr{M}_c(K), \mu(f) = 0 \text{ for each } f \in H\}.$$

It is easy to see again that this is an ideal in $\mathscr{M}_c(K)$. Analogously, for each subset L in $\mathscr{M}_c(K)$ its *orthogonal complement* in $\mathscr{C}(K)$ is defined by

$$L^\perp = \{f : f \in \mathscr{C}(K), \mu(f) = 0 \text{ for each } \mu \in L\}.$$

It follows that $\operatorname{Ann} L$ is a variety in $\mathscr{C}(K)$. The relation between annihilators and orthogonal complements of varieties and ideals is easy to describe. Indeed, we have for each variety V in $\mathscr{C}(K)$ and for each ideal I in $\mathscr{M}_c(K)$ the identities

$$\operatorname{Ann} \check{V} = (\operatorname{Ann} V)\check{}, \quad \operatorname{Ann} \check{I} = (\operatorname{Ann} I)\check{},$$

further

$$V^\perp = \operatorname{Ann} \check{V}, \quad I^\perp = \operatorname{Ann} \check{I}.$$

Hence, in the case of varieties and ideals the use of annihilators or orthogonal complements is more or less a question of taste.

It is also obvious that $V \subseteq V^{\perp\perp}$ and $I \subseteq I^{\perp\perp}$ hold for each variety V on K and for each ideal I in $\mathscr{M}_c(K)$ and similar relations hold for $\operatorname{Ann} V$ and $\operatorname{Ann} I$. Moreover, using the Hahn–Banach Theorem, it is easy to show that $V = V^{\perp\perp}$ and $V = \operatorname{Ann}(\operatorname{Ann} V)$ hold for each variety. Unfortunately, we do not have the corresponding equality for ideals, as is shown by an example in [12] in the case, when K is a group. However, if K is a discrete hypergroup, then $I = \operatorname{Ann}(\operatorname{Ann} I)$ holds for each ideal in $\mathscr{M}_c(K)$. This is also shown in [12] in the group case, and one can see immediately that the proof given there also works on hypergroups.

14.4 Exponential Polynomials

Exponential polynomials play a fundamental role in the theory of functional equations. In fact, all the functional equations mentioned above characterize functions on Abelian groups which belong to the class of exponential polynomials. Hence in order to build up a satisfactory theory of functional equations on hypergroups it is necessary to find a reasonable definition of exponential polynomials which, in the group case, coincides with the usual concept. As we mentioned above, an obvious copy of the definitions, obtained by replacing the group operation with convolution, does not work at all. Now we shortly summarize the way forward we have offered in the papers [28, 30]. For the sake of simplicity here we consider the commutative case only. Nevertheless, it will be clear for the reader that some of the methods can be extended to non-commutative hypergroups as well.

The basic idea is the use of modified difference operators. Given a hypergroup K, a continuous function $f : K \to \mathbb{C}$, and an element y in K we define the *modified difference* $\Delta_{f;y}$ as the measure

$$\Delta_{f;y} = \delta_{\check{y}} - f(y)\delta_e ,$$

where e is the identity of K. For the products of such elements we use the following notation: given a natural number n and elements $y_1, y_2, \ldots, y_{n+1}$ in K we write

$$\Delta_{f;y_1,y_2,\ldots,y_{n+1}} = \Pi_{k=1}^{n+1} \Delta_{f;y_k} ,$$

where the product means convolution. In the case f being the exponential $m = 1$ we use the simplified notation Δ_y for $\Delta_{1;y}$ and $\Delta_{y_1,y_2,\ldots,y_{n+1}}$ for $\Delta_{1;y_1,y_2,\ldots,y_{n+1}}$.

A fundamental role is played by the ideals generated by modified differences. Given the continuous function $f : K \to \mathbb{C}$ the closure of the ideal in the measure algebra generated by all modified differences of the form $\Delta_{f;y}$ with y in K will be denoted by M_f. The following theorem shows that this ideal is appropriate if and only if f is an exponential.

Theorem 14.1 *Let K be a commutative hypergroup and $f : K \to \mathbb{C}$ a function with $f(e) = 1$. Then the following statements are equivalent:*

1. *f is an exponential.*
2. *The ideal M_f is proper.*
3. *The ideal M_f is maximal.*
4. *$M_f = \operatorname{Ann} \tau(f)$.*

This theorem can be proved following the lines of [31] (see also [30]). We use this result to define generalized exponential monomials on commutative hypergroups as follows. Let K be a commutative hypergroup. The continuous function $f : K \to \mathbb{C}$ is called a *generalized exponential monomial*, if there exists an exponential m, and a natural number n such that the inclusion

$$M_m^{n+1} \subseteq \operatorname{Ann} \tau(f) \tag{14.6}$$

holds. It can be shown that if f is nonzero, then m is uniquely determined and we say that f is *associated with* m. In other words, the continuous function $f : K \to \mathbb{C}$ is a generalized exponential monomial associated with m if and only if there exists a natural number n such that

$$\Delta_{m;y_1,y_2,\ldots,y_{n+1}} * f = 0 \tag{14.7}$$

holds for each $y_1, y_2, \ldots, y_{n+1}$ in K. If f is nonzero, then the smallest n with this property is defined to be the *degree* of f. Obviously, every exponential is a generalized exponential monomial of degree zero, associated with itself. Indeed, the exponential m clearly satisfies $\Delta_{m;y} * m = 0$ for each y. To understand the difference between the group case and the case of general hypergroups we note that, for instance, in the group case every generalized exponential monomial φ of degree at most 1 associated with the given exponential m has the form

$$\varphi(x) = \big(a(x) + c\big)m(x)$$

for some complex number c, and a additive, while on hypergroups there are other functions of this type. This can be verified using the general description of exponentials and additive functions on some special hypergroups, like polynomial hypergroups, Sturm–Liouville hypergroups, etc. The interested reader will find further details in [29].

Generalized exponential monomials associated with the exponential identically 1 are called *generalized polynomials*. It is known that on Abelian groups generalized polynomials can be represented in a unique manner as the sum of the diagonalizations of symmetric multi-additive functions. To the best of our knowledge a similar result on hypergroups has not yet been found. Linear combinations of generalized exponential monomials are called generalized exponential polynomials.

An important subclass of generalized exponential monomials is formed by the ones whose variety is finite dimensional. In fact, the generalized exponential monomial f is called simply an *exponential monomial*, if $\tau(f)$ is a finite dimensional vector space. Similarly, a generalized polynomial f is called a *polynomial*, if $\tau(f)$ is a finite dimensional vector space. Accordingly, linear combinations of exponential monomials are called *exponential polynomials*. As we noted above in the group case an exponential monomial is always the product of an exponential and an ordinary polynomial of additive functions. However, in the hypergroup case we have a different situation. In fact, a complete description of all exponential monomials on commutative hypergroups is still lacking. In other words, the solution space of the functional equation (14.7) on arbitrary commutative hypergroups has not been characterized yet. Still there are some types of hypergroups on which a complete description of some of the above function classes is available (see, e.g., [15–18, 21, 24, 26, 29, 33]). In the subsequent sections we present some examples where such a description has been obtained.

14.5 Polynomial Hypergroups

An important special class of Hermitian hypergroups is closely related to orthogonal polynomials.

Let $(a_n)_{n\in\mathbb{N}}$, $(b_n)_{n\in\mathbb{N}}$, and $(c_n)_{n\in\mathbb{N}}$ be real sequences with the following properties: $c_n > 0, b_n \geq 0, a_{n+1} > 0$ for each n in $\mathbb{N}$, moreover $a_0 = b_0 = 0$ and $a_n+b_n+c_n = 1$ for each n in $\mathbb{N}$. We define the sequence of polynomials $(P_n)_{n\in\mathbb{N}}$ by $P_0(\lambda) = 1$, $P_1(\lambda) = \lambda$ and by the recursive formula

$$\lambda P_n(\lambda) = a_n P_{n-1}(\lambda) + b_n P_n(\lambda) + c_n P_{n+1}(\lambda)$$

for each $n \geq 1$ and λ in $\mathbb{R}$. The following theorem holds (see [3]).

Theorem 14.2 *If the sequence of polynomials $(P_n)_{n\in\mathbb{N}}$ satisfies the above conditions, then there exist constants $c(n,l,k)$ for each n,l,k in $\mathbb{N}$ such that*

$$P_n \cdot P_l = \sum_{k=|n-l|}^{n+l} c(n,l,k) P_k$$

holds for each n,l in $\mathbb{N}$.

The formula in the theorem is called *linearization formula* and the coefficients $c(n,l,k)$ are called *linearization coefficients*. The recursive formula for the sequence $(P_n)_{n\in\mathbb{N}}$ implies $P_n(1) = 1$ for each n in $\mathbb{N}$, hence we have

$$\sum_{k=|n-l|}^{n+l} c(n,l,k) = 1$$

for each n in $\mathbb{N}$. If the linearization is *nonnegative*, that is, the linearization coefficients are nonnegative: $c(n,l,k) \geq 0$ for each n,l,k in $\mathbb{N}$, then we can define a hypergroup structure on $\mathbb{N}$ by the following rule:

$$\delta_n * \delta_l = \sum_{k=|n-l|}^{n+l} c(n,l,k)\, \delta_k$$

for each n,l in $\mathbb{N}$, with involution as the identity mapping and with e as 0. The resulting discrete Hermitian (hence commutative) hypergroup is called *the polynomial hypergroup associated with the sequence* $(P_n)_{n\in\mathbb{N}}$. We shall denote it by $(\mathbb{N}, (P_n)_{n\in\mathbb{N}})$.

As an example we consider the hypergroup associated with the *Legendre polynomials*. The corresponding recurrence relation is

$$\lambda P_n(\lambda) = \frac{n+1}{2n+1} P_{n+1}(\lambda) + \frac{n}{2n+1} P_{n-1}(\lambda)$$

for each $n \geq 1$ and λ in $\mathbb{R}$. It can easily be seen that the linearization coefficients are nonnegative and the resulting hypergroup associated with the Legendre polynomials is the *Legendre hypergroup*.

Another interesting example for polynomial hypergroups is presented by the *Chebyshev polynomials*. The corresponding recurrence relation in the case of Chebyshev polynomials of the first kind is

$$\lambda T_n(\lambda) = \frac{1}{2} T_{n+1}(\lambda) + \frac{1}{2} T_{n-1}(\lambda)$$

for each $n \geq 1$ and λ in $\mathbb{R}$. Again, it is easy to see that the linearization coefficients are nonnegative and the resulting hypergroup associated with the Chebyshev polynomials of the first kind is the *Chebyshev hypergroup*.

The previous examples about the exponential and additive functions on the hypergroup $D(\theta)$ suggest that there is some hope to describe all exponential and additive functions on different polynomial hypergroups, too. We start with the Chebyshev hypergroup. We recall that $m : \mathbb{N} \to \mathbb{C}$ is an exponential on the Chebyshev hypergroup if and only if it satisfies

$$m(k * l) = m(k)m(l)$$

for each k, l in $\mathbb{N}$. From the linearization formula it follows easily by induction that

$$T_k(\lambda)T_l(\lambda) = \frac{1}{2} \left(T_{k+l}(\lambda) + T_{|k-l|}(\lambda)\right)$$

holds for each k, l in $\mathbb{N}$ and λ in $\mathbb{C}$. This means that for each function $f : \mathbb{N} \to \mathbb{C}$ we have

$$f(k * l) = \frac{1}{2} \Big(f(k + l) + f(|k - l|)\Big)$$

for each k, l in $\mathbb{N}$. Consequently, exponentials of the Chebyshev hypergroup are exactly the nonzero solutions of the functional equation

$$m(k + l) + m(|k - l|) = 2m(k)m(l)$$

for each k, l in $\mathbb{N}$. This functional equation is closely related to d'Alembert's functional equation and has been treated—among others—in [6] independently of hypergroups and in [21] on hypergroups. From our consideration it is clear that the functions $k \mapsto T_k(\lambda)$ satisfy this functional equation. In other words, the Chebyshev polynomials evaluated at any complex λ as functions of the subscript present exponential functions on the Chebyshev hypergroup. It turns out that this is true for every polynomial hypergroup. It turns out that the converse is also true: every exponential on a polynomial hypergroup is generated in this way. As different complex values of λ produce different exponentials, this means that the set of all exponentials of a polynomial hypergroup can be identified with the set of all complex numbers.

The following theorem presents a complete description of the exponentials on arbitrary polynomial hypergroups (see [3, 24]).

Theorem 14.3 *Let K be the polynomial hypergroup associated with the sequence of polynomials $(P_n)_{n\in\mathbb{N}}$. The function $m : \mathbb{N} \to \mathbb{C}$ is an exponential on K if and only if there exists a complex number λ such that*

$$m(k) = P_k(\lambda)$$

holds for each k in $\mathbb{N}$.

Applying this result for the Legendre hypergroup we have that the exponential functions in that case are exactly the functions $n \mapsto P_n(\lambda)$ on $\mathbb{N}$, where λ is any complex number and P_n is the n-th Legendre polynomial.

For the description of the additive functions on the Chebyshev hypergroup we know that $a : \mathbb{N} \to \mathbb{C}$ is additive on the Chebyshev hypergroup if and only if it satisfies the functional equation

$$a(k+l) + a(|k-l|) = 2a(k) + 2a(l)$$

for each k, l in $\mathbb{N}$. Surprisingly, this functional equation is closely related to the square-norm functional equation and to Apollonius Theorem (see, e.g., [8]). In fact, any solution of this functional equation has the form $a(k) = c{\cdot}k^2$ with some complex number c. This means that additive functions on the Chebyshev hypergroup are exactly the quadratic functions on $\mathbb{N}$. We can interpret this result in a somewhat surprising manner by observing that $T_n'(1) = n^2$ holds for each n in $\mathbb{N}$, where T_n' is the derivative of the n-th Chebyshev polynomial of the first kind. Consequently, additive functions of the Chebyshev hypergroup have the general form: $n \mapsto c{\cdot}T_n'(1)$ with some complex number c. This is a special case of the following remarkable result (see [24]).

Theorem 14.4 *Let K be the polynomial hypergroup associated with the sequence of polynomials $(P_n)_{n\in\mathbb{N}}$. The function $a : \mathbb{N} \to \mathbb{C}$ is an additive function on K if and only if there exists a complex number c such that*

$$a(n) = c\,P_n'(1)$$

holds for each n in $\mathbb{N}$.

Finally, we note that the study of generalized moment functions on hypergroups leads to the study of the system of functional equation (14.3). We remark that a similar system of functional equation on groupoids has been investigated and solved in [1]. The following theorem describes the generalized moment function sequences of order N in the case of polynomial hypergroups (see [17]).

Theorem 14.5 *Let K be the polynomial hypergroup associated with the sequence of polynomials $(P_n)_{n\in\mathbb{N}}$. The functions $\varphi_0, \varphi_1, \ldots, \varphi_N : K \to \mathbb{C}$ form a generalized moment function sequence of order N on K if and only if*

$$\varphi_k(n) = (P_n \circ f)^{(k)}(0)$$

holds for each n in $\mathbb{N}$ and for $k = 0, 1, \ldots, N$, where

$$f(t) = \sum_{j=0}^{N} \frac{c_j}{j!} t^j \tag{14.8}$$

for each t in $\mathbb{R}$, where c_j is a complex number ($j = 0, 1, \ldots, N$).

14.6 Stability of Additive Functions

The study of stability problems concerning functional equations started with S. Ulam's question at the Mathematics Club of the University of Wisconsin: Suppose that a group G and a metric group H are given. For any $\varepsilon > 0$, does there exist a $\delta > 0$ such that if a function $f : G \to H$ satisfies

$$d(f(xy), f(x)f(y)) < \delta$$

for all x, y in G, then a homomorphism $a : G \to H$ exists with

$$d(f(x), a(x)) < \varepsilon$$

for all x in G? These kind of questions form the material of *stability theory* and Hyers obtained the first important result on this field (see [11]). Later several mathematicians joined these investigations (see the survey papers [4, 25]) but the work of Hyers is still decisive. In fact, he proved the following theorem.

Theorem 14.6 *Let X, Y be Banach spaces and let $f : X \to Y$ be a mapping satisfying*

$$\|f(x + y) - f(x) - f(y)\| \leq \varepsilon$$

for all x, y in X. Then the limit

$$a(x) = \lim_{n\to\infty} \frac{f(2^n x)}{2^n}$$

exists for all x in X and $a : X \to Y$ is the unique additive function satisfying

$$\|f(x) - a(x)\| \leq \varepsilon$$

for all x in X.

We note that uniqueness follows immediately from the obvious fact that the difference of two additive functions is also additive, and the only bounded additive function is 0.

This pioneering result of Hyers can be expressed in the following way: *Cauchy's functional equation is stable for any pair of Banach spaces*. The function

$$(x, y) \mapsto f(x + y) - f(x) - f(y)$$

is called the *Cauchy difference* of the function f. Functions with bounded Cauchy difference are called *approximately additive mappings*. The sequence

$$\left(\frac{f(2^n x)}{2^n}\right)_{n \in \mathbb{N}}$$

is called the *Hyers–Ulam sequence*.

There are several possible ways to generalize the result of Hyers. A natural way is to generalize the domain X depending on a more general result of Rätz [19]. Here we give a corresponding result on hypergroups. Our proof has the novelty of using Banach limits (see, e.g., [5]).

Theorem 14.7 *Let K be a hypergroup with the property that for each x, y in K there exists an integer $N \geq 2$ such that for $n \geq N$ we have*

$$(x * y)^n = x^n * y^n\,. \tag{14.9}$$

Then the functional equation (14.2) *is stable for the pair* $(K, \mathbb{C})$.

We note that here powers are convolution powers which is associative by virtue of the hypergroup axiom (H1) above.

Proof By assumption, the function $f : K \to \mathbb{C}$ satisfies

$$|f(x * y) - f(x) - f(y)| \leq L \tag{14.10}$$

for each x, y in K with some positive number L. Putting $x = y$ we have

$$|f(x^2) - 2f(x)| \leq L \tag{14.11}$$

for each x in K. For x^2 in place of x this yields

$$|f(x^4) - 2f(x^2)| \leq L\,,$$

hence, by (14.11), it follows

$$|f(x^4) - 4f(x)| \leq 3L\,.$$

Repeating this argument we get by induction

$$|f(x^{2^n}) - 2^n f(x)| \leq (2^n - 1)L$$

for each x in K. Division by 2^n gives

$$\Big|\frac{f(x^{2^n})}{2^n} - f(x)\Big| \leq \big(1 - \frac{1}{2^n}\big)L, \tag{14.12}$$

which shows that the Hyers–Ulam sequence

$$\Big(\frac{f(x^{2^n})}{2^n}\Big)_{n\in\mathbb{N}}$$

is bounded. Let *LIM* denote any *Banach limit* on $\mathbb{N}$, then, by (14.12), we have that the function $a : K \to \mathbb{C}$ defined by

$$a(x) = LIM\frac{f(x^{2^n})}{2^n} \tag{14.13}$$

is well-defined for x in K, and satisfies

$$|a(x) - f(x)| \leq L$$

for each x in K. On the other hand, for each x, y in K we have

$$a(x * y) - a(x) - a(y) = LIM\Big(\frac{f\big((x * y)^{2^n}\big) - f(x^{2^n}) - f(y^{2^n})}{2^n}\Big). \tag{14.14}$$

By assumption, if n is large enough, then we have

$$|f\big((x * y)^{2^n}\big) - f(x^{2^n}) - f(y^{2^n})| = |f(x^{2^n} * y^{2^n}) - f(x^{2^n}) - f(y^{2^n})| \leq L,$$

which implies, by (14.11), that a is additive.

This theorem gives the stability of Cauchy's functional equation also in the group case, moreover, in contrast with Hyers' Theorem, we do not need the commutativity of the domain. Nevertheless, the condition of Theorem 14.7 is quite sophisticated and artificial. On non-commutative groups and semigroups the present author proposed another approach based on the concept of invariant means (see [23]). Now we show the application of this method on hypergroups.

Let K be a hypergroup and let $\mathscr{B}(K)$ denote the Banach space of all bounded complex valued functions on K equipped with the sup norm $\|.\|$. A linear functional M of the space $\mathscr{B}(K)$ is called a *right invariant mean*, if $M(1) = 1$ and $M(\tau_y f) = M(f)$ holds for each y in K and f in $\mathscr{B}(K)$. We call K *left amenable*, if there exists a left invariant mean on $\mathscr{B}(K)$. Right invariant means and right amenability are defined similarly. In the case of commutative hypergroups we simply use the terms *invariant mean* and *amenable hypergroup*. For more about invariant means, see, e.g., [9]. It turns out that wide classes of groups and even semigroups are amenable. Amenability of commutative hypergroups has been proved in [31] (see also [13]). Now we prove the stability of additive functions on right amenable hypergroups.

Theorem 14.8 *Let K be a right amenable hypergroup. Then the functional equation* (14.2) *is stable for the pair* $(K, \mathbb{C})$.

Proof Let M be a left invariant mean on K and $f : K \to \mathbb{C}$ a function satisfying

$$|f(x * y) - f(x) - f(y)| \leq L \tag{14.15}$$

for each x, y in K with some positive number L. For each y in K the function $x \mapsto f(x * y) - f(x)$ is bounded, and we define

$$a(y) = M_x\big(f(x * y) - f(x)\big)\,. \tag{14.16}$$

Here M_x denotes that M is applied to the argument as a function of x. Now we have

$$a(y * z) - a(y) - a(z) = M_x\big(f(x * y * z) - f(x * y) - f(x * z) + f(x)\big) =$$

$$M_x\big(f(x * y * z) - f(x * y)\big) - M_x\big(f(x * z) - f(x)\big) = 0\,,$$

as the argument of M in the first term is the right translate of the second term by y. It follows that a is additive. On the other hand, for each y in K we have

$$|f(y) - a(y)| = |f(y) - M_x\big(f(x * y) - f(x)\big)| = |M_x\big(f(x) + f(y) - f(x * y)\big)| \leq$$

$$M_x\big(|f(x) + f(y) - f(x * y)|\big) \leq L\,.$$

The theorem is proved.

This result extends easily to the pexiderized equation of additive functions as is shown in the following theorem.

Theorem 14.9 *Let K be a right amenable hypergroup and let $f, g, h : K \to \mathbb{C}$ be functions such that the function $(x, y) \mapsto f(x * y) - g(x) - h(y)$ is bounded. Then there exists an additive function $a : K \to \mathbb{C}$ such that $f - a$, $g - a$, and $h - a$ are bounded.*

14.7 Stability of Exponential Functions

The stability of exponential functions was first proved in [2] for real valued functions defined on linear spaces. By the results in [22] we have the following result.

Theorem 14.10 *Let S be a commutative semigroup with identity, and suppose that for the functions $f, m : S \to \mathbb{C}$ the function $x \mapsto f(x+y) - f(x)m(y)$ is bounded for each y in S. Then either f is bounded, or m is an exponential.*

This result shows the so-called *superstability* property of the exponential functional equation: the difference $f(x+y) - f(x)f(y)$ can be bounded if and only if it is either zero, or f itself is bounded. Now we study this problem on hypergroups.

Theorem 14.11 *Let K be a hypergroup and let $f, g, h : K \to \mathbb{C}$ be continuous functions. If the function*

$$x \mapsto f(x * y) - g(x)h(y)$$

is bounded for each y in K, then either f is bounded, or $h(e) \neq 0$ and $h/h(e)$ is an exponential.

Proof Suppose that f is unbounded. Then, putting $y = e$ into the above condition, $h(e) \neq 0$ follows, and we have that $f - h(e)g$ is bounded. Moreover, by assumption, we have

$$|h(e)g(x * y) - g(x)h(y)| \leq l(y),$$

for each x, y in K with some function $l : K \to \mathbb{C}$. Dividing by $h(e)^2$ it follows that the function

$$x \mapsto \frac{1}{h(e)} g(x * y) - \frac{1}{h(e)} g(x) \cdot \frac{1}{h(e)} h(y)$$

is bounded for each y in K. Theorem 11.1 in [29] implies that either g is bounded, or $h(e) \neq 0$ and $h/h(e)$ is an exponential. However, g cannot be bounded, otherwise f is bounded, too. This implies that $h(e) \neq 0$ and $h/h(e)$ is an exponential.

Obviously, this result implies the following.

Theorem 14.12 *Let K be a hypergroup and $f : K \to \mathbb{C}$ a continuous function such that the function*

$$(x, y) \mapsto f(x * y) - f(x)f(y)$$

is bounded. Then f is either bounded or an exponential.

In other words, the exponential functional equation is *superstable* on any hypergroup. In the following section we will have an application of this result for spherical functions.

14.8 Double Coset Hypergroups

In this section we exhibit another important type of hypergroups: the double coset hypergroups. The idea is that if G is a locally compact group and K is a subgroup then, in general, the left, or right, or double coset spaces with respect to K do not bear any reasonable structure. Nevertheless, if K is a compact subgroup, then a quite useful hypergroup structure can be introduced on the double coset space with respect to K. In particular, this hypergroup structure reduces to the usual group structure if K is a normal subgroup. In the subsequent paragraphs we present the details (see also [3]).

Let G be a locally compact group with identity e and K a compact subgroup with normalized Haar measure ω: $\int_K d\omega(k) = 1$. As K is unimodular ω is left and right invariant, and also inversion invariant. For each x in G we define the *double coset* of x as the set

$$KxK = \{kxl : k, l \in K\}.$$

We introduce a hypergroup structure on the set $L = G//K$ of all double cosets: the topology of L is the quotient topology, which is locally compact. The identity o is the coset $K = KeK$ itself and the involution is defined by

$$(KxK)^\vee = Kx^{-1}K\,.$$

Finally, the convolution of δ_{KxK} and δ_{KyK} is defined by

$$\delta_{KxK} * \delta_{KyK} = \int_K \delta_{KxkyK}\, d\omega(k)\,.$$

It is known that this gives a hypergroup structure on L (see [3, p. 12]), which, in general, is non-commutative. If K is a normal subgroup, then L is isomorphic to the hypergroup arising from the factor group G/K.

We note that continuous functions on L can be identified with those continuous functions on G which are K-invariant: $f(x) = f(kxl)$ for each x in G and k, l in K. Hence, for a continuous function $f : L \to \mathbb{C}$ the simplified—and somewhat loose—notation $f(x)$ can be used for the function value $f(KxK)$. Using this convention we can write for each continuous function $f : L \to \mathbb{C}$ and for each x, y in G:

$$f(x * y) = \int_K f(xky)\, d\omega(k).$$

The following theorem exhibits a close connection between exponentials on double coset hypergroups and spherical functions on locally compact groups. Following the terminology of [3] (see also [7]) we recall the concept of spherical functions. Let G be a locally compact group and K a compact subgroup with Haar measure ω. The continuous bounded K-invariant function $f : G \to \mathbb{C}$ is called a *K-spherical function* if $f(e) = 1$ and

$$\int_K f(xky)\, d\omega(k) = f(x)f(y) \tag{14.17}$$

holds for each x, y in G. A *generalized K spherical function* on G is the same as above without the boundedness hypothesis. For the sake of simplicity in this paper we use the term *spherical function* for continuous functions satisfying (14.17) without the boundedness assumption. The following theorem, which is an immediate consequence of the previous considerations gives the link between spherical functions and exponentials of double coset hypergroups.

Theorem 14.13 *Let G be a locally compact group, and $K \subseteq G$ a compact subgroup. Then the nonzero continuous complex valued function m is a K-spherical function on G if and only if it is an exponential on the double coset hypergroup $G//K$. In particular, K-spherical functions on G can be identified with the characters of $G//K$.*

By virtue of this theorem, as an application of Theorem 14.11, we obtain the following result on the superstability of functional equations related to spherical functions (see [32]).

Theorem 14.14 *Let G be a locally compact group and K a compact subgroup with normed Haar measure ω. Let $f, g, h : G \to \mathbb{C}$ be continuous K-invariant functions such that the function*

$$x \mapsto \int_K f(xky)\, d\omega(k) - g(x)h(y)$$

is bounded for each y in G. Then either f is bounded, or $h(e) \neq 0$ and $h/h(e)$ is a K-spherical function.

14.9 Superstability of Generalized Moment Functions

In this section we prove that generalized moment functions on hypergroups also have the remarkable superstability property (see also [27, 29]).

Theorem 14.15 *Let K be a hypergroup, n a nonnegative integer, and suppose that for the unbounded functions $f_k : K \to \mathbb{C}$ $(k = 0, 1, \ldots, n)$ the functions*

$$(x, y) \mapsto f_k(x * y) - \sum_{j=0}^{k} \binom{k}{j} f_j(x) f_{k-j}(y)$$

are bounded on $K \times K$. Then the sequence $(f_k)_{k \le n}$ forms a moment function sequence of order n on K.

Proof We prove the theorem for a fixed n by induction on k. For $k = 0$, by our assumption, the function

$$(x, y) \mapsto f_0(x * y) - f_0(x)f_0(y)$$

is bounded on $K \times K$. By Theorem 14.11, this implies that f_0 is an exponential on K.

Suppose now that $k \geq 1$ and we have proved that the functions f_j for $j = 0, 1, \ldots, k-1$ form a moment function sequence of order $k-1$ on K. By assumption, we have that the function

$$(x, y, z) \mapsto F(x, y, z) = f_k(x * y * z) - \sum_{j=0}^{k} \binom{k}{j} f_j(x * y)f_{k-j}(z),$$

and also the function

$$(x, y, z) \mapsto G(x, y, z) = f_k(x * y * z) - \sum_{j=0}^{k} \binom{k}{j} f_j(x)f_{k-j}(y * z)$$

is bounded on $K \times K \times K$. Then their difference

$$(x, y, z) \mapsto F(x, y, z) - G(x, y, z) =$$

$$\sum_{j=0}^{k} \binom{k}{j} f_j(x * y)f_{k-j}(z) - \sum_{j=0}^{k} \binom{k}{j} f_j(x)f_{k-j}(y * z)$$

is also bounded. By our induction hypothesis, this means that the function

$$\begin{aligned}(x, y, z) &\mapsto F(x, y, z) - G(x, y, z) = H(x, y, z) \\ &= \sum_{j=1}^{k-1} \binom{k}{j} f_j(x) \sum_{i=0}^{k-j} \binom{k-j}{i} f_i(y)f_{k-j-i}(z) - \sum_{j=1}^{k-1} \binom{k}{j} \sum_{i=0}^{j} \binom{j}{i} f_i(x)f_{j-i}(y)f_{k-j}(z) \\ &\quad + f_0(x)f_k(y * z) - f_k(x * y)f_0(z) + f_k(x)f_0(y)f_0(z) - f_0(x)f_0(y)f_k(z)\end{aligned}$$

is bounded, too. By reordering the terms in this sum we obtain

$$\begin{aligned}H(x, y, z) = {} & f_0(x)\Big[f_k(y * z) - \sum_{j=0}^{k} \binom{k}{j} f_j(y)f_{k-j}(z)\Big] \\ & - f_0(z)\Big[f_k(x * y) - \sum_{j=0}^{k} \binom{k}{j} f_j(x)f_{k-j}(y)\Big] + \sum_{j=1}^{k-1} \sum_{i=0}^{k-j-1} \binom{k}{j} \binom{k-j}{i} f_j(x)f_i(y)f_{k-j-i}(z) \\ & - \sum_{j=1}^{k-1} \sum_{i=1}^{j} \binom{k}{j} \binom{j}{i} f_i(x)f_{j-i}(y)f_{k-j}(z)\end{aligned}$$

for all x, y, z in K. We show that the two terms on the right-hand side of the last equality cancel. In the first term replacing i by $t - j$, and in the second term interchanging the sums we have

$$H(x,y,z) = f_0(x)\Big[f_k(y*z) - \sum_{j=0}^{k}\binom{k}{j}f_j(y)f_{k-j}(z)\Big]$$

$$-f_0(z)\Big[f_k(x*y) - \sum_{j=0}^{k}\binom{k}{j}f_j(x)f_{k-j}(y)\Big] + \sum_{j=1}^{k-1}\sum_{t=j}^{k-1}\binom{k}{j}\binom{k-j}{k-t}f_j(x)f_{t-j}(y)f_{k-t}(z)$$

$$-\sum_{i=1}^{k-1}\sum_{j=i}^{k-1}\binom{k}{j}\binom{j}{i}f_i(x)f_{j-i}(y)f_{k-j}(z)$$

for all x, y, z in K. In the second term we write j for i and t for j to get

$$H(x,y,z) = f_0(x)\Big[f_k(y*z) - \sum_{j=0}^{k}\binom{k}{j}f_j(y)f_{k-j}(z)\Big]$$

$$-f_0(z)\Big[f_k(x*y) - \sum_{j=0}^{k}\binom{k}{j}f_j(x)f_{k-j}(y)\Big] + \sum_{j=1}^{k-1}\sum_{t=j}^{k-1}\binom{k}{j}\binom{k-j}{k-t}f_j(x)f_{t-j}(y)f_{k-t}(z)$$

$$-\sum_{j=1}^{k-1}\sum_{t=j}^{k-1}\binom{k}{t}\binom{t}{j}f_j(x)f_{t-j}(y)f_{k-t}(z)$$

for all x, y, z in K. On the other hand, we have

$$\binom{k}{j}\binom{k-j}{k-t} = \frac{k!}{j!(k-j)!}\,\frac{(k-j)!}{(k-t)!\,(t-j)!} = \frac{k!}{t!(k-t)!}\,\frac{t!}{j!(t-j)!} = \binom{k}{t}\binom{t}{j},$$

hence the function

$$(x,y,z) \mapsto L(x,y,z) = f_0(x)\Big[f_k(y*z) - \sum_{j=0}^{k}\binom{k}{j}f_j(y)f_{k-j}(z)\Big]$$

$$-f_0(z)\Big[f_k(x*y) - \sum_{j=0}^{k}\binom{k}{j}f_j(x)f_{k-j}(y)\Big]$$

is bounded. If there are y, z in K such that

$$f_k(y*z) - \sum_{j=0}^{k}\binom{k}{j}f_j(y)f_{k-j}(z) \neq 0\,,$$

then f_0 is bounded, which is impossible. Thus we have

$$f_k(y * z) - \sum_{j=0}^{k} \binom{k}{j} f_j(y) f_{k-j}(z) = 0$$

for all y, z in K, and the proof is complete.

14.10 Stability Problems of Other Functional Equations

In the previous sections we discussed two different types of stability: one of them for additive-type equations, where the method is based on invariant means, and the other for exponential-type equations, where superstability appears, and the method is direct. The question arises concerning stability results of mixed type, where both the additive and exponential equations come into the picture and we can combine the two methods.

We mentioned above that the second equation of the system defining generalized moment function sequences is in itself interesting: we can consider it independently from the other equations. It has the form

$$f(x * y) = f(x)g(y) + g(x)f(y) . \tag{14.18}$$

It turns out that the pexiderized version

$$f(x * y) = g(x)h(y) + l(x)k(y) , \tag{14.19}$$

can be treated in the case $l = 1$. We have also considered another similar equation in [29]. In the special case of this equation when $k = 0$, we have the pexiderized exponential equation, and in the case $h = l = 1$ we obtain the pexiderized additive equation. We recall the corresponding results of [29].

Theorem 14.16 *Let K be an amenable discrete hypergroup and suppose that the functions $f, g, h, k : K \to \mathbb{C}$ are given and f is unbounded. Then the function*

$$x \mapsto f(x * y) - g(x)h(y) - k(y)$$

is bounded for each y in K if and only if we have

$$\begin{aligned} f(x) &= \varphi(x) + b_1(x) \\ g(x) &= \varphi(x) + b_2(x) \\ h(x) &= m(x) \\ k(x) &= \varphi(x) + b_3(x) \end{aligned}$$

where $m : K \to \mathbb{C}$ *is an exponential,* $b_1, b_2, b_3 : K \to \mathbb{C}$ *are bounded functions, and* $\varphi : K \to \mathbb{C}$ *satisfies the functional equation*

$$\varphi(x * y) = \varphi(x)m(y) + \varphi(y) \tag{14.20}$$

for each x, y *in* K*, further, if* b_2 *is nonzero, then* m *is bounded.*

We can see here that we have superstability with respect to h and stability with respect to the other three functions. Another special feature is that we have the stability result without knowing the general solution of the corresponding functional equation

$$f(x * y) = g(x)h(y) + k(y)\,. \tag{14.21}$$

At this point we mention the open problem concerning the stability of the sine equation (14.18). Another related stability result in [29] is the following.

Theorem 14.17 *Let* K *be a discrete commutative hypergroup and suppose that functions* $f, g, h, k, l : K \to \mathbb{C}$ *are unbounded. Then the function*

$$x \mapsto f(x * y) - g(x)h(y) - k(x) - l(y) \tag{14.22}$$

is bounded for each y *in* K *if and only if either*

$$f(x) = \frac{\lambda}{2}\, a(x)^2 + d_0 a(x) + a_0(x) + b_1(x) \tag{14.23}$$

$$g(x) = a(x) + c_0, \qquad h(x) = \lambda a(x) + d_0$$

$$k(x) = \frac{\lambda}{2}\, a(x)^2 + a_0(x) + b_2(x)$$

$$l(x) = \frac{\lambda}{2}\, a(x)^2 + (d_0 - \lambda c_0)a(x) + a_0(x) + b_3(x)\,,$$

or

$$f(x) = cd[m(x) - 1] + a(x) + b_1(x) \tag{14.24}$$

$$g(x) = c[m(x) - 1] + c_0,\ h(x) = d[m(x) - 1] + d_0$$

$$k(x) = c(d - d_0)[m(x) - 1] + a(x) + b_2(x)$$

$$l(x) = d(c - c_0)[m(x) - 1] + a(x) + b_3(x)\,,$$

where $m : K \to \mathbb{C}$ *is an exponential,* $a, a_0 : K \to \mathbb{C}$ *are additive functions,* $b_1, b_2, b_3 : K \to \mathbb{C}$ *are bounded functions, and* λ, c, d, c_0, d_0 *are complex numbers.*

The proof is again a combination of direct methods and of the invariant mean technique. We note that—in contrast with the previous theorem—commutativity is used instead of amenability.

Acknowledgements Research was supported by OTKA Grant No. K111651.

References

1. Aczél, J.: Functions of binomial type mapping groupoids into rings. Math. Z. **154**(2), 115–124 (1977)
2. Baker, J., Lawrence, J., Zorzitto, F.: The stability of the equation $f(x + y) = f(x)f(y)$. Proc. Am. Math. Soc. **74**(2), 242–246 (1979)
3. Bloom, W.R., Heyer, H.: Harmonic Analysis of Probability Measures on Hypergroups. de Gruyter Studies in Mathematics, vol. 20. Walter de Gruyter, Berlin (1995)
4. Cabello Sánchez, F., Castillo, J.M.F.: Banach space techniques underpinning a theory for nearly additive mappings. Dissertationes Math. (Rozprawy Mat.) **404**, 73 pp. (2002)
5. Conway, J.B.: A Course in Functional Analysis. Springer, New York (1990)
6. Davison, T.M.K.: D'Alembert's functional equation and Chebyshev polynomials. Ann. Acad. Paed. Cracov. Studia Math. **4**, 31–38 (2001)
7. Dieudonné, J.: Treatise on Analysis, vol. VI. Academic [Harcourt Brace Jovanovich, Publishers], New York (1978)
8. Godfrey, C., Siddons, A.W.: Modern Geometry. University Press, Cambridge (1908)
9. Greenleaf, F.P.: Invariant means on topological groups and their applications. Van Nostrand Mathematical Studies, vol. 16. Van Nostrand Reinhold, New York (1969)
10. Hewitt, E., Ross, K.A.: Abstract Harmonic Analysis, Volume I: Structure of Topological Groups. Integration Theory, Group Representations. Die Grundlehren der mathematischen Wissenschaften, vol. 115. Academic, New York (1963)
11. Hyers, D.H.: On the stability of the linear functional equation. Proc. Natl. Acad. Sci. USA **27**, 222–224 (1941)
12. Laczkovich, M., Székelyhidi, L.: Spectral synthesis on discrete abelian groups. Math. Proc. Camb. Philos. Soc. **143**(1), 103–120 (2007)
13. Lau, A.T.M.: Analysis on a class of Banach algebras with applications to harmonic analysis on locally compact groups and semigroups. Fundam. Math. **118**(3), 161–175 (1983)
14. Michael, E.: Topologies on spaces of subsets. Trans. Am. Math. Soc. **71**, 152–182 (1951)
15. Orosz, Á.: Sine and cosine equation on discrete polynomial hypergroups. Aequationes Math. **72**(3), 225–233 (2006)
16. Orosz, Á., Székelyhidi, L.: Moment functions on polynomial hypergroups in several variables. Publ. Math. Debrecen **65**(3–4), 429–438 (2004)
17. Orosz, Á., Székelyhidi, L.: Moment functions on polynomial hypergroups. Arch. Math. (Basel) **85**(2), 141–150 (2005)
18. Orosz, Á., Székelyhidi, L.: Moment functions on Sturm-Liouville hypergroups. Ann. Univ. Sci. Budapest. Sect. Comput. **29**, 141–156 (2008)
19. Rätz, J.: On approximately additive mappings. In: General Inequalities, 2 (Proceedings of the Second International Conference, Oberwolfach, 1978), pp. 233–251. Birkhäuser, Basel (1980)
20. Ross, K.A.: Hypergroups and signed hypergroups. In: Harmonic Analysis and Hypergroups. Trends in Mathematics (Delhi, 1995), pp. 77–91. Birkhäuser Boston, Boston (1998)
21. Roukbi, A., Zeglami, D.: d'Alembert's functional equations on hypergroups. Adv. Pure Appl. Math. **2**(2), 147–166 (2011)
22. Székelyhidi, L.: On a theorem of Baker, Lawrence and Zorzitto. Proc. Am. Math. Soc. **84**(1), 95–96 (1982)

23. Székelyhidi, L.: Remark 17. In: Report of Meeting: The Twenty-second International Symposium on Functional Equations, December 16 – December 22, 1984, Oberwolfach, Germany. Aequationes Math. **29(1)**, 62–111 (1985)
24. Székelyhidi, L.: Functional equations on hypergroups. In: Functional Equations, Inequalities and Applications, pp. 167–181. Kluwer Academic Publishers, Dordrecht (2000)
25. Székelyhidi, L.: Ulam's problem, Hyers's solution—and to where they led. In: Functional Equations and Inequalities, pp. 259–285. Kluwer Academic Publishers, Dordrecht (2000)
26. Székelyhidi, L.: Functional equations on Sturm-Liouville hypergroups. Math. Pannon. **17**(2), 169–182 (2006)
27. Székelyhidi, L.: Superstability of moment functions on hypergroups. Nonlinear Funct. Anal. Appl. **11**(5), 815–821 (2006)
28. Székelyhidi, L.: Exponential polynomials on commutative hypergroups. Arch. Math. (Basel) **101**(4), 341–347 (2013)
29. Székelyhidi, L.: Functional Equations on Hypergroups. World Scientific, Hackensack (2013)
30. Székelyhidi, L.: Characterization of exponential polynomials on commutative hypergroups. Ann. Funct. Anal. **5**(2), 53–60 (2014)
31. Székelyhidi, L.: Harmonic and Spectral Analysis. World Scientific, Hackensack (2014)
32. Székelyhidi, L.: Superstability of functional equations related to spherical functions. Open Math. **15**, 427–432 (2017)
33. Székelyhidi, L., Vajday, L.: Functional equations on the $SU(2)$-hypergroup. Math. Pannon. **23(2)**, 187–193 (2012)

Chapter 15
Stability of Systems of General Functional Equations in the Compact-Open Topology

Pavol Zlatoš

Abstract We introduce a fairly general concept of functional equation for k-tuples of functions $f_1, \ldots, f_k: X \to Y$ between arbitrary sets. The homomorphy equations for mappings between groups and other algebraic systems, as well as various types of functional equations and recursion formulas occurring in mathematical analysis or combinatorics, respectively, become special cases (of systems) of such equations. Assuming that X is a locally compact and Y is a completely regular topological space, we show that systems of such functional equations, with parameters satisfying rather a modest continuity condition, are stable in the following intuitive sense: Every k-tuple of "sufficiently continuous," "reasonably bounded" functions $X \to Y$ satisfying the given system with a "sufficient precision" on a "big enough" compact set is already "arbitrarily close" on an "arbitrarily big" compact set to a k-tuple of continuous functions solving the system. The result is derived as a consequence of certain intuitively appealing "almost-near" principle using the relation of infinitesimal nearness formulated in terms of nonstandard analysis.

Keywords System of functional equations • Continuous solution • Stability • Locally compact • Completely regular • Uniformity • Nonstandard analysis

Mathematics Subject Classification (2010) Primary 39B82; Secondary 39B72, 54D45, 54E15, 54J05

P. Zlatoš (✉)
Faculty of Mathematics, Physics and Informatics, Comenius University,
Mlynská dolina, 842 48 Bratislava, Slovakia
e-mail: zlatos@fmph.uniba.sk

J. Brzdęk et al. (eds.), *Developments in Functional Equations and Related Topics*, Springer Optimization and Its Applications 124,
DOI 10.1007/978-3-319-61732-9_15

15.1 Introduction

The study of stability of functional equations in the spirit of Ulam started with examining the stability of additive functions and more generally of homomorphisms between metrizable topological groups, cf. [3, 13, 14, 19, 20, 27, 28]. Since that time it has developed to an established topic in mathematical and functional analysis and extended to a variety of (systems of) functional equations—see, e.g., [6, 9, 15, 21, 22, 26]. However, in most cases the stability issue was considered (explicitly or implicitly) either within the topology of uniform convergence or within the (strong) topology given by a norm on some functional space. On the other hand, especially when dealing with spaces of continuous functions defined on a locally compact space, the compact-open topology (i.e., the topology of uniform convergence on compact sets) is the most natural one. The systematic study of such local stability on compacts and its relation to the "usual" global or uniform stability was commenced by the author for homomorphisms between topological groups in [30] and extended to homomorphisms between topological universal algebras in [31]; cf. also [18, 24].

In the present paper we introduce a fairly general concept of functional equation for k-tuples of functions $f_1, \ldots, f_k \colon X \to Y$ between arbitrary sets. Then the homomorphy equations for mappings between groups and other algebraic systems, as well as various types of functional equations occurring in mathematical analysis (like, e.g., the sine and cosine addition formulas) or various recursion formulas occurring in combinatorics become just special cases (of systems) of such equations. Assuming that X is a locally compact and Y is a completely regular (i.e., uniformizable) topological space, we will show that systems of such functional equations, with functional parameters satisfying rather a modest continuity condition, are stable in the following intuitive sense, which will be made precise in the final Section 15.4 (Theorems 15.2, 15.3): Every k-tuple of "sufficiently continuous," "reasonably bounded" functions $X \to Y$ satisfying the given system with a "sufficient precision" on a "big enough" compact set is already "arbitrarily close" on an "arbitrarily big" compact set to a k-tuple of continuous functions solving the system. The result is a generalization comprising several former results by the author and his collaborators [24, 25, 29–31], as special cases. It is derived as a consequence of certain intuitively appealing stability or "almost-near" principle (in the sense of [2, 5]) using the relation of infinitesimal nearness formulated in terms of nonstandard analysis in Section 15.3 (Theorem 15.1, Corollary 15.2), generalizing a more specific principle of this kind from [24].

15.2 General Form of Functional Equations

Let X, Y be arbitrary nonempty sets and $k, m, n \geq 1$, $p \geq 0$ be integers. A k-tuple of functions $\boldsymbol{f} = (f_1, \ldots, f_k), f_i \colon X \to Y$, is viewed as a single function $\boldsymbol{f} \colon X \to Y^k$. Further, let $\boldsymbol{\alpha} = (\alpha_1, \ldots, \alpha_m)$ be an m-tuple of p-ary operations $\alpha_j \colon X^p \to X$

(if $p = 0$, a nullary operation α on X is simply a constant $\alpha \in X$). We use the tensor product notation to denote the function $\boldsymbol{f} \otimes \boldsymbol{\alpha}: X^p \to Y^{k\times m}$ assigning to every p-tuple $\boldsymbol{x} = (x_1, \ldots, x_p) \in X^p$ the $k \times m$ matrix

$$(\boldsymbol{f} \otimes \boldsymbol{\alpha})(\boldsymbol{x}) = \big((f_i \circ \alpha_j)(\boldsymbol{x})\big) = \begin{pmatrix} f_1\big(\alpha_1(\boldsymbol{x})\big) \ldots f_1\big(\alpha_m(\boldsymbol{x})\big) \\ \vdots \quad \ddots \quad \vdots \\ f_k\big(\alpha_1(\boldsymbol{x})\big) \ldots f_k\big(\alpha_m(\boldsymbol{x})\big) \end{pmatrix}.$$

In the trivial case when $k = m = 1$ we can identify $\boldsymbol{f} = f, \boldsymbol{\alpha} = \alpha$; then $\boldsymbol{f} \otimes \boldsymbol{\alpha}$ is just the composition of functions $f \circ \alpha: X^p \to Y$. If $m = 1$ and $\alpha(x) = x$ is the identity Id_X on X, then $\boldsymbol{f} \otimes \alpha = (f_1, \ldots, f_k) = \boldsymbol{f}$. If $m = p$ and $\boldsymbol{\alpha} = \boldsymbol{\pi} = (\pi_1, \ldots, \pi_m)$, where $\pi_j: X^m \to X$ is the jth projection, i.e., $\pi_j(x_1, \ldots, x_m) = x_j$, then $(\boldsymbol{f} \otimes \boldsymbol{\pi})(\boldsymbol{x}) = \big(f_i(x_j)\big) \in Y^{k\times m}$. In general, the function $\boldsymbol{f} \otimes \boldsymbol{\alpha}$ can be identified with the matrix of composed functions $f_i \circ \alpha_j: X^p \to Y$ $(i \le k, j \le m)$.

Additionally, if $F: Y^{k\times m} \to Y$ is a $(k \times m)$-ary operation on Y, then $F(\boldsymbol{f} \otimes \boldsymbol{\alpha}) = F \circ (\boldsymbol{f} \otimes \boldsymbol{\alpha}): X^p \to Y$ denotes the function given by

$$F(\boldsymbol{f} \otimes \boldsymbol{\alpha})(\boldsymbol{x}) = F\big((\boldsymbol{f} \otimes \boldsymbol{\alpha})(\boldsymbol{x})\big),$$

for $\boldsymbol{x} \in X^p$. More generally, for any mapping $F: Y^{k\times m} \times X^p \to Y$ we denote by $\widetilde{F}(\boldsymbol{f} \otimes \boldsymbol{\alpha}): X^p \to Y$ the function given by

$$\widetilde{F}(\boldsymbol{f} \otimes \boldsymbol{\alpha})(\boldsymbol{x}) = F\big((\boldsymbol{f} \otimes \boldsymbol{\alpha})(\boldsymbol{x}), \boldsymbol{x}\big),$$

for $\boldsymbol{x} \in X^p$. Further on (except for some Examples) we will study exclusively the latter more general case which includes the former one, when the mapping F does not depend on $\boldsymbol{x}$, i.e., when $F(\boldsymbol{A}, \boldsymbol{x}) = F(\boldsymbol{A}, \boldsymbol{x}')$ for any matrix $\boldsymbol{A} \in Y^{k\times m}$ and all $\boldsymbol{x}, \boldsymbol{x}' \in X^p$.

A *general functional equation*, briefly a *GFE, of type* (k, m, n, p), with $k, m, n \ge 1, p \ge 0$, is a functional equation of the form

$$\widetilde{F}(\boldsymbol{f} \otimes \boldsymbol{\alpha}) = \widetilde{G}(\boldsymbol{f} \otimes \boldsymbol{\beta}), \tag{15.1}$$

where $\boldsymbol{f} = (f_1, \ldots, f_k)$ is a k-tuple of *functional variables* or "unknown" functions $f_i: X \to Y$, $\boldsymbol{\alpha} = (\alpha_1, \ldots, \alpha_m)$ is an m-tuple and $\boldsymbol{\beta} = (\beta_1, \ldots, \beta_n)$ is an n-tuple of p-ary operations on the set X, and, finally, $F: Y^{k\times m} \times X^p \to Y$ and $G: Y^{k\times n} \times X^p \to Y$ are any mappings. The operations (mappings) α_i, β_j, F, and G are called the *functional coefficients* or *parameters* of the equation. A k-tuple of functions $\boldsymbol{f} = (f_1, \ldots, f_k): X \to Y^k$ *satisfies* the GFE (15.1), or it is a *solution* of it, if the functions $\widetilde{F}(\boldsymbol{f} \otimes \boldsymbol{\alpha})$, $\widetilde{G}(\boldsymbol{f} \otimes \boldsymbol{\beta})$ coincide, i.e., if

$$F\big((\boldsymbol{f} \otimes \boldsymbol{\alpha})(\boldsymbol{x}), \boldsymbol{x}\big) = G\big((\boldsymbol{f} \otimes \boldsymbol{\beta})(\boldsymbol{x}), \boldsymbol{x}\big),$$

for all $\boldsymbol{x} \in X^p$. More generally, $\boldsymbol{f}$ *satisfies* the GFE (15.1) *on a set* $S \subseteq X^p$ if the above equation holds for each $\boldsymbol{x} \in S$; we say that $\boldsymbol{f}$ *satisfies* the GFE (15.1) *on a set* $A \subseteq X$ if it satisfies (15.1) on the set $A^p \subseteq X^p$.

A system of GFEs

$$\widetilde{F}_\lambda(\boldsymbol{f} \otimes \boldsymbol{\alpha}_\lambda) = \widetilde{G}_\lambda(\boldsymbol{f} \otimes \boldsymbol{\beta}_\lambda) \qquad (\lambda \in \Lambda)\,, \tag{15.2}$$

with (finite or infinite) index set $\Lambda \neq \emptyset$, consist of GFEs of particular types $(k, m_\lambda, n_\lambda, p_\lambda)$ (with k fixed and m_λ, n_λ, p_λ depending on $\lambda \in \Lambda$). Then $f = (f_1, \ldots, f_k)$ is a *solution of the system* if $\boldsymbol{f}$ satisfies all the equations in it. Satisfaction of the system on some set $A \subseteq X$ is defined in the obvious way.

We do not maintain that the (systems of) GFEs of the form just defined cover all the (systems of) functional equations one can meet, as such a claim would be too ambitious and, obviously, not founded well enough. In particular, functional equations dealing with compositions of functional variables $f_i \circ f_j$ or with iterated compositions like $f, f^2 = f \circ f, f^3 = f \circ f \circ f$, etc., do not fall under this scheme. On the other hand, as indicated by the examples below, they still comprise a large and representative variety of (systems of) functional equations studied so far.

Let us start with three closely related examples of algebraic nature.

Example 15.1 Let $(X, *)$, $(Y, \star)$ be two groupoids, i.e., algebraic structures with arbitrary binary operations $*$, $\star$ on the sets X and Y, respectively. Let $\alpha: X^2 \to X$ be the operation $\alpha(x_1, x_2) = x_1 * x_2$ on X, π_1, $\pi_2: X^2 \to X$ be the projections on the first and the second variable, respectively, $F = \mathrm{Id}_Y: Y \to Y$ be the identity mapping and $G: Y^2 \to Y$ be the operation $G(y_1, y_2) = y_1 \star y_2$ on Y. Then the GFE

$$F(\boldsymbol{f} \otimes \boldsymbol{\alpha}) = G(\boldsymbol{f} \otimes \boldsymbol{\pi})$$

of type $(1, 1, 2, 2)$, with $\boldsymbol{f} = f: X \to Y$, $\boldsymbol{\alpha} = \alpha$ and $\boldsymbol{\pi} = (\pi_1, \pi_2)$, which rewrites as

$$f \circ \alpha = G\big(f \otimes (\pi_1, \pi_2)\big)\,,$$

simply means that

$$f(x_1 * x_2) = f(x_1) \star f(x_2)$$

for all $x_2, x_2 \in X$. In other words, a function f satisfies the above GFE if and only if it is a homomorphism $f: (X, *) \to (Y, \star)$.

If both $(X, *)$, $(Y, \star)$ coincide with the additive group $(\mathbb{R}, +)$ of reals, we get the Cauchy functional equation

$$f(x + y) = f(x) + f(y)\,.$$

If $(X, *) = (\mathbb{R}, +)$ and $(Y, \star)$ is the multiplicative group $(\mathbb{R}^+, \cdot)$ of positive reals, we obtain the equation

$$f(x + y) = f(x)f(y)\,,$$

characterizing exponential functions. If both $(X, *)$, $(Y, \star)$ denote the set $\mathbb{R}$ with the arithmetical mean $x * y = x \star y = (x + y)/2$, we have Jensen's functional equation

$$f\left(\frac{x+y}{2}\right) = \frac{f(x) + f(y)}{2}.$$

And the list could be continued indefinitely.

Example 15.2 More generally, let Λ be a set of operation symbols with finite arities p_λ ($\lambda \in \Lambda$), and $\mathfrak{X} = (X, \alpha_\lambda)_{\lambda\in\Lambda}$, $\mathfrak{Y} = (Y, G_\lambda)_{\lambda\in\Lambda}$ be two universal algebras of signature $(p_\lambda)_{\lambda\in\Lambda}$, i.e., $\alpha_\lambda = \lambda^{\mathfrak{X}}: X^{p_\lambda} \to X$, $G_\lambda = \lambda^{\mathfrak{Y}}: Y^{p_\lambda} \to Y$ are p_λ-ary operations on the sets X, Y, respectively, corresponding to the symbol $\lambda \in \Lambda$, cf. [10]. A function $f: X \to Y$ is called a *homomorphism* from $\mathfrak{X}$ to $\mathfrak{Y}$, briefly $f: \mathfrak{X} \to \mathfrak{Y}$, if for each $\lambda \in \Lambda$ and any p_λ-tuple $\boldsymbol{x} = (x_1, \ldots, x_{p_\lambda}) \in X^{p_\lambda}$ we have

$$f\big(\alpha_\lambda(x_1, \ldots, x_{p_\lambda})\big) = G_\lambda\big(f(x_1), \ldots, f(x_{p_\lambda})\big),$$

(for nullary operation symbols $\lambda \in \Lambda$ this simply means that $f(\alpha_\lambda) = G_\lambda$). Similarly as in the previous Example 15.1, we see immediately that this is the case if and only if f satisfies the system of GFEs

$$f \circ \alpha_\lambda = G_\lambda\big(f \otimes (\pi_1, \ldots, \pi_{p_\lambda})\big) \qquad (\lambda \in \Lambda),$$

of types $(1, 1, p_\lambda, p_\lambda)$, where $\pi_j: X^{p_\lambda} \to X$, $\pi_j(\boldsymbol{x}) = x_j$, is the jth projection for $j \leq p_\lambda$.

Example 15.3 Let $(\Lambda, +, \cdot, 0, 1)$, be a ring with unit $1 \neq 0$. A (left) Λ-module X is an abelian group $(X, +)$ with scalar multiplication $\Lambda \times X \to X$, sending each pair $(\lambda, x) \in \Lambda \times X$ to the scalar multiple $\lambda x \in X$, satisfying the usual axioms. Then each scalar $\lambda \in \Lambda$ can be regarded as an endomorphism $\lambda^X: X \to X$ of the abelian group $(X, +)$, and the assignment $\lambda \mapsto \lambda^X$ becomes a homomorphism of rings $(\Lambda, +, \cdot, 0, 1) \to \big(\mathrm{End}(X, +), +, \circ, 0, \mathrm{Id}_X\big)$, cf. [12]. In particular, if Λ is a field, then a Λ-module is just a vector space over Λ.

A homomorphism of Λ-modules X, Y is a mapping $f: X \to Y$, preserving the addition and scalar multiplication, i.e., satisfying

$$f(x + y) = f(x) + f(y),$$
$$f(\lambda x) = \lambda f(x)$$

for any $x, y \in X$, $\lambda \in \Lambda$. If Λ is a field, then this is the usual definition of a linear mapping between the vector spaces X, Y.

Regarding $\Lambda^+ = \{+\} \cup \Lambda$ as a set of operation symbols (+ binary, and each $\lambda \in \Lambda$ unary), every Λ-module is simply a universal algebra $\mathfrak{X} = (X, +, \lambda)_{\lambda\in\Lambda}$, satisfying the Λ-module axioms, and a Λ-module homomorphism is a homomorphism of such algebras. Now, the previous Example 15.2 applies, i.e., $f: X \to Y$ is a Λ-module homomorphism if and only if it satisfies the system of GFEs consisting of

$$f \circ \alpha = G\big(f \otimes (\pi_1, \pi_2)\big),$$

where α is the addition in X and G is the addition in Y, and

$$f \circ \lambda^X = \lambda^Y \circ f \qquad (\lambda \in \Lambda).$$

We continue with two examples of more analytic character.

Example 15.4 Let $\alpha: \mathbb{R}^2 \to \mathbb{R}$ be the addition on $\mathbb{R}$, $F_1 = \pi_1$, $F_2 = \pi_2: \mathbb{R}^2 \to \mathbb{R}$ denote the projections, and the functions $G_1, G_2: \mathbb{R}^{2\times 2} \to \mathbb{R}$ be given by

$$G_1 \begin{pmatrix} a_{11} & a_{12} \\ a_{21} & a_{22} \end{pmatrix} = \operatorname{Per} \begin{pmatrix} a_{11} & a_{12} \\ a_{21} & a_{22} \end{pmatrix} = a_{11}a_{22} + a_{21}a_{12},$$

$$G_2 \begin{pmatrix} a_{11} & a_{12} \\ a_{21} & a_{22} \end{pmatrix} = \operatorname{Det} \begin{pmatrix} a_{21} & a_{12} \\ a_{11} & a_{22} \end{pmatrix} = a_{21}a_{22} - a_{11}a_{12},$$

(notice the reversed order of elements in the first column of the determinant). Then the system of the following two GFEs, both of type $(2, 1, 2, 2)$, in the couple of functional variables $\boldsymbol{f} = (f_1, f_2)$, standing for the sine and cosine, respectively,

$$\pi_1(\boldsymbol{f} \otimes \alpha) = G_1\big(\boldsymbol{f} \otimes (\pi_1, \pi_2)\big),$$

$$\pi_2(\boldsymbol{f} \otimes \alpha) = G_2\big(\boldsymbol{f} \otimes (\pi_1, \pi_2)\big),$$

is nothing else but the well-known sine and cosine addition formulas

$$\sin(x + y) = \sin x \cos y + \cos x \sin y,$$

$$\cos(x + y) = \cos x \cos y - \sin x \sin y.$$

Example 15.5 Let $\sigma: \mathbb{C} \to \mathbb{C}$ denote the shift $\sigma(x) = x + 1$ and $G: \mathbb{C} \times \mathbb{C} \to \mathbb{C}$ be the multiplication $G(y, x) = yx$ on $\mathbb{C}$. Then the GFE

$$f \circ \sigma = \widetilde{G}(f)$$

of type $(1, 1, 1, 1)$ is the functional equation

$$f(x + 1) = f(x)\, x,$$

satisfied by the Euler function Γ on the open complex half plane $\{x \in \mathbb{C} \mid \operatorname{Re} x > 0\}$.

We conclude with two examples dealing with recursion in one and two variables.

Example 15.6 Let $\big(f(x)\big)_{x\in\mathbb{N}}$ be a sequence of elements of a set A, i.e., a function $f: \mathbb{N} \to A$, satisfying the recursion

$$f(x + n) = G\big(f(x), \ldots, f(x + n - 1)\big)$$

for a fixed $n \geq 1$ and an n-ary operation $G: A^n \to A$ given in advance. For each $j \in \mathbb{N}$ we denote by $\sigma^j: \mathbb{N} \to \mathbb{N}$ the shift $\sigma^j(x) = x + j$. Then the above recursion formula takes the form of the GFE

$$f \circ \sigma^n = G(f \otimes (\sigma^0, \ldots, \sigma^{n-1}))$$

of type $(1, 1, n, 1)$. The more general recursion formula

$$f(x+n) = G(f(x), \ldots, f(x+n-1), x),$$

where $G: A^n \times \mathbb{N} \to A$, takes the form of the GFE of type $(1, 1, n, 1)$

$$f \circ \sigma^n = \widetilde{G}(f \otimes (\sigma^0, \ldots, \sigma^{n-1})).$$

Example 15.7 Let A be a set and $G: A^3 \times \mathbb{N}^2 \to A$ be an arbitrary mapping. Consider the following recursion formula:

$$f(x+1, y+1) = G(f(x,y), f(x+1,y), f(x,y+1), x, y),$$

expressing the value of a function $f: \mathbb{N}^2 \to A$ at $(x+1, y+1)$ in terms of its values at the preceding neighbors (x, y), $(x+1, y)$, $(x, y+1)$, and the position (x, y) itself. The notorious recursion formulas

$$\binom{n+1}{k+1} = \binom{n}{k} + \binom{n}{k+1},$$

$$c(k+1, l+1) = c(k+1, l) + c(k, l+1),$$

$$s(n+1, k+1) = s(n,k) - n\, s(n, k+1),$$

$$S(n+1, k+1) = S(n,k) + (k+1) S(n, k+1),$$

for binomial coefficients (both in the usual form or for $c(k, l) = \binom{k+l}{k}$), as well as for Stirling numbers of the first and the second kind, respectively, are just some special cases of such functional equations for functions $f: \mathbb{N}^2 \to \mathbb{Z}$.

Let $\sigma_1, \sigma_2: \mathbb{N}^2 \to \mathbb{N}^2$ denote the shifts in the first and the second variable, respectively, i.e., $\sigma_1(x, y) = (x+1, y)$, $\sigma_2(x, y) = (x, y+1)$, and $\sigma_{12} = \sigma_1 \circ \sigma_2 = \sigma_2 \circ \sigma_1: \mathbb{N}^2 \to \mathbb{N}^2$ be the double shift, i.e., $\sigma_{12}(x, y) = (x+1, y+1)$. Then the original recursion formula can be written as the GFE

$$f \circ \sigma_{12} = \widetilde{G}(f \otimes (\sigma_0, \sigma_1, \sigma_2)),$$

with $\sigma_0 = \mathrm{Id}_{\mathbb{N}^2}: \mathbb{N}^2 \to \mathbb{N}^2$ denoting the identity. The generalization to recursion formulas for functions $f: \mathbb{N}^n \to A$ with $n \geq 2$ variables is straightforward.

15.3 Infinitesimal Nearness and S-Continuity

In this section we modify the short introduction to the nonstandard approach to continuity of mappings between topological groups from [24] to the more general situation of mappings between completely regular topological spaces. We use [8] as a reference source for general topology. In order to simplify our terminology, we assume that all (standard) topological or uniform spaces dealt with are Hausdorff.

The reader is assumed to have some basic acquaintance with nonstandard analysis in an extent covered either by the original Robinson's monograph [23] or, e.g., by Albeverio et al. [1], or Davis [7], or Arkeryd et al. [4], mainly in the parts [11] and [17]. In particular, some knowledge of the nonstandard approach to topology, based on the equivalence relation of infinitesimal nearness, is desirable.

Our exposition takes place in a nonstandard universe which is an elementary extension $^*\mathbf{V}$ of a superstructure $\mathbf{V}$ over some set of individuals containing at least all (classical) complex numbers and the elements of the universal algebras or topological spaces dealt with. In particular, this means that every standard universal algebra $\mathfrak{A} = (A, F_\lambda)_{\lambda \in \Lambda}$ is embedded into its nonstandard extension $^*\mathfrak{A} = (^*A, {}^*F_\lambda)_{\lambda \in \Lambda}$ via the canonic elementary embedding $a \mapsto {}^*a$, and identified with its image under *, in such a way that for any formula $\Phi(v_1, \dots, v_n)$ of the first-order language built upon the operation symbols $\lambda \in \Lambda$ and any $a_1, \dots, a_n \in A$ we have

$$\Phi(a_1, \dots, a_n) \text{ holds in } \mathfrak{A} \quad \text{if and only if} \quad {}^*\Phi(a_1, \dots, a_n) \text{ holds in } {}^*\mathfrak{A},$$

where $^*\Phi$ is the formula obtained from Φ by replacing each operation $F_\lambda \colon A^{p_\lambda} \to A$ by its extension $^*F_\lambda \colon {}^*A^{p_\lambda} \to {}^*A$. This rule is referred to as the *transfer principle.* However, this principle applies to any tuples of functions $\boldsymbol{f} = (f_1, \dots, f_k) \colon X \to Y^k$ and their nonstandard extensions $^*\boldsymbol{f} = ({}^*f_1, \dots, {}^*f_k) \colon {}^*X \to {}^*Y^k$, as well.

Objects belonging to the original universe are called *standard* and objects belonging to its nonstandard extension are called *internal.* Taking the advantage of the relation between the universes of standard and internal objects, we cannot avoid the so-called *external sets*, i.e., sets of internal objects, which themselves are not necessarily internal.

We assume that our nonstandard universe is *κ-saturated* for some sufficiently big uncountable cardinal κ, which will be specified later on. This is to say that any system of less than κ internal sets with the finite intersection property has itself nonempty intersection. Informally, we refer to this situation by the phrase that our nonstandard universe is *sufficiently saturated*. In a similar vein, a set of *admissible size* means a set of cardinality $< \kappa$.

If $(X, \mathscr{T})$ is a topological space, then the topology $\mathscr{T}$ (i.e., the system of open sets in X) gives rise to two different topologies on its nonstandard extension $^*X$.

The *Q-topology* is given by the base $^*\mathscr{T}$; it is Hausdorff if and only if the original topology $\mathscr{T}$ on X is Hausdorff. This topology plays rather an auxiliary role in our accounts.

The *S-topology* is given by the base $\{{}^*A \mid A \in \mathscr{T}\}$. Obviously, the S-topology is coarser than the Q-topology and it is not Hausdorff, unless $(X, \mathscr{T})$ is discrete.

We will systematically take advantage of the fact that if $(X, \mathscr{T})$ is a (Hausdorff) completely regular space, whose topology is induced by a uniformity $\mathscr{U}$ on X, then, in a sufficiently saturated nonstandard universe, the S-topology is fully determined by a single external equivalence relation

$$x \approx y \Leftrightarrow \forall U \in \mathscr{U} : (x, y) \in {}^*U ,$$

called the *relation of infinitesimal nearness* on ${}^*X$. At the same time the system $\{{}^*U \mid U \in \mathscr{U}\}$ is a base of the *S-uniformity* on ${}^*X$. Uniform continuity with respect to it is referred to as the *uniform S-continuity*.

The external set of all elements indiscernible from $x \in {}^*X$ is called the *monad* of x, i.e.,

$$\mathrm{Mon}(x) = \{y \in {}^*X \mid y \approx x\} .$$

An element $x \in {}^*X$ is called *nearstandard* if $x \approx x_0$ for some $x_0 \in X$. The (external) set of all nearstandard elements in ${}^*X$ is denoted by $\mathrm{Ns}({}^*X)$, i.e.,

$$\mathrm{Ns}({}^*X) = \bigcup_{x \in X} \mathrm{Mon}(x) .$$

For $x \in \mathrm{Ns}({}^*X)$ we denote by ${}^\circ x$ the unique element $x_0 \in X$ infinitesimally close to x, called the *standard part* or *shadow* of x.

For the rest of this section $(X, \mathscr{T}_X)$ and $(Y, \mathscr{T}_Y)$ denote some completely regular topological spaces, whose topologies are induced by some uniformities $\mathscr{U}_X$, $\mathscr{U}_Y$, respectively, and ${}^*X$, ${}^*Y$ are their canonical extensions in a sufficiently saturated nonstandard universe; more precisely, we assume that our nonstandard universe is κ-saturated for some cardinal κ bigger than the cardinalities of some bases of the uniformities $\mathscr{U}_X$, $\mathscr{U}_Y$.

While the Q-continuity of internal functions $f: {}^*X \to {}^*Y$ is just the *continuity, their S-continuity can be characterized in the following intuitively appealing way in the spirit of the original infinitesimal calculus (below, we denote the relations of infinitesimal nearness on ${}^*X$, ${}^*Y$ by $\approx_X$, $\approx_Y$, respectively):

Proposition 15.1 *Let $f: {}^*X \to {}^*Y$ be an internal function. Then*

(a) *f is S-continuous in a point $x_0 \in {}^*X$ if and only if*

$$\forall x \in {}^*X : x \approx_X x_0 \Rightarrow f(x) \approx_Y f(x_0) ;$$

(b) *f is S-continuous on a set $A \subseteq {}^*X$ (i.e., f is S-continuous in every point $a \in A$) if and only if*

$$\forall a \in A \ \forall x \in {}^*X : x \approx_X y \Rightarrow f(x) \approx_Y f(y) ;$$

(c) *if* $A \subseteq {}^*X$ *is an intersection of admissibly many internal sets, then* f *is S-continuous on A if and only if f is uniformly S-continuous on A.*

In view of (a) and (b), S-continuity of an internal function $f\colon {}^*X \to {}^*Y$ can be alternatively *defined* as preservation of the relation of infinitesimal nearness by f. In particular, for the canonic extension ${}^*f\colon {}^*X \to {}^*Y$ of a standard function $f\colon X \to Y$ we have the following criteria (notice the subtle difference between (b) and (c)).

Corollary 15.1 *Let* $f\colon X \to Y$ *be a function. Then*

(a) f *is continuous in a point* $x_0 \in X$ *if and only if*

$$\forall\, x \in {}^*X : x \approx_X x_0 \;\Rightarrow\; {}^*f(x) \approx_Y f(x_0)\,;$$

(b) f *is continuous on a set* $A \subseteq X$ *(i.e.,* f *is continuous in every point* $a \in A$*) if and only if*

$$\forall\, a \in A \;\; \forall\, x \in {}^*X : x \approx_X a \;\Rightarrow\; {}^*f(x) \approx_Y f(a)\,;$$

(c) f *is uniformly continuous on a set* $A \subseteq X$ *if and only if*

$$\forall\, x, y \in {}^*A : x \approx_X y \;\Rightarrow\; {}^*f(x) \approx_Y {}^*f(y)\,.$$

Notice that under the assumption of (b), ${}^*f$ is Q-continuous on ${}^*A$, as well.

An internal function $f\colon {}^*X \to {}^*Y$ is called *nearstandard* if $f(x) \in \mathrm{Ns}({}^*Y)$ for each $x \in X$. Let us remark that this is indeed equivalent to f be a nearstandard point in the nonstandard extension ${}^*\big(Y^X\big)$ of the Tikhonov product $Y^X = \{f \mid f\colon X \to Y\}$. Any nearstandard function $f\colon {}^*X \to {}^*Y$ gives rise to a function ${}^\circ\! f\colon X \to Y$ given by

$$({}^\circ\! f)(x) = {}^\circ(f(x))\,,$$

for $x \in X$, called the *standard part* of f. If f is additionally S-continuous on $\mathrm{Ns}({}^*X)$, then its standard part can be extended to a map ${}^\circ\! f\colon \mathrm{Ns}({}^*X) \to Y$ (denoted in the same way), such that

$${}^\circ\! f(x) = {}^\circ\! f({}^\circ x) = {}^\circ(f(x))$$

for any $x \in \mathrm{Ns}({}^*X)$. The situation can be depicted by the following commutative diagram:

$$\begin{array}{ccc} \mathrm{Ns}({}^*X) & \xrightarrow{\;\;f\;\;} & \mathrm{Ns}({}^*Y) \\ {\scriptstyle\circ}\big\downarrow & & \big\downarrow{\scriptstyle\circ} \\ X & \xrightarrow[\;{}^\circ\! f\;]{} & Y \end{array}$$

A function $f: {}^*X \to {}^*Y$ is called *NS-continuous* if it is S-continuous on $\mathrm{Ns}({}^*X)$. Now we have the following supplement to Proposition 15.1 and its Corollary 15.1.

Proposition 15.2 *Let $f: {}^*X \to {}^*Y$ be a nearstandard internal function. Then the following implications hold:*

(a) *if f is NS-continuous, then its standard part ${}^\circ\! f: X \to Y$ is continuous and ${}^*({}^\circ\! f)(x) \approx_Y f(x)$ for $x \in \mathrm{Ns}({}^*X)$;*
(b) *if f is S-continuous on some internal set $A \supseteq \mathrm{Ns}({}^*X)$, then its standard part ${}^\circ\! f: X \to Y$ is uniformly continuous.*

Notice that the function ${}^*({}^\circ\! f)$ is also Q-continuous. However, even if f were S-continuous on the whole of ${}^*X$, the second conclusion in (a) still cannot be strengthened to ${}^*({}^\circ\! f)(x) \approx_Y f(x)$ for all $x \in {}^*X$.

Proof We will prove just the first statement in (a); then the second statement easily follows and (b) can be proved in a similar way.

Assume that f is NS-continuous and denote $g = {}^\circ\! f: X \to Y$ its standard part. In order to prove the continuity of g, pick an arbitrary $x_0 \in X$ and $V \in \mathscr{U}_Y$. Let $W \in \mathscr{U}_Y$ be symmetric, such that $W^3 = W \circ W \circ W \subseteq V$. As f is internal and NS-continuous, it is also continuous in x_0 with respect to the S-topology on ${}^*X$, hence there is a $U \in \mathscr{U}_X$ such that $(x, x_0) \in {}^*U$ implies $\big(f(x), f(x_0)\big) \in {}^*W$ for any $x \in {}^*X$. In particular, for $x \in X$ such that $(x, x_0) \in U$, we have $g(x) \approx_Y f(x)$, $\big(f(x), f(x_0)\big) \in {}^*U$, as well as $f(x_0) \approx_Y g(x_0)$, hence $\big(g(x), g(x_0)\big) \in {}^*W^3 \subseteq {}^*V$. Since $g(x), g(x_0) \in Y$, by transfer principle $\big(g(x), g(x_0)\big) \in V$.

Let us conclude this section with a remark that the introduced continuity notions can be easily generalized to tuples of functions $\boldsymbol{f} = (f_1, \ldots, f_k): X \to Y^k$. As well known, $\boldsymbol{f}$ has whatever standard continuity property if and only if all the functions f_i have this property. The relation of infinitesimal nearness $\approx_Y$ can be extended to ${}^*Y^k$ by

$$y \approx_Y z \Leftrightarrow y_1 \approx_Y z_1 \;\&\; \ldots \;\&\; y_k \approx_Y z_k\,,$$

(similarly, $\approx_X$ can be extended to ${}^*X^p$). Then an internal function $\boldsymbol{f}: {}^*X \to {}^*Y^k$ is nearstandard if and only if all the functions f_i are nearstandard; $\boldsymbol{f}$ has anyone of the S-continuity properties if and only if all the functions f_i have the corresponding property. If $\boldsymbol{f}$ is nearstandard, then the k-tuple ${}^\circ\boldsymbol{f} = \big({}^\circ\! f_1, \ldots, {}^\circ\! f_k\big)$ of functions ${}^\circ\! f_i: X \to Y$ is called the *standard part* of $\boldsymbol{f}$.

15.4 An Infinitesimal "Almost-Near" Principle for Systems of General Functional Equations

Let $(X, \mathscr{T}_X)$, $(Y, \mathscr{T}_Y)$ be two completely regular topological spaces with topologies induced by some uniformities $\mathscr{U}_X$, $\mathscr{U}_Y$, respectively. If there is no danger of confusion, we omit the subscripts X, Y in the notation of the relations of infinitesimal nearness $\approx_X$, $\approx_Y$ on ${}^*X$, ${}^*Y$, respectively.

Consider the GFE (15.1) for a k-tuple of functional variables $\boldsymbol{f} = (f_1, \ldots, f_k)$. Embedding the situation into some nonstandard universe we say that an internal function $\boldsymbol{f} = (f_1, \ldots, f_k)\colon {}^*X \to {}^*Y^k$, *almost satisfies* Equation (15.1) *on* $\mathrm{Ns}({}^*X)$ if

$$^*\widetilde{F}(\boldsymbol{f} \otimes {}^*\boldsymbol{\alpha})(\boldsymbol{x}) \approx {}^*\widetilde{G}(\boldsymbol{f} \otimes {}^*\boldsymbol{\beta})(\boldsymbol{x})$$

for all $\boldsymbol{x} = (x_1 \ldots, x_p) \in \mathrm{Ns}({}^*X^p)$. Similarly, $\boldsymbol{f}$ *almost satisfies* the system of GFEs (15.2) *on* $\mathrm{Ns}({}^*X)$ if it almost satisfies on $\mathrm{Ns}({}^*X)$ every equation in it. (Notice that, due to the transfer principle, ${}^*(\widetilde{F}) = \widetilde{{}^*F}$, and similarly for G, hence the notation ${}^*\widetilde{F}$, ${}^*\widetilde{G}$ is unambiguous.)

Theorem 15.1 *Let the mappings* $F\colon Y^{k\times m} \times X^p \to Y$, $G\colon Y^{k\times n} \times X^p \to Y$ *be continuous in the "matrix" variables* $y_{ij} \in Y$ *for all* $i \le k$ *and* $j \le m, n$, *respectively. If a nearstandard internal function* $\boldsymbol{f} = (f_1, \ldots, f_k)\colon {}^*X \to {}^*Y^k$ *almost satisfies the GFE* (15.1) *on* $\mathrm{Ns}({}^*X)$, *then its standard part* $^\circ f = \left({}^\circ f_1, \ldots, {}^\circ f_k\right)$ *is a solution of the GFE* (15.1).

Proof Take an arbitrary $\boldsymbol{x} = (x_1, \ldots, x_p) \in X^p$. We have

$$^*\widetilde{F}(\boldsymbol{f} \otimes {}^*\boldsymbol{\alpha})(\boldsymbol{x}) \approx {}^*\widetilde{G}(\boldsymbol{f} \otimes {}^*\boldsymbol{\beta})(\boldsymbol{x})\,.$$

As $\boldsymbol{x}$ is standard, ${}^*\boldsymbol{\alpha}(\boldsymbol{x}) = \boldsymbol{\alpha}(\boldsymbol{x})$ is standard, as well, hence $f_i\big(\alpha_j(\boldsymbol{x})\big) \approx {}^\circ f_i\big(\alpha_j(\boldsymbol{x})\big)$ for any $i \le k, j \le m$, and, as ${}^*F$ is *NS*-continuous in the matrix variables y_{ij},

$$\begin{aligned}
^*\widetilde{F}(\boldsymbol{f} \otimes {}^*\boldsymbol{\alpha})(\boldsymbol{x}) &= {}^*F\big((\boldsymbol{f} \otimes {}^*\boldsymbol{\alpha})(\boldsymbol{x}), \boldsymbol{x}\big) = {}^*F\big((\boldsymbol{f} \otimes \boldsymbol{\alpha})(\boldsymbol{x}), \boldsymbol{x}\big) \\
&\approx {}^*F\big(({}^\circ f \otimes \boldsymbol{\alpha})(\boldsymbol{x}), \boldsymbol{x}\big) = F\big(({}^\circ f \otimes \boldsymbol{\alpha})(\boldsymbol{x}), \boldsymbol{x}\big) = \widetilde{F}({}^\circ f \otimes \boldsymbol{\alpha})(\boldsymbol{x})\,.
\end{aligned}$$

Similarly we can get

$$^*\widetilde{G}(\boldsymbol{f} \otimes {}^*\boldsymbol{\beta})(\boldsymbol{x}) \approx \widetilde{G}({}^\circ f \otimes \boldsymbol{\beta})(\boldsymbol{x})\,.$$

Therefore,

$$\widetilde{F}({}^\circ f \otimes \boldsymbol{\alpha})(\boldsymbol{x}) \approx \widetilde{G}({}^\circ f \otimes \boldsymbol{\beta})(\boldsymbol{x})\,,$$

and, as both the expressions are standard,

$$\widetilde{F}({}^\circ f \otimes \boldsymbol{\alpha})(\boldsymbol{x}) = \widetilde{G}({}^\circ f \otimes \boldsymbol{\beta})(\boldsymbol{x})\,,$$

i.e., ${}^\circ f$ is a solution of the GFE (15.1).

From Theorem 15.1 and Proposition 15.2 (b) we readily obtain the following consequence generalizing Theorem 2.2 from [24], dealing just with the homomorphy equation in topological groups.

Corollary 15.2 *Assume that F, G are continuous in the matrix variables y_{ij}. Then for every nearstandard NS-continuous internal function $\boldsymbol{f} = (f_1, \ldots, f_k)\colon {}^*X \to {}^*Y^k$ which almost satisfies the system of GFEs (15.2) on $\mathrm{Ns}({}^*X)$, there is a continuous solution $\boldsymbol{\varphi} = (\varphi_1, \ldots, \varphi_k)$ of the system, such that $\boldsymbol{\varphi}(x) \approx \boldsymbol{f}(x)$ for each $x \in X$.*

15.5 Stability of Systems of General Functional Equations

In order to formulate a standard version of the just established nonstandard stability principle, we need to introduce some notions—cf. [24, 30, 31].

Definition 15.1 Let $(X, \mathscr{T}_X)$ be a topological space and $(Y, \mathscr{U}_Y)$ be a uniform space.

(a) A $(\mathscr{T}_X, \mathscr{U}_Y)$ *continuity scale* is a mapping $\Gamma\colon X \times \mathscr{B} \to \mathscr{T}_X$, such that $\mathscr{B}$ is a base of the uniformity $\mathscr{U}_Y$ and $\Gamma(x, V)$ is a neighborhood of x in $(X, \mathscr{T}_X)$, satisfying

$$V \subseteq W \;\Rightarrow\; \Gamma(x, V) \subseteq \Gamma(x, W)$$

for any $x \in X$, and $V, W \in \mathscr{B}$.

(b) Given a continuity scale $\Gamma\colon X \times \mathscr{B} \to \mathscr{T}_X$, a function $f\colon X \to Y$ is called *Γ-continuous in a point $x_0 \in X$*, or *continuous in x_0 with respect to Γ*, if

$$x \in \Gamma(x_0, V) \;\Rightarrow\; \big(f(x), f(x_0)\big) \in V$$

for each $x \in X$; f is *Γ-continuous on a set $A \subseteq X$* if it is Γ-continuous in each point $a \in A$; it is *Γ-continuous* if it is Γ-continuous on X.

(c) Given a continuity scale $\Gamma\colon X \times \mathscr{B} \to \mathscr{T}_X$ and an entourage $U \in \mathscr{B}$, a function $f\colon X \to Y$ is *(Γ, U)-precontinuous in a point $x_0 \in X$* if

$$x \in \Gamma(x_0, V) \;\Rightarrow\; \big(f(x), f(x_0)\big) \in V$$

for any $V \in \mathscr{B}$, such that $U \subseteq V$, and each $x \in X$; (Γ, U)-precontinuity on a set $A \subseteq X$ and on X are defined in the obvious way.

(d) If $(X, \mathscr{U}_X)$ is a uniform space, too, then a $(\mathscr{U}_X, \mathscr{U}_Y)$ *uniform continuity scale* is a mapping $\Gamma\colon \mathscr{B} \to \mathscr{U}_X$ such that $\mathscr{B}$ is some base of the uniformity $\mathscr{U}_Y$ and

$$V \subseteq W \;\Rightarrow\; \Gamma(V) \subseteq \Gamma(W)$$

for any $V, W \in \mathscr{B}$.

(e) Given a uniform continuity scale $\Gamma\colon \mathscr{B} \to \mathscr{U}_X$, a function $f\colon X \to Y$ is *uniformly Γ-continuous on a set $A \subseteq X$* if

$$(x, y) \in \Gamma(V) \;\Rightarrow\; \big(f(x), f(y)\big) \in V$$

for any $x, y \in A$; f is *uniformly Γ-continuous* if it is uniformly Γ-continuous on X.

(f) Given a uniform continuity scale $\Gamma: \mathscr{B} \to \mathscr{U}_X$ and an entourage $U \in \mathscr{B}$, a function $f: X \to Y$ is *uniformly (Γ, U)-precontinuous on a set $A \subseteq X$* if

$$(x, y) \in \Gamma(V) \Rightarrow \big(f(x), f(y)\big) \in V$$

for any $V \in \mathscr{B}$, such that $U \subseteq V$ and all $x, y \in A$; f is *uniformly (Γ, U)-precontinuous* if it is (Γ, U)-precontinuous on X.

Obviously, if a function $f: X \to Y$ is Γ-continuous with respect to some continuity scale Γ, then it is continuous. Conversely, if f is continuous, then, given any base $\mathscr{B}$ of $\mathscr{U}_Y$, the assignment

$$\Gamma(x_0, V) = \left\{x \in X \,\middle|\, \big(f(x), f(x_0)\big) \in V\right\},$$

for $x_0 \in X$, $V \in \mathscr{B}$, defines a $(\mathscr{T}_X, \mathscr{U}_Y)$ continuity scale $\Gamma: X \times \mathscr{B} \to \mathscr{T}_X$, and, of course, f is continuous with respect to it.

The other way round, f is Γ-continuous if and only if it is (Γ, U)-precontinuous for all $U \in \mathscr{B}$. Thus each particular condition of (Γ, U)-precontinuity for an entourage $U \in \mathscr{U}_Y$ can be regarded as an approximate continuity property. Informally, f is "almost Γ-continuous" if it is (Γ, U)-precontinuous for a "sufficiently small" $U \in \mathscr{B}$. The relation between the uniform versions of these notions is similar.

If (X, d), (Y, e) are metric spaces, then a (d, e)-continuity scale is just a mapping $\gamma: X \times \mathbb{R}^+ \to \mathbb{R}^+$ such that $\gamma(x, \epsilon) \leq \gamma(x, \epsilon')$ for any $x \in X$, $\epsilon' \geq \epsilon > 0$. Then a function $f: X \to Y$ is γ-continuous in $x_0 \in X$ if

$$d(x, x_0) < \gamma(x_0, \epsilon) \Rightarrow e\big(f(x), f(x_0)\big) < \epsilon$$

for all $\epsilon > 0$ and $x \in X$. A uniform (d, e)-continuity scale is an isotone mapping $\gamma: \mathbb{R}^+ \to \mathbb{R}^+$. A function $f: X \to Y$ is uniformly γ-continuous if

$$d(x, y) < \gamma(\epsilon) \Rightarrow e\big(f(x), f(y)\big) < \epsilon$$

for all $\epsilon > 0$ and $x, y \in X$.

Definition 15.2 Let X, Y be arbitrary sets.

(a) A *bounding relation* from X to Y is any binary relation $R \subseteq X \times Y$ such that all its stalks $R[x] = \{y \in Y \mid (x, y) \in R\}$, for $x \in X$, are nonempty.

(b) Given a bounding relation $R \subseteq X \times Y$, a function $f: X \to Y$ is *R-bounded on a set $A \subseteq X$* if $f(a) \in R[a]$ for each $a \in A$; f is *R-bounded* if it is R-bounded on X, i.e., if $f \subseteq R$.

(c) A bounding relation $R \subseteq X \times Y$ is *stalkwise finite* if all its stalks $R[x]$ are finite. If, additionally, $(Y, \mathscr{T}_Y)$ is a topological space, then R is called *stalkwise compact* if all its stalks $R[x]$ are compact.

Definition 15.3 Let X be any set, $(Y, \mathscr{U}_Y)$ be a uniform space and $V \in \mathscr{U}_Y$.

(a) Two functions $f, g: X \to Y$ are *V-close on a set $A \subseteq X$* if $\big(f(a), g(a)\big) \in V$ for all $a \in A$.
(b) A k-tuple $\boldsymbol{f} = (f_1, \ldots, f_k)$ of functions $f_i: X \to Y$ is a *V-solution* of the GFE (15.1) *on a set $S \subseteq X^p$* if

$$\big(\widetilde{F}(\boldsymbol{f} \otimes \boldsymbol{\alpha})(\boldsymbol{x}), \widetilde{G}(\boldsymbol{f} \otimes \boldsymbol{\beta})(\boldsymbol{x})\big) \in V$$

for all $\boldsymbol{x} \in S$; $\boldsymbol{f}$ is a *V-solution* the GFE (15.1) *on a set $A \subseteq X$* if it is its V-solution on A^p; $\boldsymbol{f}$ is a *V-solution* of the system of GFEs (15.2) on $A \subseteq X$ if it is a V-solution of every equation in the system on A.

For brevity's sake we say that a function $\boldsymbol{f} = (f_1, \ldots, f_k): X \to Y^k$ has any of the just introduced Γ-continuity properties if and only if each particular function f_i has the corresponding property. Similarly, $\boldsymbol{f}$ is R-bounded (on a set $A \subseteq X$) if and only each function f_i is R-bounded. We say that two such functions $\boldsymbol{f}, \boldsymbol{g}: X \to Y^k$ are V-close on $A \subseteq X$ if f_i, g_i are V-close on A for each $i \leq k$.

The system of all nonempty compact sets of a topological space $(X, \mathscr{T}_X)$ is denoted by $\mathscr{K}(X)$.

Theorem 15.2 *Let $(X, \mathscr{T}_X)$ be a locally compact topological space, $(Y, \mathscr{U}_Y)$ be a uniform space, $\Gamma: X \times \mathscr{B} \to \mathscr{T}_X$ be a $(\mathscr{T}_X, \mathscr{U}_Y)$ continuity scale, and $R \subseteq X \times Y$ be a stalkwise compact bounding relation. Assume that all the functional coefficients $F_\lambda: Y^{k \times m_\lambda} \times X^{p_\lambda} \to Y$, $G_\lambda: Y^{k \times n_\lambda} \times X^{p_\lambda} \to Y$ in the system of GFEs* (15.2) *are continuous in the matrix variables y_{ij}. Then for each pair $D \in \mathscr{K}(X)$, $V \in \mathscr{U}_Y$ there exists a pair $C \in \mathscr{K}(X)$, $U \in \mathscr{U}_Y$ such that $D \subseteq C$ and the following implication holds true:*

If a U-solution $\boldsymbol{f} = (f_1, \ldots, f_k): X \to Y^k$ of the system (15.2) *on C is both (Γ, U)-precontinuous and R-bounded on C, then there exists a continuous solution $\boldsymbol{\varphi} = (\varphi_1, \ldots, \varphi_k)$ of the system, such that $\boldsymbol{f}, \boldsymbol{\varphi}$ are V-close on D.*

Proof Let $(X, \mathscr{T}_X)$, $(Y, \mathscr{U}_Y)$, $\Gamma: X \times \mathscr{B} \to \mathscr{T}_X$, $R \subseteq X \times Y$, as well as the system of GFEs (15.2) satisfy the assumptions of the theorem. Then $(X, \mathscr{T}_X)$ is completely regular, as well, hence its topology is induced by some uniformity $\mathscr{U}_X$. Admit, in order to obtain a contradiction, that there is a pair $D \in \mathscr{K}(X)$, $V \in \mathscr{U}_Y$ for which the conclusion of the theorem fails. For each pair $C \in \mathscr{K}(X)$, $U \in \mathscr{B}$ such that $C \supseteq D$ we denote by $\mathscr{F}(C, U)$ the set of all U-solutions $\boldsymbol{f} = (f_1, \ldots, f_k): X \to Y^k$ of the system of GFEs (15.2) on C which are both (Γ, U)-precontinuous and R-bounded on C, nonetheless, $\boldsymbol{f}$ is not V-close on D to any continuous solution $\boldsymbol{\varphi} = (\varphi_1, \ldots, \varphi_k)$ of the system (15.2). According to our assumption, all the sets $\mathscr{F}(C, U)$ are nonempty, and, for all $C, C' \in \mathscr{K}(X)$, $U, U' \in \mathscr{B}$, we obviously have

$$D \subseteq C \subseteq C' \;\&\; U' \subseteq U \;\Rightarrow\; \mathscr{F}(C', U') \subseteq \mathscr{F}(C, U)\,.$$

Let us embed the situation into a sufficiently saturated nonstandard universe. More precisely, we assume that it is κ-saturated for some uncountable cardinal κ

such that card $\mathscr{B} < \kappa$, as well as card $\mathscr{C} < \kappa$ for some cofinal subset $\mathscr{C} \subseteq \mathscr{K}(X)$ such that $D \subseteq C$ for each $C \in \mathscr{C}$. Then

$$\bigcap_{C \in \mathscr{C},\, U \in \mathscr{B}} {}^*\mathscr{F}(C, U) \neq \emptyset\,.$$

Let $\boldsymbol{f} = (f_1, \ldots, f_k)$ belong to this intersection. Then $\boldsymbol{f}\colon {}^*X \to {}^*Y^k$ is an internal function, for all $U \in \mathscr{U}_Y$, $C \in \mathscr{C}$, $\boldsymbol{f}$ is ${}^*(\Gamma, U)$-precontinuous and ${}^*R$-bounded on ${}^*C$ and it satisfies

$$\left({}^*\widetilde{F}_\lambda(\boldsymbol{f} \otimes {}^*\boldsymbol{\alpha}_\lambda)(\boldsymbol{x}),\ {}^*\widetilde{G}_\lambda(\boldsymbol{f} \otimes {}^*\boldsymbol{\beta}_\lambda)(\boldsymbol{x})\right) \in {}^*U$$

for any $\lambda \in \Lambda$ and $\boldsymbol{x} \in {}^*C^{p_\lambda}$. Since X is locally compact, $\mathrm{Ns}({}^*X) = \bigcup_{C \in \mathscr{C}} {}^*C$. It follows that $\boldsymbol{f}$ is *NS*-continuous and almost satisfies the system (15.2) on $\mathrm{Ns}({}^*X)$. Finally, $\boldsymbol{f}(x) \in \left({}^*R[x]\right)^k$ for any $C \in \mathscr{C}$ and $x \in {}^*C$. As $R[x]$ is compact for $x \in X$, in that case we have $\boldsymbol{f}(x) \in \left({}^*R[x]\right)^k \subseteq \mathrm{Ns}\left({}^*Y^k\right)$. Thus $\boldsymbol{f}$ is nearstandard. According to Corollary 15.2, there is a continuous solution $\boldsymbol{\varphi} = (\varphi_1, \ldots, \varphi_k)$ of the system (15.2), such that $\boldsymbol{f}(x) \approx \boldsymbol{\varphi}(x)$ each $x \in X$. On the other hand, ${}^*\boldsymbol{\varphi}$ is Q-continuous (i.e., *continuous), hence $\boldsymbol{f}$ and ${}^*\boldsymbol{\varphi}$ cannot be ${}^*V$-close on ${}^*D$. Thus there are an $i \leq k$ and an $x \in {}^*D$ such that $\left(f_i(x), {}^*\varphi_i(x)\right) \notin {}^*V$. However, as D is compact, ${}^*D \subseteq \mathrm{Ns}({}^*X)$. Since both f_i and ${}^*\varphi_i$ are *NS*-continuous, taking an $x_0 \in X$ such that $x \approx x_0$, we obtain

$${}^*\varphi_i(x) \approx \varphi_i(x_0) \approx f_i(x_0) \approx f_i(x)\,,$$

i.e., a contradiction.

Like in Theorem 15.2, we assume in the next three Corollaries that all the mappings F_λ, G_λ in the system of GFEs (15.2) are continuous in the matrix variables y_{ij} (but, for brevity's sake, we do not mention that explicitly). In the fourth Corollary 15.6 this assumption is superfluous as it is satisfied automatically.

If $(Y, \mathscr{U}_Y)$ is compact, then $R = X \times Y$ is a stalkwise compact bounding relation such that every function $\boldsymbol{f}\colon X \to Y^k$ is R-bounded. This makes possible to avoid mentioning any bounding relation in the formulation of Theorem 15.2.

Corollary 15.3 *Let $(X, \mathscr{T}_X)$ be a locally compact topological space, $(Y, \mathscr{U}_Y)$ be a compact uniform space, and $\Gamma\colon X \times \mathscr{B} \to \mathscr{T}_X$ be a $(\mathscr{T}_X, \mathscr{U}_Y)$ continuity scale. Then for each pair $D \in \mathscr{K}(X)$, $V \in \mathscr{U}_Y$ there is a pair $C \in \mathscr{K}(X)$, $U \in \mathscr{U}_Y$ such that $D \subseteq C$ and the following implication holds true:*

If a U-solution $\boldsymbol{f} = (f_1, \ldots, f_k)\colon X \to Y^k$ of the system of GFEs (15.2) *on C is (Γ, U)-precontinuous on C, then there exists a continuous solution $\boldsymbol{\varphi} = (\varphi_1, \ldots, \varphi_k)$ of the system, such that $\boldsymbol{f}$, $\boldsymbol{\varphi}$ are V-close on D.*

If $(X, \mathscr{T}_X)$ is compact, then its topology is induced by a unique uniformity $\mathscr{U}_X$ and, at the same time, it is enough to control the continuity of functions $\boldsymbol{f}\colon X \to Y^k$ by means of a uniform continuity scale. Choosing $D = X$ we get the following *global* version of Theorem 15.2.

Corollary 15.4 *Let $(X, \mathscr{U}_X)$ be a compact and $(Y, \mathscr{U}_Y)$ be an arbitrary uniform space, $\Gamma: \mathscr{B} \to \mathscr{U}_X$ be a $(\mathscr{U}_X, \mathscr{U}_Y)$ uniform continuity scale and $R \subseteq X \times Y$ be a stalkwise compact bounding relation. Then for each $V \in \mathscr{U}_Y$ there is a $U \in \mathscr{U}_Y$ such that the following implication holds true:*

If $\boldsymbol{f} = (f_1, \ldots, f_k): X \to Y^k$ is a uniformly (Γ, U)-precontinuous and R-bounded U-solution of the system of GFEs (15.2)*, then there exists a (uniformly) continuous solution $\boldsymbol{\varphi} = (\varphi_1, \ldots, \varphi_k)$ of the system, such that $\boldsymbol{f}, \boldsymbol{\varphi}$ are V-close on X.*

Under the assumptions of both Corollaries 15.3 and 15.4 we have

Corollary 15.5 *Let $(X, \mathscr{U}_X)$, $(Y, \mathscr{U}_Y)$ be compact uniform spaces and $\Gamma: \mathscr{B} \to \mathscr{U}_X$ be a $(\mathscr{U}_X, \mathscr{U}_Y)$ uniform continuity scale. Then for each $V \in \mathscr{U}_Y$ there is a $U \in \mathscr{U}_Y$ such that the following implication holds true:*

If $\boldsymbol{f} = (f_1, \ldots, f_k): X \to Y^k$ is a uniformly (Γ, U)-precontinuous U-solution of the system of GFEs (15.2)*, then there exists a (uniformly) continuous solution $\boldsymbol{\varphi} = (\varphi_1, \ldots, \varphi_k)$ of the system, such that $\boldsymbol{f}, \boldsymbol{\varphi}$ are V-close on X.*

The interested reader can easily formulate the metric versions of Theorem 15.2, as well as of Corollaries 15.3–15.5.

Endowing both the sets X, Y with discrete topologies (uniformities), all the functions $X \to Y$ become (uniformly) continuous. Then compact subsets of X are just the finite ones and, similarly, a stalkwise compact bounding relation $R \subseteq X \times Y$ is a stalkwise finite one. In that case, choosing $U = \mathrm{Id}_Y$ in Theorem 15.2, we obtain the following result on extendability of functions satisfying a system of GFEs (15.2) on some finite set to its (global) solutions.

Corollary 15.6 *Let X and Y be arbitrary sets and $R \subseteq X \times Y$ be a stalkwise finite bounding relation. Then for each finite set $D \subseteq X$ there is a finite set $C \subseteq X$ such that $D \subseteq C$ and for every R-bounded partial solution $\boldsymbol{f} = (f_1, \ldots, f_k): X \to Y^k$ of the system of GFEs* (15.2) *on C there exists a solution $\boldsymbol{\varphi} = (\varphi_1, \ldots, \varphi_k)$ of the system, such that $\boldsymbol{\varphi}(x) = \boldsymbol{f}(x)$ for all $x \in D$.*

If the arity numbers p_λ in the system of GFEs (15.2) have a common upper bound p, then all the particular equations in the system can be considered as being of types $(k, m_\lambda, n_\lambda, p)$. In such a case, given a $U \in \mathscr{U}_Y$, we say that a function $\boldsymbol{f}: X \to Y^k$ is a *U-solution* of the system (15.2) *on a set $S \subseteq X^p$* if it is a U-solution of each its particular equation on S. Then we have the following variant of Theorem 15.2. Its proof can be obtained by slight modifications of the proof of Theorem 15.2 and is left to the reader.

Theorem 15.3 *Let $(X, \mathscr{T}_X)$ be any topological space, $(Y, \mathscr{U}_Y)$ be a uniform space, $\Gamma: X \times \mathscr{B} \to \mathscr{T}_X$ be a $(\mathscr{T}_X, \mathscr{U}_Y)$ continuity scale, and $R \subseteq X \times Y^k$ be a stalkwise compact bounding relation. Assume that all the equations in the system of GFEs* (15.2) *have the same arity $p_\lambda = p$, S is a locally compact subspace of X^p and each of the maps $F_\lambda: Y^{k \times m_\lambda} \times X^p \to Y$, $G_\lambda: Y^{k \times n_\lambda} \times X^p \to Y$ is continuous in the matrix variables y_{ij}. Then for each pair $D \in \mathscr{K}(X)$, $V \in \mathscr{U}_Y$, such that $D^p \subseteq S$, there is a triple $C \in \mathscr{K}(X)$, $K \in \mathscr{K}\big(X^p\big)$, $U \in \mathscr{U}_Y$, such that $D \subseteq C$, $D^p \subseteq K \subseteq S$ and the following implication holds true:*

If a U-solution $\boldsymbol{f} = (f_1, \dots, f_k): X \to Y^k$ *of the system* (15.2) *on K is both* (Γ, U)*-precontinuous and R-bounded on C, then there exists a continuous solution* $\boldsymbol{\varphi} = (\varphi_1, \dots, \varphi_k)$ *of the system on S, such that* $\boldsymbol{f}, \boldsymbol{\varphi}$ *are V-close on D.*

Formulation of the corresponding modified versions of Corollaries 15.3–15.6 is left to the reader, as well.

Comparing the "local" stability Theorems 15.2, 15.3 and Corollaries 15.3, 15.6 with "global" Corollaries 15.4, 15.5 and other global stability results we see that while global stability deals with approximation of functions $\boldsymbol{f} = (f_1, \dots, f_k): X \to Y^k$ by continuous solutions $\boldsymbol{\varphi} = (\varphi_1, \dots \varphi_k)$ of the given (system of) functional equation(s) on the whole space X, local stability deals with approximate extension (and if Y is discrete, then right by extension) of restrictions $\boldsymbol{f} \restriction D = (f_1 \restriction D, \dots, f_k \restriction D)$ of such functions to some (in the present setting compact) subset $D \subseteq X$ to continuous solutions of the (system of) functional equation(s).

The interested reader can find a brief discussion of the role of nonstandard analysis in establishing our results as well of the possibility to replace it by some standard methods in the final part of [24].

Final Remark The general form of functional equations introduced in Section 15.1 was designed with the aim to prove the stability Theorems 15.2, 15.3 for all of them in a uniform way. I expected that in order to achieve this goal it will be necessary to assume that all the functional coefficients $F_\lambda, G_\lambda, \boldsymbol{\alpha}_\lambda, \boldsymbol{\beta}_\lambda$ are continuous (in all their variables). Having succeeded just with the continuity of F_λ and G_λ in the "matrix" variables $y_{ij} \in Y$, only, without requiring their continuity in the remaining variables $x_i \in X$, and, at the same time, without any continuity assumption on the tuples of operations $\boldsymbol{\alpha}_\lambda, \boldsymbol{\beta}_\lambda$, was then a true surprise for me.

A revision of the results established in [24, 25, 29–31] from such a point of view reveals that in most of them some continuity assumptions can be omitted. For instance, Theorem 3 from [30] (as well as Theorem 2.6 from [24]) on stability of continuous homomorphisms from a locally compact topological group G into any topological group H remains true *without assuming that G is a topological group*. It suffices that G be both a group and a locally compact topological space. Similarly, Theorem 3.1 from [31] on stability of continuous homomorphisms from a locally compact topological algebra $\mathfrak{A}$ into a completely regular topological algebra $\mathfrak{B}$ remains true for any universal algebra $\mathfrak{A}$ endowed with a locally compact (Hausdorff) topology, without assuming continuity of the operations in $\mathfrak{A}$.

Theorems 15.2, 15.3 also show that both the above-mentioned results admit a generalization in yet another direction, for the former one stated already in Theorem 2.6 in [24]. Namely for a mapping $f: G \to H$ or $f: A \to B$ in order to be close to a continuous homomorphism it is *not* necessary to assume that it is Γ-continuous with respect to the given continuity scale Γ (as both the above-mentioned theorems in [30] and [31] do); it is enough that f be (Γ, U)-precontinuous for a sufficiently small entourage U.

On the other hand, as shown by several counterexamples in [25] and [30], even in those weaker results one cannot manage without the control of the examined functions by means of some continuity scale and a stalkwise compact bounding relation. The more interesting are then the stability results not requiring the continuity scale and/or the bounding relation in their formulation. This is, e.g., the case of the global stability result for homomorphisms from amenable groups into the group of unitary operators on a Hilbert space in [16] (covering many more specific results proved both before and afterwards), as well as of the local stability result for homomorphisms from amenable groups into the unit circle $\mathbb{T}$ in [30].

Acknowledgements Research supported by grants no. 1/0608/13 and 1/0333/17 of the Slovak grant agency VEGA.

References

1. Albeverio, S., Fenstad, J.E., Høegh-Krohn, R., Lindstrøm, T.: Nonstandard Methods in Stochastic Analysis and Mathematical Physics. Academic, London (1986)
2. Anderson, R.M.: "Almost" implies "near." Trans. Am. Math. Soc. **196**, 229–237 (1986)
3. Aoki, T.: On the stability of the linear transformation in Banach spaces. J. Math. Soc. Jpn. **2**, 64–66 (1950)
4. Arkeryd, L.O., Cutland, N.J., Henson, C.W. (eds.): Nonstandard Analysis, Theory and Applications. Kluwer Academic Publishers, Dordrecht (1997)
5. Boualem, H., Brouzet, R.: On what is the almost-near principle. Am. Math. Mon. **119**(5), 381–393 (2012)
6. Czerwik, S.: Functional Equations and Inequalities in Several Variables. World Scientific, Singapore (2002)
7. Davis, M.: Applied Nonstandard Analysis. Wiley, New York (1977)
8. Engelking, R.: General Topology. PWN — Polish Scientific Publishers, Warszawa (1977)
9. Forti, G.L.: Hyers-Ulam stability of functional equations in several variables. Aequationes Math. **50**, 143–190 (1995)
10. Grätzer, G.: Universal Algebra. Van Nostrand, Princeton (1968)
11. Henson, C.W.: Foundations of nonstandard analysis: a gentle introduction to nonstandard extensions. In: Arkeryd, L.O., Cutland, N.J., Henson, C.W. (eds.) Nonstandard Analysis, Theory and Applications, pp. 1–49. Kluwer Academic Publishers, Dordrecht (1997)
12. Hilton, P.J., Stammbach, U.: A Course in Homological Algebra. Springer, New York (1970)
13. Hyers, D.H.: On the stability of the linear functional equation. Proc. Natl. Acad. Sci. USA **27**, 222–224 (1941)
14. Hyers, T.H., Rassias, T.M.: Approximate homomorphisms. Aequationes Math. **44**, 125–153 (1992)
15. Hyers, D.H., Isac, G., Rassias, T.M.: Stability of Functional Equations in Several Variables. Birkhäuser Verlag, Basel (1998)
16. Kazhdan, D.: On ϵ-representations. Isr. J. Math. **43**, 315–323 (1982)
17. Loeb, P.A.: Nonstandard analysis and topology. In: Arkeryd, L.O., Cutland, N.J., Henson, C.W. (eds.) Nonstandard Analysis, Theory and Applications, pp. 77–89. Kluwer Academic Publishers, Dordrecht (1997)
18. Mačaj, M., Zlatoš, P.: Approximate extension of partial ϵ-characters of abelian groups to characters with application to integral point lattices. Indag. Math. **16**, 237–250 (2005)
19. Mauldin, R.D.: The Scottish Book. Birkhäuser Verlag, Boston (1981)
20. Rassias, T.M.: On the stability of the linear mapping in Banach spaces. Proc. Am. Math. Soc. **72**, 297–300 (1978)
21. Rassias, T.M.: On the stability of functional equations and a problem of Ulam. Acta Appl. Math. **62**, 23–130 (2000)

22. Rassias, T.M. (ed.): Functional Equations and Inequalities. Mathematics and Its Applications, vol. 518. Kluwer Academic Publishers, Dordrecht (2000)
23. Robinson A.: Non-Standard Analysis (revised edn.). Princeton University Press, Princeton (1996)
24. Sládek, F., Zlatoš, P.: A local stability principle for continuous group homomorphisms in nonstandard setting. Aequationes Math. **89**, 991–1001 (2015)
25. Špakula, J., Zlatoš, P.: Almost homomorphisms of compact groups. Ill. J. Math. **48**, 1183–1189 (2004)
26. Székelyhidi, L.: Ulam's problem, Hyers's solution — and where they led. In: Rassias, T.M. (ed.) Functional Equations and Inequalities. Mathematics and Its Applications, vol. 518, pp. 259–285. Kluwer, Dordrecht (2000)
27. Ulam, S.M.: A Collection of Mathematical Problems. Interscience Publications, New York (1961)
28. Ulam, S.M.: Problems in Modern Mathematics. Wiley, New York (1964)
29. Zlatoš, P.: Stability of homomorphisms between compact algebras. Acta Univ. M. Belii. Ser. Math. **15**, 73–78 (2009)
30. Zlatoš, P.: Stability of group homomorphisms in the compact-open topology. J. Log. Anal. **2:3**, 1–15 (2010)
31. Zlatoš, P.: Stability of homomorphisms in the compact-open topology. Algebra Univers. **64**, 203–212 (2010)

MIX
Papier aus verantwortungsvollen Quellen
Paper from responsible sources
FSC® C105338

If you have any concerns about our products,
you can contact us on
ProductSafety@springernature.com

In case Publisher is established outside the EU,
the EU authorized representative is:
Springer Nature Customer Service Center GmbH
Europaplatz 3, 69115 Heidelberg, Germany

Printed by Libri Plureos GmbH
in Hamburg, Germany